Quantum Foundations

Quantum Foundations

90 Years of Uncertainty

Special Issue Editors

Pedro W. Lamberti
Gustavo M. Bosyk
Sebastian Fortin
Federico Holik

MDPI • Basel • Beijing • Wuhan • Barcelona • Belgrade

MDPI

Special Issue Editors
Pedro W. Lamberti
Universidad Nacional de Córdoba & CONICET
Argentina

Gustavo M. Bosyk
Instituto de Física La Plata
Argentina

Sebastian Fortin
Universidad de Buenos Aires
Argentina

Federico Holik
Instituto de Física La Plata
Argentina

Editorial Office
MDPI
St. Alban-Anlage 66
4052 Basel, Switzerland

This is a reprint of articles from the Special Issue published online in the open access journal *Entropy* (ISSN 1099-4300) from 2018 to 2019 (available at: https://www.mdpi.com/journal/entropy/special_issues/90_Years_Uncertainty)

For citation purposes, cite each article independently as indicated on the article page online and as indicated below:

LastName, A.A.; LastName, B.B.; LastName, C.C. Article Title. *Journal Name* **Year**, *Article Number*, Page Range.

ISBN 978-3-03897-754-4 (Pbk)
ISBN 978-3-03897-755-1 (PDF)

Contents

About the Special Issue Editors

Pedro W. Lamberti, PhD: Dr Pedro Lamberti is a Full Professor of Theoretical Physics in the Faculty of Mathematics, Astronomy, Physics and Computation at The National University of Córdoba in Argentina. He is also a member of the National Council of Science and Technology (CONICET, Argentina). His research focuses on studying the definition of quantum correlations by using the notion of distances between quantum states. He also studies the dynamical properties of time series by using information theory tools. He obtained a PhD in Physics in the area of General Relativity Theory from the National University of Córdoba (in 1990).

Gustavo M. Bosyk, PhD: Dr Gustavo Bosyk has been a Researcher at Instituto de Física La Plata—CONICET (Argentina) since 2016. He studied Physics at the University of Buenos Aires (diploma awarded in 2010). He received his PhD in Physics (in 2014) from Universidad Nacional de La Plata (Argentina). His scientific interests focus on the broad field of quantum information and quantum foundations, with a particular focus on entanglement theory, quantum correlations, uncertainty relations, quantum coding, quantum resource theories, information theoretic measures, and majorization. He is the author of more than 20 peer-reviewed publications on these topics and he has participated in several international workshops and conferences on the field.

Sebastian Fortin, PhD: Dr Sebastian Fortin has a degree in Physics from the University of Buenos Aires and a PhD in Epistemology and History of Science from the National University of Tres de Febrero (Argentina). He is an Assistant Researcher at CONICET and a First-Class Professor Assistant at the Physics Department of the Faculty of Exact and Natural Sciences at the University of Buenos Aires. He specializes in the philosophy of physics, particularly quantum mechanics. His research focuses on studying the classical limit based on the decoherence phenomena, quantum information, interpretation of quantum mechanics, the modal-Hamiltonian interpretation, nonunitary evolutions, irreversibility, and quantum logic.

Federico Holik, PhD: Dr Federico Holik is a Research Fellow of the National Scientific and Technical Research Council in Argentina. Dr Holik studied at the University of Buenos Aires (Argentina) and held postdoctoral positions at Instituto de Física La Plata (Argentina) and Université Paris Diderot (France). His research focuses on quantum information theory, the foundations of quantum mechanics, the interpretation of quantum probabilities, and the study of the logical, algebraic, and geometrical aspects of the quantum formalism.

entropy

MDPI

Editorial

Special Issue "Quantum Foundations: 90 Years of Uncertainty"

Gustavo M. Bosyk [1]**, Sebastian Fortin** [2]**, Pedro W. Lamberti** [3] **and Federico Holik** [1,*]

[1]   Instituto de Física La Plata , UNLP, CONICET, Facultad de Ciencias Exactas, 1900 La Plata, Argentina; gbosyk@gmail.com

[2]   CONICET, Departamento de Física, Universidad de Buenos Aires, Buenos Aires, C1053 CABA, Argentina; sebastian.fortin@gmail.com

[3]   Facultad de Matemática, Astronomía, Física y Computación, Universidad Nacional de Córdoba & CONICET, Córdoba, Argentina; pwlamberti@gmail.com

*   Correspondence: holik@fisica.unlp.edu.ar or olentiev2@gmail.com

Received: 2 February 2019; Accepted: 6 February 2019; Published: 8 February 2019

Keywords: foundations of quantum mechanics; uncertainty relations; bell inequalities; entropy; quantum computing

---

The VII Conference on Quantum Foundations: 90 years of uncertainty was held during November 29th to December 1st, in 2017, at the Facultad de Matemática, Astronomía, Física y Computación, Córdoba, Argentina. It gathered experts in the foundations of quantum mechanics from different countries around the world, interested in promoting a multidisciplinary approach to the fundamental questions of quantum theory and its applications, by taking in consideration not only the physical, but also the philosophical and mathematical aspects of the theory. By those days, 90 years had passed since the seminal paper of Werner Heisenberg [1], describing the reciprocal uncertainty relation between position and momentum in the quantum realm. But the intriguing questions about the interpretation of those relations in connection to the general problems of the interpretation of the quantum formalism, still remain. This was reflected in the vivid discussions that were posed during the Conference.

This special issue captures the main aspects of this debate in connection with other fundamental questions of quantum theory and its applications, by incorporating a selected list of contributions that we now present below.

In the paper "Evaluating the Maximal Violation of the Original Bell Inequality by Two-Qudit States Exhibiting Perfect Correlations/Anticorrelations", by Andrei Y. Khrennikov and Elena R. Loubenets [2], a general class of symmetric two-qubit states with perfect correlations or anticorrelations between Alice and Bob was introduced. It was proved that, for all states belonging to this class, the maximal violation of the original Bell inequality is upper bounded by a factor $\frac{3}{2}$ and the two-qubit states where this quantum upper bound is attained were given. This is a step forward for solving the problem of finding the quantum upper bound for the original Bell inequality. The experimental implications of these results were also discussed.

In the paper "Revisiting Entanglement within the Bohmian Approach to Quantum Mechanics", by Claudia Zander and Angel Ricardo Plastino [3], the concept of entanglement was discussed in the framework of the Bohmian approach to quantum mechanics. Using this approach, two partial measures for the amount of entanglement corresponding to a pure state of a pair of quantum particles were constructed. These measures were then put in connection with the notion of total entanglement—that relies on the linear entropy of the single-particle reduced density matrix—which was shown to be equal to their sum. A clear interpretation of the introduced measures was given in terms of the ontology of Bohmian dynamics.

In the paper "New Forms of Quantum Value Indefiniteness Suggest that Incompatible Views on Contexts Are Epistemic", by Karl Svozil [4], the problem of quantum probabilities and quantum contextuality was addressed. Quantum logics used in extensions of the Kochen–Specker theorem were discussed. The study of these logics and the structure of the probabilistic states that can be built using them, lead the author to suggest a natural interpretation for the quantum formalism. According to this view, quantum systems can be completely characterized by a unique context and a "true" proposition within this context; this situation defines the ontic state of the quantum system. It was argued that, unless there is a total match between preparation and measurement contexts, information about the former from the latter cannot be ontic, but epistemic.

In the paper "Adiabatic Quantum Computation Applied to Deep Learning Networks", by Jeremy Liu et al. [5], the task of training deep learning networks was addressed. This was done by exploring the possibility of using quantum devices. The authors do this by focusing on a restricted form of adiabatic quantum computation known as quantum annealing, performed by a D-Wave processor. They propose a particular network topology that can be trained to classify MNIST and neutrino detection data. They compared their quantum annealing approach with other extant alternatives, and showed that the quantum approach can find good network parameters in a reasonable time, despite increased network topology complexity.

In the paper "Entropic Uncertainty Relations for Successive Measurements in the Presence of a Minimal Length" by Alexey E. Rastegin [6], the generalized uncertainty principle for successive measurements in the presence of a minimal length was discussed. Uncertainties were described by appealing to generalized entropies of both the Rényi and Tsallis types. The specific features of measurements of observables with continuous spectra were taken into account. It was first shown that, since uncertainty relations formulated in terms of Shannon entropies involve a state-dependent correction term, they will be different, in general, from preparation uncertainty relations. Next, it was shown that state-independent uncertainty relations can be obtained in terms of Rényi and Tsallis entropies. These have the same lower bounds as in the preparation scenario and were shown to depend on the acceptance function of apparatuses in momentum measurements.

In the paper "Quantization and Bifurcation beyond Square-Integrable Wavefunctions", by Ciann–Dong Yang and Chung–Hsuan Kuo [7], nonsquare-integrable (NSI) solutions of the Schrödinger equation are discussed. These solutions are ruled out in the majority of the formulations of quantum mechanics, due to problems with the conservation of probability. Contrarily, in this paper, a quantum-trajectory approach to energy quantization that includes the possibility of nonsquare-integrable solutions of the Schrödinger equation was considered. It was shown that both, normalized and unnormalized wavefunctions contribute to energy quantization. While square-integrable wavefunctions help to locate the bifurcation points at which energy has a step jump, it turns out that the non square-integrable ones form the flat parts of the stair-like distribution of the quantized energies. The synchronicity between the energy quantization process and the center-saddle bifurcation process was also discussed, in connection to the nonsquare-integrable wave functions.

In the paper "Gudder's Theorem and the Born Rule", by Francisco De Zela [8], the Born probability rule was discussed. The author proves that it can be derived from Gudder's theorem [9]. In doing so, the author tried to identify the fundamental underlying assumptions that lead to a probability rule such as Born's. It was then argued that Born's rule applies to both the classical and the quantum domains.

In the paper "Uncertainty Relation Based on Wigner–Yanase–Dyson Skew Information with Quantum Memory" by Jun Li and Shao–Ming Fei [10], uncertainty relations based on Wigner–Yanase–Dyson skew information with quantum memory were studied. The authors derive uncertainty inequalities in product and summation forms. The lower bounds of these inequalities were found and were shown to contain two terms. One of them is related to the degree of compatibility of two measurements. The other one is connected to the quantum correlation between the measured system and the quantum memory.

In the review paper "Uncertainty Relations for Coarse-Grained Measurements: An Overview", by Fabricio Toscano et al. [11], the problem of uncertainty relations tailored specifically to coarse-grained measurement of continuous quantum observables was addressed, including both theoretical and experimental aspects. These inequalities have applications in detection of quantum correlations and security requirements in quantum cryptography. In order to deal with continuous variable systems, measurements are coarse grained, but the coarse-grained observables do not necessarily obey the same uncertainty relations as the original ones. This leads to the study of coarse-grained uncertainty relations associated to continuous variable quantum systems. This review focused on such uncertainty relations as well as their applications in quantum information theory.

We hope that the selected papers will be of interest for the community of physicists and philosophers working on the foundations of quantum mechanics.

Acknowledgments: We acknowledge all authors for their contributions, all participants of the VII Conference on Quantum Foundations at Córodba (Argentina), as well as the anonymous reviewers of the articles here, and editorial staff of Entropy.

Conflicts of Interest: The authors declare no conflict of interest.

References

1. Heisenberg, W. Über den anschaulichen Inhalt der quantentheoretischen Kinematik und Mechanik. *Z. Phys.* **1927**, *43*, 172–198. [CrossRef]
2. Khrennikov, A.Y.; Loubenets, E.R. Evaluating the Maximal Violation of the Original Bell Inequality by Two-Qudit States Exhibiting Perfect Correlations/Anticorrelations. *Entropy* **2018**, *20*, 829. [CrossRef]
3. Zander, C.; Plastino, A.R. Revisiting Entanglement within the Bohmian Approach to Quantum Mechanics. *Entropy* **2018**, *20*, 473. [CrossRef]
4. Svozil, K. New Forms of Quantum Value Indefiniteness Suggest That Incompatible Views on Contexts Are Epistemic. *Entropy* **2018**, *20*, 406. [CrossRef]
5. Liu, J.; Spedalieri, F.M.; Yao, K.-T.; Potok, T.E.; Schuman, C.; Young, S.; Patton, R.; Rose, G.S.; Chamka, G. Adiabatic Quantum Computation Applied to Deep Learning Networks. *Entropy* **2018**, *20*, 380. [CrossRef]
6. Rastegin, A.E. Entropic Uncertainty Relations for Successive Measurements in the Presence of a Minimal Length. *Entropy* **2018**, *20*, 354. [CrossRef]
7. Yang, C.-D.; Kuo, C.-H. Quantization and Bifurcation beyond Square-Integrable Wavefunctions. *Entropy* **2018**, *20*, 327. [CrossRef]
8. De Zela, F. Gudder's Theorem and the Born Rule. *Entropy* **2018**, *20*, 158. [CrossRef]
9. Gudder, S.P. *Stochastic Methods in Quantum Mechanics*; North-Holland: New York, NY, USA, 1979.
10. Li, J.; Fei, S.-M. Uncertainty Relation Based on Wigner–Yanase–Dyson Skew Information with Quantum Memory. *Entropy* **2018**, *20*, 132. [CrossRef]
11. Toscano, F.; Tasca, D.S.; Rudnicki, Ł.; Walborn, S.P. Uncertainty Relations for Coarse–Grained Measurements: An Overview. *Entropy* **2018**, *20*, 454. [CrossRef]

entropy

MDPI

Article

Evaluating the Maximal Violation of the Original Bell Inequality by Two-Qudit States Exhibiting Perfect Correlations/Anticorrelations

Andrei Y. Khrennikov [1,2,]* **and Elena R. Loubenets** [2]

[1]   International Center for Mathematical Modeling, Linnaeus University, 35195 Vaxjo, Sweden
[2]   Applied Mathematics Department, National Research University Higher School of Economics,
        101000 Moscow, Russia; elena.loubenets@hse.ru
*   Correspondence: Andrei.Khrennikov@lnu.se; Tel.: +46-725-941-531

Received: 25 August 2018; Accepted: 22 October 2018; Published: 29 October 2018

Abstract: We introduce the general class of symmetric two-qubit states guaranteeing the perfect correlation or anticorrelation of Alice and Bob outcomes whenever some spin observable is measured at both sites. We prove that, for all states from this class, the maximal violation of the original Bell inequality is upper bounded by $\frac{3}{2}$ and specify the two-qubit states where this quantum upper bound is attained. The case of two-qutrit states is more complicated. Here, for all two-qutrit states, we obtain the same upper bound $\frac{3}{2}$ for violation of the original Bell inequality under Alice and Bob spin measurements, but we have not yet been able to show that this quantum upper bound is the least one. We discuss experimental consequences of our mathematical study.

Keywords: original Bell inequality; perfect correlation/anticorrelation; qudit states; quantum bound; measure of classicality

---

1. Introduction

The recent loophole free experiments [1–3] demonstrated violations of classical bounds for the wide class of the Bell-type inequalities which derivations are not based on perfect (anti-) correlations, for example, the Clauser–Horne–Shimony–Holt (CHSH) inequality [4] and its further various generalizations [5–14]. These experiments have very high value for foundations of quantum mechanics (QM) and interrelation between QM and hidden variable models, see, for example, [15–22] for recent debates.

However, John Bell started his voyage beyond QM not with such inequalities, but with the original Bell inequality [23,24] the derivation of which is based on perfect anticorrelations—the condition which is explicitly related to the Einstein–Podolsky–Rosen (EPR) argument [25].

At the time of the derivation of the original Bell inequality, the experimental technology was not so advanced and preparation of sufficiently clean ensembles of singlet states was practically dificult. Therefore, Bell enthusiastically supported the proposal of Clauser, Horne, Shimony, and Holt, which is based on a new scheme (without exploring perfect correlations) and the CHSH inequality [4].

The tremendous technological success of recent years, especially, in preparation of the two-qubit singlet state and high efficiency detection, makes the original Bell's project at least less dificult. This novel situation attracted again attention to the original Bell inequality [26]. We also point to related theoretical studies on the original Bell inequality which were done during the previous years, see [27–31]. In [29,31], it is, for example, shown that, unlike the CHSH inequality, the original Bell inequality distinguishes between classicality and quantum separability.

Finally, we point to a practically unknown paper of Pitowsky [32] where he claims that by violating the original Bell inequality and its generalizations it would be possible to approach a higher degree of nonclassicality than for the CHSH-like inequalities.

This claim is built upon the fact that, for the CHSH inequality $|\mathcal{B}^{\text{CHSH}}_{clas}| \leq 2$, the fraction $F^{(\rho_d)}_{\text{CHSH}}$ of the quantum (Tsirelson) upper bound [33,34] $2\sqrt{2}$ to the classical one is equal to $F^{(\rho_d)}_{\text{CHSH}} = \sqrt{2}$ for a bipartite state ρ_d of an arbitrary dimension $d \geq 2$, whereas, for the original Bell inequality, the fraction $F^{(\rho_{singlet})}_{\text{OB}}$ of the quantum upper bound for the two-qubit singlet ($d = 2$) to the classical bound (equal to one see in Section 2) is given by [26,32]

$$F^{(\rho_{singlet})}_{\text{OB}} = \frac{3}{2} > \sqrt{2} = F^{(\rho_d)}_{\text{CHSH}}, \quad \forall d \geq 2. \tag{1}$$

The rigorous mathematical proof of the least upper bound $\frac{3}{2}$ on the violation of the original Bell inequality by the two-qubit singlet was presented in the article [26] written under the influence of Pitowsky's paper [32]. In both papers—References [26,32], the considerations were restricted only to the two-qubit singlet case.

However, for the violation $F^{(\rho_d)}_{\text{OB}}$ of the original Bell inequality by a two-qudit state ρ_d exhibiting perfect correlations/anticorrelations, the CHSH inequality implies for all $d \geq 2$ the upper bound $(2\sqrt{2} - 1)$ (see in Section 3) and the latter upper bound is more than the least upper bound $\frac{3}{2}$ proved [26,32] for the two-qubit singlet.

We stress that quantum nonlocality is not equivalent [35] to quantum entanglement and that larger violations of Bell inequalities can be reached [36] by states with less entanglement. Therefore, the proof [26] that, for the two-qubit singlet state (which is maximally entangled), the least upper bound on violation of the original Bell inequality is equal to $\frac{3}{2}$ does not automatically mean that $\frac{3}{2}$ is the least upper bound on violation of the original Bell inequality for all two-qubit states. Moreover, the proof of the least upper bound $\frac{3}{2}$ on violation of the original Bell inequality by the singlet state has no any consequence for quantifying violation of this inequality by a two-qudit state of an arbitrary dimension $d \geq 2$.

In the present paper, we rigorously prove that under Alice and Bob spin measurements, the least upper bound $\frac{3}{2}$ on the violation of the original Bell inequality holds for all two-qubit and all two-qutrit states exhibiting perfect correlations/anticorrelations. In the sequel to this article, we intend to prove that, quite similarly to the CHSH case where the least upper bound $\sqrt{2}$ on quantum violations holds for all dimensions $d \geq 2$, under the condition on perfect correlations/anticorrelations, the least upper bound $\frac{3}{2}$ on quantum violations of the original Bell inequality holds for all $d \geq 2$ (see in Section 6).

In Section 2 (Preliminaries), we present the condition [31] on perfect correlations or anticorrelations for joint probabilities and prove, under this condition, the validity of the original Bell inequality in the local hidden variable (LHV) frame. This general condition is true for any number of outcomes at each site and reduces to the Bell's perfect correlation/anticorrelation condition [23] on the correlation function only in case of Alice and Bob outcomes ± 1.

In Section 3, we analyse violation of the original Bell inequality by a two-qudit quantum state and show that, for all dimensions of a two-qudit state exhibiting perfect correlations/anticorrelations and any three qudit observables, the maximal violation of the original Bell inequality cannot exceed the value $(2\sqrt{2} - 1)$.

In Section 4, we introduce (Proposition 2) the general class of symmetric two-qubit density operators which guarantee perfect correlation or anticorrelation of Alice and Bob outcomes whenever some (the same) spin observable is measured at both sites. We prove (Theorem 1) that, for all states from this class, the maximal violation of the original Bell inequality is upper bounded by $\frac{3}{2}$ and specify the two-qubit states for which this quantum upper bound is attained.

In Section 5, we consider Alice and Bob spin measurements on two-qutrit states. This case is more complicated. Here, we are also able to prove the upper bound $\frac{3}{2}$ for all spin measurements on an arbitrary two-qutrit state, but we have not yet been able to find two-qutrit states for which this upper bound is attained. In future, we plan to study this problem as well as to consider spaces of higher dimensions.

In Secton 6, we summarize the main results and stress that description of general density operators ensuring perfect correlations or anti-correlations for spin or polarization observables may simplify performance of a hypothetical experiment on violation of the original Bell inequality. In principle, experimenters need not prepare an ensemble of systems in the singlet state since, by Proposition 2 and Theorem 1, for such experiments, a variety of two-qubit states, pure and mixed, can be used and it might be easier to prepare some of such states.

2. Preliminaries: Derivation of the Original Bell Inequality in a General Case

Both Bell's proofs [23,24] of the original Bell inequality in a local hidden variable (LHV) frame are essentially built up on two assumptions: a dichotomic character of Alice's and Bob's measurements plus the perfect correlation or anticorrelation of their outcomes for a definite pair of their local settings. Specifically, the latter assumption is abbreviated in quantum information as the condition on perfect correlations or anticorrelations.

In this section, we present the proof [31] of the original Bell inequality in the LHV frame for any numbers of Alice and Bob outcomes in $[-1,1]$ and under the condition which is more general than the one introduced by Bell.

Consider an arbitrary bipartite correlation scenario with two measurement settings $a_i, b_k, i, k = 1, 2$, and any numbers of discrete outcomes $\lambda_a, \lambda_b \in [-1,1]$ at Alice and Bob sites, respectively. This bipartite scenario is described by four joint measurements (a_i, b_k), $i, k = 1, 2$, with joint probability distributions $P_{(a_i, b_k)}$ of outcomes in $[-1,1]^2$. Notation $P_{(a_i, b_k)}(\lambda_a, \lambda_b)$ means the joint probability of the event that, under a measurement (a_i, b_k), Alice observes an outcome λ_a while Bob—an outcome λ_b. For the general framework on the probabilistic description of an arbitrary N-partite correlation scenario with any numbers of measurement settings and any spectral type of outcomes at each site, discrete or continuous, see [37].

For a joint measurement (a_i, b_k), we denote by

$$\langle \lambda_{a_i} \rangle = \sum_{\lambda_a, \lambda_b \in [-1,1]} \lambda_a P_{(a_i, b_k)}(\lambda_a, \lambda_b), \quad \langle \lambda_{b_k} \rangle = \sum_{\lambda_a, \lambda_b \in [-1,1]} \lambda_b P_{(a_i, b_k)}(\lambda_a, \lambda_b) \tag{2}$$

the averages of outcomes, observed by Alice and Bob, and by

$$\langle \lambda_{a_i} \lambda_{b_k} \rangle = \sum_{\lambda_a, \lambda_b \in [-1,1]} \lambda_a \lambda_b P_{(a_i, b_k)}(\lambda_a, \lambda_b) \tag{3}$$

the average of the product $\lambda_a \lambda_b$ of their outcomes.

Let, under a joint measurement (a_i, b_k), Alice and Bob outcomes satisfy the conditions that either the event

$$\{\lambda_a = \lambda_b\} := \left\{ (\lambda_a, \lambda_b) \in [-1,1]^2 \mid \lambda_a = \lambda_b \right\} \tag{4}$$

or the event

$$\{\lambda_a = -\lambda_b \neq 0\} := \left\{ (\lambda_a, \lambda_b) \in [-1,1]^2 \mid \lambda_a = -\lambda_b \neq 0 \right\} \tag{5}$$

are observed with certainty, that is [31]:

$$P_{(a_i, b_k)}(\{\lambda_a = \lambda_b\}) = \sum_{\lambda_a = \lambda_b} P_{(a_i, b_k)}(\lambda_a, \lambda_b) = 1 \tag{6}$$

or

$$P_{(a_i, b_k)}(\{\lambda_a = -\lambda_b \neq 0\}) = \sum_{\lambda_a = -\lambda_b \neq 0} P_{(a_i, b_k)}(\lambda_a, \lambda_b) = 1, \tag{7}$$

respectively.

To demonstrate that, under conditions (6) or (7) on probabilities, outcomes of Alice and Bob are perfectly correlated or anticorrelated, consider, for example, the plus sign case (6). From (6) it follows that, for arbitrary $\lambda_a \neq \lambda_b$, the joint probability

$$P_{(a_i,b_k)}(\lambda_a, \lambda_b)|_{\lambda_a \neq \lambda_b} = 0. \tag{8}$$

Hence, under a joint measurement (a_i, b_k), the marginal probabilities at Alice and Bob sites are given by

$$
\begin{aligned}
P_{a_i}(\lambda_a) &= \sum_{\lambda_b} P_{(a_i,b_k)}(\lambda_a, \lambda_b) = P_{(a_i,b_k)}(\lambda_a, \lambda_b)|_{\lambda_b = \lambda_a}, \quad \forall \lambda_a, \\
P_{b_k}(\lambda_b) &= \sum_{\lambda_a} P_{(a_i,b_k)}(\lambda_a, \lambda_b) = P_{(a_i,b_k)}(\lambda_a, \lambda_b)|_{\lambda_a = \lambda_b}, \quad \forall \lambda_b.
\end{aligned}
\tag{9}
$$

Therefore, under this joint measurement, at Alice and Bob sites the marginal probability distributions of observed outcomes $\lambda \in [-1, 1]$ coincide $P_{a_i}(\lambda) = P_{b_k}(\lambda)$ and, given, for example, that Alice observes an outcome $\lambda_a = \lambda_0$, Bob observes the outcome $\lambda_b = \lambda_0$ with certainty, i.e., the conditional probability $P_{b_k}(\lambda_b = \lambda_0 \mid \lambda_a = \lambda_0) = 1, \forall \lambda_0$. Also, under condition (6), the Pearson correlation coefficient γ_{cor}, considered in statistics, is given by

$$\gamma_{cor} = \frac{\sum_{\lambda_a, \lambda_b} (\lambda_a - \langle \lambda_a \rangle)(\lambda_b - \langle \lambda_b \rangle) P_{(a_i,b_k)}(\lambda_a, \lambda_b)}{\sqrt{\sum_{\lambda_a} (\lambda_a - \langle \lambda_a \rangle)^2 P_{a_i}(\lambda_a)} \sqrt{\sum_{\lambda_b} (\lambda_b - \langle \lambda_b \rangle)^2 P_{b_k}(\lambda_b)}} = 1. \tag{10}$$

Therefore, under the plus sign condition (6), Alice and Bob outcomes are perfectly correlated also in the meaning generally accepted in statistics.

The minus sign case (7) is considered quite similarly and results in the relation $P_{a_i}(\lambda) = P_{b_k}(-\lambda)$, $\forall \lambda \in [-1, 1]$, for marginal distributions of Alice and Bob, the relation $P_{b_k}(\lambda_b = -\lambda_0 \mid \lambda_a = \lambda_0) = 1$, $\forall \lambda_0$, for the conditional probability and the Pearson correlation coefficient $\gamma_{cor} = -1$. All this means the perfect anticorrelation of Alice and Bob outcomes.

For a joint measurement with outcomes ± 1, the general conditions (6), (7) are equivalently represented by the condition on the product expectation

$$\langle \lambda_a \lambda_b \rangle = \pm 1. \tag{11}$$

respectively, introduced originally in Bell [23]. However, for any number of outcomes in $[-1, 1]$ at both sites, Alice and Bob outcomes may be correlated or anticorrelated in the sense of (6) or (7), respectively, but their product expectation $\langle \lambda_a \lambda_b \rangle \neq \pm 1$.

Thus, under a bipartite scenario with any number of different outcomes in $[-1, 1]$, relations (6) and (7) introduced in [31], constitute the general condition on perfect correlation or anticorrelation of outcomes observed by Alice and Bob. This general perfect correlations/anticorrelations condition reduces to the Bell one (11) only in a dichotomic case with $\lambda_a, \lambda_b = \pm 1$.

Let a 2×2-setting correlation scenario with joint measurements $(a_i, b_k,)$, $i, k = 1, 2$ and outcomes $\lambda_{a_i}, \lambda_{b_k} \in [-1, 1]$ admit a local hidden variable (LHV) model for joint probabilities, for details, see Section 4 in [37], that is, all joint distributions $P_{(a_i,b_k)}$, $i, k = 1, 2$, admit the representation

$$P_{(a_i,b_k)}(\lambda_a, \lambda_b) = \int_\Omega P_{a_i}(\lambda_a | \omega) P_{b_k}(\lambda_b | \omega) \, \nu(d\omega), \quad \forall \lambda_{a_i}, \lambda_{b_k}, \tag{12}$$

via a single probability distribution ν of some variables $\omega \in \Omega$ and conditional probability distributions $P_{a_i}(\cdot \mid \omega)$, $P_{b_k}(\cdot \mid \omega)$ of outcomes at Alice's and Bob's sites. The latter conditional probabilities are usually referred to as "local" in the sense that each of them depends only on a measurement setting at the corresponding site.

Then all scenario product expectations $\langle \lambda_{a_i} \lambda_{b_k} \rangle$, $i, k = 1, 2$, admit the LHV representation

$$\langle \lambda_{a_i} \lambda_{b_k} \rangle = \int_\Omega f_{a_i}(\omega) \, f_{b_k}(\omega) \, \nu(d\omega) \tag{13}$$

with

$$f_{a_i}(\omega) := \sum_{\lambda_a \in [-1,1]} \lambda_a P_{a_i}(\lambda_a | \omega) \in [-1,1], \quad f_{b_k}(\omega) := \sum_{\lambda_b \in [-1,1]} \lambda_b P_{b_k}(\lambda_b | \omega) \in [-1,1]. \tag{14}$$

If an LHV model (12) for joint probabilities is deterministic [37,38], then the values of functions f_{a_i}, f_{b_k}, $i, k = 1, 2$, constitute outcomes under Alice and Bob corresponding measurements with settings a_i and b_k, respectively. However, in a stochastic LHV model [37,38], functions f_{a_i}, f_{b_k} may take any values in $[-1, 1]$ even in a dichotomic case.

On the other side, if, for a scenario admitting an LHV model (12) and having outcomes λ_{a_i}, $\lambda_{b_k} = \pm 1$, the Bell perfect correlation/anticorrelation restriction $\langle \lambda_{a_{i_0}} \lambda_{b_{k_0}} \rangle = \pm 1$ is fulfilled under some joint measurement (a_{i_0}, b_{k_0}), then, in this LHV model, the corresponding functions $f_{a_{i_0}}$, $f_{b_{k_0}}$ take only two values ± 1 and, moreover, $f_{a_{i_0}}(\omega) = \pm f_{b_{k_0}}(\omega)$, ν-almost everywhere (a.e.) on Ω.

We have the following statement [31] (see Appendix, for the proof).

Proposition 1. *Let, under a 2×2-setting correlation scenario with joint measurements $(a_i, b_k,)$, $i, k = 1, 2$ and any number of outcomes λ_{a_i}, λ_{b_k} in $[-1, 1]$, Alice's and Bob's outcomes under the joint measurement (a_2, b_1) be perfectly correlated or anticorrelated:*

$$P_{(a_2, b_1)}(\{\lambda_a = \lambda_b\}) = 1 \tag{15}$$

or

$$P_{(a_2, b_1)}(\{\lambda_a = -\lambda_b \neq 0\}) = 1 \tag{16}$$

If this scenario admits an LHV model (12), then its product expectations satisfy the original Bell inequality:

$$\left| \langle \lambda_{a_1} \lambda_{b_1} \rangle - \langle \lambda_{a_1} \lambda_{b_2} \rangle \right| \pm \langle \lambda_{a_2} \lambda_{b_2} \rangle \leq 1, \tag{17}$$

in its perfect correlation (plus sign) or perfect anticorrelation (minus sign) forms, respectively.

We stress that, for the validity of the original Bell inequality (17) in the LHV frame, it is suffice for condition (15) or condition (16) on perfect correlations or anticorrelations be fulfilled only under a joint measurement (a_2, b_1).

Furthermore, it was proved in [31] that, in the LHV frame, the original Bell inequality (17) holds under the LHV condition which is more general than conditions (15), (16) on perfect correlation/anticorrelations, does not imply for the LHV functions (14) relations $f_{a_2}(\omega) = \pm f_{b_1}(\omega)$, ν-a.e. on Ω and incorporates conditions (15), (16) on perfect correlation/anticorrelations only as particular cases.

For many bipartite quantum states admitting 2×2-setting LHV models, specifically, this general sufficient condition in [31] ensures [30,31,39] the validity of the perfect correlation form of the original Bell inequality for Alice and Bob measurements for any three qudit quantum observables $X_{a_1}, X_{a_2} = X_{b_1}, X_{b_2}$ with operator norms ≤ 1. Satisfying the perfect correlation form of the original Bell inequality (17), these states do not need to exhibit perfect correlations and may even have a negative correlation function (see relation (61) in [31]) whenever the same quantum observable $X_{a_2} = X_{b_1}$ is measured at both sites.

For example, all two-qudit Werner state [35]

$$W_{d,\Phi} = \frac{1+\Phi}{2}\frac{P_d^{(+)}}{r_d^{(+)}} + \frac{1-\Phi}{2}\frac{P_d^{(-)}}{r_d^{(-)}}, \quad \Phi \in [-1,1], \tag{18}$$

on $\mathbb{C}^d \otimes \mathbb{C}^d$, $d \geq 3$, separable ($\Phi \in [0,1]$) or nonseparable ($\Phi \in [-1,0)$), and all separable two-qubit Werner stated $W_{2,\Phi}(\Phi)$, $\Phi \in [0,1]$, satisfy the general sufficient condition, introduced in [31], and do not violate the perfect correlation form of the original Bell inequality (17) for any three quantum observables X_{a_1}, $X_{a_2} = X_{b_1}$, X_{b_2} but do not exhibit perfect correlations whenever the same observable $X_{a_2} = X_{b_1}$ is measured at both sites. In (18), $P_d^{(\pm)}$ are the orthogonal projections onto the symmetric and antisymmetric subspaces of $\mathbb{C}^d \otimes \mathbb{C}^d$ with dimensions $r_d^{(\pm)} = \text{tr}[P_d^{(\pm)}] = \frac{d(d\pm 1)}{2}$, respectively.

3. Quantum Violation

Consider Alice and Bob projective measurements of quantum qudit observable X_{a_1}, $X_{a_2} = X_{b_1}$, X_{b_2} in an arbitrary two-qudit state ρ on $\mathbb{C}^d \otimes \mathbb{C}^d$.

In this case, Alice and Bob outcomes coincide with eigenvalues λ_a, λ_b of these observables and restriction $\lambda_a, \lambda_b \in [-1,1]$ implies the restriction on operators norms $\|X_{a_i}\|, \|X_{b_k}\| \leq 1$. The joint probability $P_{(a_i,b_k)}(\lambda_a, \lambda_b)$ that, under a joint measurement (a_i, b_k), Alice observes an outcome λ_a, while Bob—and outcome λ_b is given by

$$\text{tr}[\rho\{P_{X_{a_i}}(\lambda_a) \otimes P_{X_{b_k}}(\lambda_b)\}] \tag{19}$$

where $P_{X_{a_i}}(\lambda_a)$, $P_{X_{b_k}}(\lambda_b)$, $i, k = 1, 2$, are the spectral projections of observables X_{a_i} and X_{b_k}, corresponding to eigenvalues λ_a and λ_b, respectively. The averages in (2), (3) take the form

$$\langle \lambda_{a_i} \rangle = \text{tr}[\rho X_{a_i}], \quad \langle \lambda_{b_k} \rangle = \text{tr}[\rho X_{b_k}], \quad \langle \lambda_{a_i}\lambda_{b_k} \rangle = \text{tr}[\rho\{X_{a_i} \otimes X_{b_k}\}], \, i, k = 1, 2 \tag{20}$$

The general conditions (15), (16) on perfect correlations or anticorrelations of Alice and Bob outcomes under a joint measurement (a_2, b_1) reduce to

$$\sum_{\lambda_a = \lambda_b} \text{tr}[\rho\{P_{X_{b_1}}(\lambda_a) \otimes P_{X_{b_1}}(\lambda_b)\}] = 1, \tag{21}$$

$$\sum_{\lambda_a = -\lambda_b \neq 0} \text{tr}[\rho\{P_{X_{b_1}}(\lambda_a) \otimes P_{X_{b_1}}(\lambda_b)\}] = 1, \tag{22}$$

respectively, and for observables with eigenvalues ± 1, these conditions are equivalent to

$$\text{tr}[\rho\{X_{b_1} \otimes X_{b_1}\}] = \pm 1. \tag{23}$$

Thus, under the considered quantum scenario, the left hand-side $W_{\rho_d}^{(\pm)}$ of the original Bell inequality (17) takes the form

$$W_\rho^{(\pm)}(X_a, X_{b_1}, X_{b_2}) = | \text{tr}[\rho\{X_a \otimes X_{b_1}\}] - \text{tr}[\rho\{X_a \otimes X_{b_2}\}] | \pm \text{tr}[\rho\{X_{b_1} \otimes X_{b_2}\}], \tag{24}$$

where, for short, we changed the index notation $a_1 \rightarrow a$, and the general condition on perfect correlations/anticorrelations of Alice and Bob outcomes under a joint measurement (b_1, b_1) is given by (21)/(22).

It is, however, well known that the two-qubit singlet state $\rho_{singlet}$ satisfies the perfect anticorrelation (minus sign) condition (in the form (23)) whenever the same qubit observable X_b with eigenvalues ± 1 is measured at both sites but, depending on a choice of qubit observables

X_a, X_{b_1}, X_{b_2}, this state may, however, violate [23,24] the perfect anticorrelation form of the original Bell inequality (17).

As it has been proven in [26,32], for the singlet $\rho_{singlet}$, the maximal value of the left hand-side (24) of the original Bell inequality (17) over qubit observables with eigenvalues ± 1 is equal to $\frac{3}{2}$.

This value is beyond the well-known Tsirelson [33,34] maximal value $\sqrt{2}$ for the quantum violation parameter $\left| \mathcal{B}_{quant}^{CHSH} \right| / \left| \mathcal{B}_{lhv}^{CHSH} \right|$ of the Clauser–Horne–Shimony–Holt (CHSH) inequality [4] $\left| \mathcal{B}_{lhv}^{CHSH} \right| \leq 2$ and, moreover, beyond the least upper bound $\sqrt{2}$ on the quantum violation parameter $\left| \mathcal{B}_{quant} \right| / \left| \mathcal{B}_{lhv} \right|$ for all unconditional Bell functionals $\mathcal{B}(\cdot)$ for two settings and two outcomes per site [40–43].

On the other side, the Tsirelson bound $2\sqrt{2}$ on the quantum violation of the CHSH inequality [4] holds for a bipartite quantum state of an arbitrary dimension. For different choices of signs, this implies

$$
\begin{aligned}
\mathrm{tr}[\rho\{X_a \otimes X_{b_1}\}] - \mathrm{tr}[\rho\{X_a \otimes X_{b_2}\}] + \mathrm{tr}[\rho\{X_{b_1} \otimes X_{b_1}\}] + \mathrm{tr}[\rho\{X_{b_1} \otimes X_{b_2}\}] &\leq 2\sqrt{2} \\
\mathrm{tr}[\rho\{X_a \otimes X_{b_1}\}] - \mathrm{tr}[\rho\{X_a \otimes X_{b_2}\}] - \mathrm{tr}[\rho\{X_{b_1} \otimes X_{b_1}\}] - \mathrm{tr}[\rho\{X_{b_1} \otimes X_{b_2}\}] &\leq 2\sqrt{2} \\
-\mathrm{tr}[\rho\{X_a \otimes X_{b_1}\}] + \mathrm{tr}[\rho\{X_a \otimes X_{b_2}\}] + \mathrm{tr}[\rho\{X_{b_1} \otimes X_{b_1}\}] + \mathrm{tr}[\rho\{X_{b_1} \otimes X_{b_2}\}] &\leq 2\sqrt{2} \\
-\mathrm{tr}[\rho\{X_a \otimes X_{b_1}\}] + \mathrm{tr}[\rho\{X_a \otimes X_{b_2}\}] - \mathrm{tr}[\rho\{X_{b_1} \otimes X_{b_1}\}] - \mathrm{tr}[\rho\{X_{b_1} \otimes X_{b_2}\}] &\leq 2\sqrt{2}
\end{aligned}
\tag{25}
$$

Combining the first line with the third one, for a two-qudit state exhibiting perfect correlations (condition (21)), we get the following upper bound

$$
\begin{aligned}
W_\rho^{(+)}(X_a, X_{b_1}, X_{b_2})|_{perfect} &= \left| \mathrm{tr}[\rho\{X_a \otimes X_{b_1}\}] - \mathrm{tr}[\rho\{X_a \otimes X_{b_2}\}] + \mathrm{tr}[\rho\{X_{b_1} \otimes X_{b_2}\}] \right| \\
&\leq 2\sqrt{2} - \left| \mathrm{tr}[\rho\{X_{b_1} \otimes X_{b_1}\}] \right|
\end{aligned}
\tag{26}
$$

on the left-hand side of the original Bell inequality. Similarly, combining the second line with the fourth one under condition (22) on perfect anticorrelations, we derive

$$
\begin{aligned}
W_\rho^{(-)}(X_a, X_{b_1}, X_{b_2})|_{perfect} &= \left| \mathrm{tr}[\rho\{X_a \otimes X_{b_1}\}] - \mathrm{tr}[\rho\{X_a \otimes X_{b_2}\}] - \mathrm{tr}[\rho\{X_{b_1} \otimes X_{b_2}\}] \right| \\
&\leq 2\sqrt{2} - \left| \mathrm{tr}[\rho\{X_{b_1} \otimes X_{b_1}\}] \right|
\end{aligned}
\tag{27}
$$

Thus, for an arbitrary two-qudit state exhibiting perfect correlation/anticorrelations whenever the same quantum observable X_{b_1} is measured at both sites we have

$$
\begin{aligned}
W_\rho^{(\pm)}(X_a, X_{b_1}, X_{b_2})|_{perfect} &= \left| \mathrm{tr}[\rho\{X_a \otimes X_{b_1}\}] - \mathrm{tr}[\rho\{X_a \otimes X_{b_2}\}] \right| \pm \mathrm{tr}[\rho\{X_{b_1} \otimes X_{b_2}\}] \right| \\
&\leq 2\sqrt{2} - \left| \mathrm{tr}[\rho\{X_{b_1} \otimes X_{b_1}\}] \right|
\end{aligned}
\tag{28}
$$

If observable X_{b_1} has only eigenvalues ± 1, then conditions (21), (22) reduce to the Bell condition (23) and the upper bound (28) takes the form

$$
W_\rho^{(\pm)}(X_a, X_{b_1}, X_{b_2})|_{perfect} \leq 2\sqrt{2} - 1
\tag{29}
$$

and holds for a two-qudit state ρ of an arbitrary dimension $d \geq 2$. For $d = 2$, this upper bound is more than the maximal value $\frac{3}{2}$ proved [26,32] for the two-qubit singlet.

Therefore, in the following section, we proceed to analyze the maximal value which the left-hand of $W_\rho^{(\pm)}(X_a, X_{b_1}, X_{b_2})|_{perfect}$ over all qubit observables X_a, X_{b_1}, X_{b_2} with eigenvalues ± 1 and all two-qubit states ρ, satisfying the perfect correlation/anticorrelation condition (23).

4. Two-Qubit Case

Consider the violation of the original Bell inequality (17) by a two-qubit state exhibiting perfect correlations/anticorrelations whenever the same qubit quantum observable with eigenvalues ± 1 is projectively measured at both sites.

We further consider only symmetric two-qubit states ρ (identical quantum particles), that is, states on $\mathbb{C}^2 \otimes \mathbb{C}^2$ which do not change under the permutation of the Hilbert spaces $\mathbb{C}^2$ in the tensor product $\mathbb{C}^2 \otimes \mathbb{C}^2$, and, for simplicity, change index notations $b_1 \to r$, $b_2 \to c$ in (24).

For $d = 2$, a generic qubit observable X on $\mathbb{C}^2$ admits the representation

$$X = \alpha \mathbb{I}_{\mathbb{C}^2} + r \cdot \sigma, \tag{30}$$

$$r \cdot \sigma = r_1 \sigma_1 + r_2 \sigma_2 + r_3 \sigma_3 \tag{31}$$

where $\alpha = \frac{1}{2} \text{tr}[X]$, $r = (r_1, r_2, r_3)$ is a vector in $\mathbb{R}^3$ with components

$$r_1 = \frac{1}{2} \text{tr}[X\sigma_1], \quad r_2 = \frac{1}{2} \text{tr}[X\sigma_2], \quad r_3 = \frac{1}{2} \text{tr}[X\sigma_3], \tag{32}$$

and

$$\sigma_1 = |e_1\rangle\langle e_2| + |e_2\rangle\langle e_1|, \quad \sigma_2 = i(|e_2\rangle\langle e_1| - |e_1\rangle\langle e_2|), \quad \sigma_3 = |e_1\rangle\langle e_1| - |e_2\rangle\langle e_2| \tag{33}$$

are self-adjoint operators on $\mathbb{C}^2$ with eigenvalues ± 1, represented in the standard orthonormal basis $\{e_1, e_2\}$ in $\mathbb{C}^2$ by the Pauli matrices

$$\sigma_1 = \begin{pmatrix} 0 & 1 \\ 1 & 0 \end{pmatrix}, \ \sigma_2 = \begin{pmatrix} 0 & -i \\ i & 0 \end{pmatrix}, \ \sigma_3 = \begin{pmatrix} 1 & 0 \\ 0 & -1 \end{pmatrix}. \tag{34}$$

Every qubit observable with eigenvalues ± 1 is represented in (30) by some unit vector $\|r\| = 1$ and constitutes projection $\sigma_r := r \cdot \sigma$ of the qubit spin along a unit vector (direction) r in $\mathbb{R}^3$.

Therefore, for Alice and Bob measurements of qubit observables with eigenvalues ± 1, the left-hand side (24) of the original Bell inequality takes the form

$$W_\rho^{(\pm)}(\sigma_a, \sigma_r, \sigma_c) = |\, \text{tr}[\rho\{\sigma_a \otimes \sigma_r\}] - \text{tr}[\rho\{\sigma_a \otimes \sigma_c\}]\,| \pm \text{tr}[\rho\{\sigma_r \otimes \sigma_c\}] \tag{35}$$

where a, r, c are unit vectors in $\mathbb{R}^3$ and the relation

$$\text{tr}[\rho\{\sigma_r \otimes \sigma_r\}] = \pm 1 \tag{36}$$

constitutes the perfect correlation/anticorrelation of Alice and Bob outcomes whenever the same spin observable σ_r—the projection of qubit spin along the same direction r in $\mathbb{R}^3$—is measured at both sites.

Substituting representation (31) into (35) and (36), we rewrite these relations via scalar products of vectors in $\mathbb{R}^3$:

$$W_\rho^{(\pm)}(\sigma_a, \sigma_r, \sigma_c) = \left|(a, T^{(\rho)}r) - (a, T^{(\rho)}c)\right| \pm (r, T^{(\rho)}c), \tag{37}$$

$$(r, T^{(\rho)}r) = \pm 1, \tag{38}$$

where $(a, T^{(\rho)}r) := \sum_{i,j} T_{ij}^{(\rho)} a_i r_j$ and $T^{(\rho)}$ is the linear operator on $\mathbb{R}^3$, defined in the canonical basis in $\mathbb{R}^3$ by the matrix with real elements

$$T_{ij}^{(\rho)} := \text{tr}[\rho\{\sigma_i \otimes \sigma_j\}, \ i, j = 1, 2, 3, \tag{39}$$

This correlation matrix is symmetric (since ρ is symmetric), has eigenvalues λ_m, $m = 1, 2, 3$, where all $|\lambda_m| \leq 1$, and is similar by its form to the matrix considered in [44].

Let us first analyze when an arbitrary symmetric two-qubit state ρ may satisfy condition (38). By decomposing a unit vector $r = \sum_m \beta_m v_m$, $\sum_m \beta_m^2 = 1$, in the orthonormal basis $\{v_j, j = 1, 2, 3\}$ of eigenvectors of $T^{(\rho)}$, we rewrite condition (38) in the form

$$\sum_m \beta_m^2 (\lambda_m \mp 1) = 0. \tag{40}$$

Since all eigenvalues $|\lambda_m| \leq 1$, relation (40) implies the following statement.

Proposition 2. *A symmetric two-qubit state ρ exhibits perfect correlation/anticorrelations*

$$\mathrm{tr}[\rho\{\sigma_r \otimes \sigma_r\}] = \pm 1 \tag{41}$$

if and only if its correlation matrix $T^{(\rho)}$ has at least one eigenvalue equal to ± 1, respectively. In this case:

(1) if only one of eigenvalues of $T^{(\rho)}$ is equal to ± 1, say $\lambda_{m_0} = \pm 1$, then ρ satisfies the perfect correlation/anticorrelation condition (41), respectively, only for the unit vector $r = v_{m_0}$;

(2) if $T^{(\rho)}$ has two eigenvalues equal to ± 1, say $\lambda_{m_1}, \lambda_{m_2} = \pm 1$, then ρ satisfies the perfect correlation/anticorrelation condition (41), respectively for every unit vector $r = \beta_{m_1} v_{m_1} + \beta_{m_2} v_{m_2}$, $\beta_{m_1}^2 + \beta_{m_2}^2 = 1$ in the plane determined by the eigenvectors $\{v_{m_1}, v_{m_2}\}$ of $T^{(\rho)}$;

(3) if all three eigenvalues of $T^{(\rho)}$ are equal to ± 1, then ρ satisfies the perfect correlation/anticorrelation condition (41), respectively, for any unit vector r in $\mathbb{R}^3$.

For the two-qubit Bell states

$$\phi_{(\pm)} = \frac{1}{\sqrt{2}}\left(e_1 \otimes e_1 \pm e_2 \otimes e_2\right), \quad \psi_{(\pm)} = \frac{1}{\sqrt{2}}\left(e_1 \otimes e_2 \pm e_2 \otimes e_1\right), \tag{42}$$

we have

$$
\begin{aligned}
T^{(\phi_+)} &= \begin{pmatrix} 1 & 0 & 0 \\ 0 & -1 & 0 \\ 0 & 0 & 1 \end{pmatrix}, \quad T^{(\phi_-)} = \begin{pmatrix} -1 & 0 & 0 \\ 0 & 1 & 0 \\ 0 & 0 & 1 \end{pmatrix} \\
T^{(\psi_+)} &= \begin{pmatrix} 1 & 0 & 0 \\ 0 & 1 & 0 \\ 0 & 0 & -1 \end{pmatrix}, \quad T^{(\psi_-)} = \begin{pmatrix} -1 & 0 & 0 \\ 0 & -1 & 0 \\ 0 & 0 & -1 \end{pmatrix}
\end{aligned}
\tag{43}
$$

and this implies.

Corollary 1. *(1) The Bell state ϕ_+ exhibits perfect anticorrelations under spin measurements at both sites along the coordinate axis Y and perfect correlations under spin measurements at both sites along the same arbitrary direction in the coordinate plane XZ;*

(2) The Bell state ϕ_- exhibits perfect anticorrelations under spin measurements at both sites along the coordinate axis X and perfect correlations—under spin measurements at both sites along the same arbitrary direction in the coordinate plane YZ;

(3) The Bell state ψ_+ exhibits perfect anticorrelations under measurements at both sites of spin projections along the coordinate axis Z and perfect correlations—under spin measurements at both along the same arbitrary direction in the coordinate plane XY;

(4) The Bell state (singlet) ψ_- exhibits perfect anticorrelations under spin measurements at both sites along the same arbitrary direction in $\mathbb{R}^3$.

Let us now analyze the maximal value of the left-hand side (37) of the original Bell inequality for a two-qubit state ρ exhibiting perfect correlations/anticorrelations (38).

Under condition $\|a\| = 1$, the maximum of $W_\rho^{(\pm)}(\sigma_a, \sigma_r, \sigma_c)$ over a is reached on the unit vector

$$a = \pm \frac{T^{(\rho)}(r-c)}{\left\|T^{(\rho)}(r-c)\right\|} \tag{44}$$

and is given by

$$\left\|T^{(\rho)}(r-c)\right\| \pm (r, T^{(\rho)}c). \tag{45}$$

Expanding vectors $r = \sum_m \beta_m v_m$, $\sum \beta_m^2 = 1$, $c = \sum_m \gamma_m v_m$, $\sum_m \gamma_m^2 = 1$, in terms of the orthonormal eigenvectors $\{v_m\}$ of $T^{(\rho)}$, we rewrite (45) in the form

$$\sqrt{\sum_{m=1,2,3} \lambda_m^2 (\beta_m - \gamma_m)^2} \pm \sum_{m=1,2,3} \lambda_m \beta_m \gamma_m, \tag{46}$$

where, due to perfect correlations/anticorrelations condition (38), the coefficients β_m are specified in Proposition 2.

Consider the maximum of expression (46) over coefficients γ_m. By Proposition 2, expression (46) reduces to

$$\sqrt{\sum_{\lambda_m^2 = 1}(\beta_m - \gamma_m)^2 + \sum_{\lambda_m^2 \neq 1} \lambda_m^2 \gamma_m^2} + \sum_{\lambda_m^2 = 1} \beta_m \gamma_m$$

$$= \sqrt{2(1 - \sum_{\lambda_m^2 = 1} \beta_m \gamma_m) - \sum_{\lambda_m^2 \neq 1}(1 - \lambda_m^2)\gamma_m^2} + \sum_{\lambda_m^2 = 1} \beta_m \gamma_m \tag{47}$$

since $\sum_{\lambda_m^2 = 1} \beta_m^2 = 1$. From (47) it follows that, for all choices of a direction r—coefficients β_m in (47) specified in Proposition 2, we have

$$\sup_{a,c} W_\rho^{(\pm)}(\sigma_a, \sigma_r, \sigma_c)|_{perfect} \leq \max_{z \in [-1,1]} \left(\sqrt{2(1-z)} + z\right) = \frac{3}{2} \tag{48}$$

where the upper bound $\frac{3}{2}$ is, for example, reached on every Bell state where all eigenvalues of the correlation matrices $\lambda_m \in \{-1, 1\}$, $m = 1, 2, 3$.

Also, if a two-qubit state, exhibiting perfect correlations/anticorrelations (see Proposition 2), has the correlation matrix with at least two eigenvalues, say $\lambda_{m_1}, \lambda_{m_2}$, with $|\lambda_{m_1}|, |\lambda_{m_2}| = 1$, then the upper bound $\frac{3}{2}$ is reached on the unit vector c which is in the plane of eigenvectors v_{m_1}, v_{m_2} corresponding to these eigenvalues (vector r is in this plane, see Proposition 2) and satisfies condition $c \cdot r = \sum_{\lambda_m^2 = 1} \beta_m \gamma_m = \frac{1}{2}$, that is, at angle $\pi/3$ to vector r.

Thus, we have proved the following new result.

Theorem 1. *Let ρ be a symmetric two-qubit states on $\mathbb{C}^2 \otimes \mathbb{C}^2$ exhibiting perfect correlations/anticorrelations whenever the same qubit observable σ_r is measured at both sites. Then the maximal value of the left-hand side $W_\rho^{(\pm)}(\sigma_a, \sigma_r, \sigma_c)$ of the original Bell inequality is given by*

$$\max_{\rho,a,r,c} W_\rho^{(\pm)}(\sigma_a, \sigma_r, \sigma_c)|_{perfect} = \frac{3}{2} \tag{49}$$

and is reached on symmetric two-qubit states discussed in lines after Equation (48).

We stress that this maximal value is less than the upper bound (29) following from the CHSH inequality.

5. Two-Qutrit Case

Consider now the violation of the original Bell inequality under Alice and Bob spin measurements on a symmetric two-qutrit state ρ on $\mathbb{C}^3 \otimes \mathbb{C}^3$, exhibiting perfect correlations or anticorrelations.

For Alice and Bob spin measurements in a two-qutrit state ρ, the left-hand side (24) of the original Bell inequality and the condition on perfect correlations/anticorrelations take the forms

$$
\begin{aligned}
W_\rho^{(\pm)}(S_a, S_r, S_c) &= \left| \operatorname{tr}[\rho\{S_a \otimes S_r\}] - \operatorname{tr}[\rho\{S_a \otimes S_c\}] \right| \pm \operatorname{tr}[\rho\{S_r \otimes S_c\}], \quad (50) \\
\operatorname{tr}[\rho\{S_r \otimes S_r\}] &= \pm 1, \quad (51)
\end{aligned}
$$

where a, r, c are unit vectors in $\mathbb{R}^3$ and

$$
S_r = r \cdot S = r_1 S_1 + r_2 S_2 + r_3 S_3, \quad S = (S_1, S_2, S_3), \quad (52)
$$

is the qutrit observable with eigenvalues $\{1, 0, -1\}$, describing projection of qutrit spin along a unit vector r in $\mathbb{R}^3$.

Note that if a two-qutrit state ρ exhibits perfect correlations/anticorrelations (51) under measurements in this state at both sites of spin projection along a direction r, the probability of event that either Alice or Bob observe at their site the outcome $\lambda = 0$ is equal to zero.

In the standard orthonormal basis $\{e_1, e_2, e_3\}$ in $\mathbb{C}^3$ these operators have the following matrix representations:

$$
S_1 = \frac{1}{\sqrt{2}} \begin{pmatrix} 0 & 1 & 0 \\ 1 & 0 & 1 \\ 0 & 1 & 0 \end{pmatrix}, \quad
S_2 = \frac{1}{\sqrt{2}} \begin{pmatrix} 0 & -i & 0 \\ i & 0 & -i \\ 0 & i & 0 \end{pmatrix}, \quad
S_3 = \begin{pmatrix} 1 & 0 & 0 \\ 0 & 0 & 0 \\ 0 & 0 & -1 \end{pmatrix} \quad (53)
$$

and

$$
S_r = \begin{pmatrix} r_3 & \frac{r_1 - i r_2}{\sqrt{2}} & 0 \\ \frac{r_1 + i r_2}{\sqrt{2}} & 0 & \frac{r_1 - i r_2}{\sqrt{2}} \\ 0 & \frac{r_1 + i r_2}{\sqrt{2}} & -r_3 \end{pmatrix} \quad (54)
$$

In view of (52), quite similarly to our techniques in Section 4 we introduce for a symmetric two-qutrit state ρ the correlation matrix $Z^{(\rho)}$ with real elements

$$
Z_{ij}^{(\rho)} = \operatorname{tr}[\rho\{S_i \otimes S_j\}], \quad (55)
$$

which is symmetric, diagonalized and has eigenvalues $|\lambda_m| \leq 1$, and this allows us to rewrite (50), (51) in the form:

$$
\begin{aligned}
W_\rho^{(\pm)}(S_a, S_r, S_c) &= \left| (a, Z^{(\rho)} r) - (a, Z^{(\rho)} c) \right| \pm (r, Z^{(\rho)} c), \\
(r, Z^{(\rho)} r) &= \pm 1.
\end{aligned} \quad (56)
$$

These expressions are quite the same by their form to expressions (37), (38) for a two-qubit state. By using the same techniques as in a qubit case, we derive

$$
\sup_{a,c} W_\rho^{(\pm)}(S_a, S_r, S_c)|_{perfect} \leq \frac{3}{2}. \quad (57)
$$

We, however, do not know whether under the considered measurements this supremum is reached.

Theorem 2. *Let ρ be a symmetric two-qutrit states on $\mathbb{C}^3 \otimes \mathbb{C}^3$ exhibiting perfect correlations/anticorrelations whenever spin projection S_r along a direction r is measured at both sites. Then, under Alice and Bob spin* measurements on these two-qutrit states, the maximal value of the left-hand side $W_\rho^{(\pm)}(S_a, S_r, S_c)$ of the original Bell inequality (17) is upper bounded as

$$\sup_{\rho, a, r, c} W_\rho^{(\pm)}(S_a, S_r, S_c)|_{perfectBell} \leq \frac{3}{2}. \tag{58}$$

This two-qutrit upper bound is less than the upper bound (29) following from the CHSH inequality.

6. Conclusions

As was pointed out in the Introduction, the recent tremendous developments in quantum technologies make experiments to test the original Bell inequality at least less difficult. This stimulates interest in novel theoretical, foundational, and mathematical studies on this inequality. In particular, it is important to find the quantum bound, the analog of the Tsirelson bound, for the original Bell inequality. It was well-known that in the two-qubit singlet case this bound equals 3/2, see, e.g., [26,32]. A year ago, I. Basieva and A. Khrennikov came with the conjecture [45] that the same upper bound holds in case of arbitrary two-qudit states and qudit observables coupled by perfect correlations/anticorrelations. The question of quantum upper bound for the original Bell inequality became actual in connection with studies on quantum-like modeling of psychological behavior, see related paper [46].

In the present article, we have proven this conjecture for all two-qubit states and all traceless qubit observables and all two-qubit states and spin qutrit observables. This is the first step towards justifying this conjecture for an arbitrary two-qudit case, and the authors of the present paper plan to continue studies on this problem. Since in the multi-dimensional case the analytical expressions are very complex, it may be useful to try to perform preliminary numerical study, cf. [47]. We also point to technique for evaluation of the quantum upper bound which was elaborated in [48,49] and tested on the CHSH-like inequalities. In principle, this technique can be applied to the original Bell inequality.

Author Contributions: Conceptualization, A.Y.K. and E.R.L.; Methodology, A.Y.K. and E.R.L.; Validation, A.Y.K. and E.R.L.; Formal Analysis, A.Y.K.; Investigation, A.Y.K. and E.R.L.; Writing—Original Draft Preparation, E.R.L.; Writing—Review & Editing, A.Y.K. and E.R.L.

Funding: E.R. Loubenets was supported within the framework of the Academic Fund Program at the National Research University Higher School of Economics (HSE) in 2018-2019 (grant N 18-01-0064) and by the Russian Academic Excellence Project "5-100".

Conflicts of Interest: The authors declare no conflict of interest.

Appendix A

Consider the proof of Proposition 1.

Let, for a joint measurement (a_2, b_1), the perfect anticorrelation (16) be fulfilled and this scenario admit an LHV model (12). This and (14) imply:

$$0 \leq \int_\Omega |f_{a_2}(\omega) + f_{b_1}(\omega)| \, \nu(d\omega)$$

$$= \int_\Omega \left| \sum_{\lambda_a, \lambda_b} (\lambda_a + \lambda_b) \, P_{a_2}(\lambda_a|\omega) P_{b_1}(\lambda_b|\omega) \right| \nu(d\omega)$$

$$\leq \int_\Omega \sum_{\lambda_a, \lambda_b} |\lambda_a + \lambda_b| \, P_{a_2}(\lambda_a|\omega) P_{b_1}(\lambda_b|\omega) \nu(d\omega) \leq 2 \sum_{\lambda_a \neq -\lambda_b} P_{(a_2, b_1)}(\lambda_a, \lambda_b) = 0.$$

Thus, under condition (16) on scenario joint probabilities, the LHV functions $f_{a_2}(\omega) = -f_{b_1}(\omega)$, ν-*a.e.* on Ω. Quite similarly, for the case of perfect correlations (15) we derive $f_{a_2}(\omega) = f_{b_1}(\omega)$, ν-*a.e.* on Ω. These relations and the number inequality

$$|x - y| \leq 1 - xy, \quad \forall \, x, y \in [-1, 1],$$

give:

$$\left| \langle \lambda_{a_1} \lambda_{b_1} \rangle - \langle \lambda_{a_1} \lambda_{b_2} \rangle \right| \pm \langle \lambda_{a_2} \lambda_{b_2} \rangle$$

$$= \left| \int_{\Omega} f_{a_1}(\omega) f_{b_1}(\omega) - f_{a_1}(\omega) f_{b_2}(\omega) \, \nu(\mathrm{d}\omega) \right| \pm \int_{\Omega} f_{a_2}(\omega) f_{b_2}(\omega) \nu(\mathrm{d}\omega)$$

$$\leq \int_{\Omega} \left| (f_{b_1}(\omega) - f_{b_2}(\omega)) \right| \nu(\mathrm{d}\omega) \pm \int_{\Omega} f_{a_2}(\omega) f_{b_2}(\omega) \, \nu(\mathrm{d}\omega) \leq 1.$$

This proves the statement.

References and Notes

1. Hensen, B.; Bernien, H.; Dréau, A.E.; Reiserer, A.; Kalb, N.; Blok, M.S.; Ruitenberg, J.; Vermeulen, R.F.; Schouten, R.N.; Abellán, C.; et al. Experimental loophole-free violation of a Bell inequality using entangled electron spins separated by 1.3 km. *Nature* **2015**, *526*, 682–686. [CrossRef] [PubMed]
2. Giustina, M.; Versteegh, M.A.; Wengerowsky, S.; Handsteiner, J.; Hochrainer, A.; Phelan, K.; Steinlechner, F.; Kofler, J.; Larsson, J.Å.; Abellán, C.; et al. A significant-loophole-free test of Bell's theorem with entangled photons. *Phys. Rev. Lett.* **2015**, *115*, 250401, arXiv:quant-ph1511.03190. [CrossRef] [PubMed]
3. Shalm, L.K.; Meyer-Scott, E.; Christensen, B.G.; Bierhorst, P.; Wayne, M.A.; Stevens, M.J.; Gerrits, T.; Glancy, S.; Hamel, D.R.; Allman, M.S.; et al. A strong loophole-free test of local realism. *Phys. Rev. Lett.* **2015**, *115*, 250402, arXiv:quant-ph1511.03189. [CrossRef] [PubMed]
4. Clauser, J.F.; Horne, M.A.; Shimony, A.; Holt, R.A. Proposed experiment to test local hidden-variable theories. *Phys. Rev. Lett.* **1969**, *23*, 880–884. [CrossRef]
5. Clauser, J.F.; Horne, M.A. Experimental consequences of objective local theories. *Phys. Rev. D* **1974**, *10*, 526–535. [CrossRef]
6. Clauser, J.F.; Shimony, A. Bell's theorem. Experimental tests and implications. *Rep. Prog. Phys.* **1978**, *41*, 1881–1927. [CrossRef]
7. Eberhard, P.H. Background level and counter efficiencies required for a loophole-free Einstein–Podolsky–Rosen experiment. *Phys. Rev. A* **1993**, *47*, 477–750. [CrossRef]
8. Khrennikov, A.; Ramelow, S.; Ursin, R.; Wittmann, B.; Kofler, J.; Basieva, I. On the equivalence of the Clauser–Horne and Eberhard inequality based tests. *Phys. Scr.* **2014**, *163*, 014019. [CrossRef]
9. Aspect, A.; Dalibard, J.; Roger, G. Experimental test of Bell's inequalities using time-varying analyzers. *Phys. Rev. Lett.* **1982**, *49*, 1805–1807. [CrossRef]
10. Aspect, A. Three Experimental Tests of Bell Inequalities by the Measurement of Polarization Correlations between Photons. Ph.D. Thesis, Universiti Paris-Sud, Orsay, France, 1983.
11. Weihs, G.; Jennewein, T.; Simon, C.; Weinfurter, R.; Zeilinger, A. Violation of Bell's inequality under strict Einstein locality conditions. *Phys. Rev. Lett.* **1998**, *81*, 5039–5043. [CrossRef]
12. Rowe, M.A.; Kielpinski, D.; Meyer, V.; Sackett, C.A.; Itano, W.M.; Monroe, C.; Wineland, D.J. Experimental violation of a Bell's inequality with efficient detection. *Nature* **2001**, *409*, 791–794. [CrossRef] [PubMed]
13. Giustina, M.; Mech, A.; Ramelow, S.; Wittmann, B.; Kofler, J.; Beyer, J.; Lita, A.; Calkins, B.; Gerrits, T.; Woo Nam, S.; et al. Bell violation using entangled photons without the fair-sampling assumption. *Nature* **2013**, *497*, 227–230. [CrossRef] [PubMed]
14. Christensen, B.G.; McCusker, K.T.; Altepeter, J.; Calkins, B.; Gerrits, T.; Lita, A.; Miller, A.; Shalm, L.K.; Zhang, Y.; Nam, S.W.; et al. Detection-Loophole-Free Test of Quantum Nonlocality, and Application. *Phys. Rev. Lett.* **2013**, *111*, 1304–1306. [CrossRef] [PubMed]
15. Aspect, A. Closing the door on Einstein and Bohr's quantum debate. *Physics* **2015**, *8*, 123. [CrossRef]
16. Wiseman, H. Quantum physics: Death by experiment for local realism. *Nature* **2015**, *526*, 649–650. [CrossRef] [PubMed]

17. Kupczynski, M. Can Einstein with Bohr debate on quantum mechanics be closed? *Philos. Trans. R. Soc. A* **2017**, *375*, 2016039, arXiv:1603.00266.
18. Khrennikov, A. After Bell. *Prog. Phys.* **2017**, *65*, 1600014. [CrossRef]
19. Khrennikov, A. Bohr against Bell: Complementarity versus nonlocality. *Open Phys.* **2017**, *15*, 734–738. [CrossRef]
20. La Cour, B.R. Local hidden-variable model for a recent experimental test of quantum nonlocality and local contextuality. *Phys. Lett. A* **2017**, *381*, 2230–2234. [CrossRef]
21. Zhou, Z.Y.; Zhu, Z.H.; Liu, S.L.; Li, Y.H.; Shi, S.; Ding, D.S.; Chen, L.X.; Gao, W.; Guo, G.C.; Shi, B.S. Quantum twisted double-slits experiments: confirming wavefunctions physical reality. *Sci. Bull.* **2017**, *62*, 1185–1192. [CrossRef]
22. Long, G.; Qin, W.; Yang, Z.; Li, J.L. Realistic interpretation of quantum mechanics and encounter-delayed-choice experiment. *Sci. China Phys. Mech. Astron.* **2018**, *61*, 030311. [CrossRef]
23. Bell, J.S. On the Einstein Podolsky Rosen paradox. *Physics* **1964**, *1*, 195–200. [CrossRef]
24. Bell, J.S. *Speakable and Unspeakable in Quantum Mechanics*, 2nd ed.; Cambridge University Press: Cambridge, UK, 2004.
25. Einstein, A.; Podolsky, B.; Rosen, N. Can quantum-mechanical description of physical reality be considered complete? *Phys. Rev.* **1935**, *47*, 777–780. [CrossRef]
26. Khrennikov, A.; Basieva, I. Towards experiments to test violation of the original Bell inequality. *Entropy* **2018**, *20*, 280. [CrossRef]
27. Ryff, L.C. Bell and Greenberger, Horne, and Zeilinger theorems revisited. *Am. J. Phys.* **1997**, *12*, 1197–1199. [CrossRef]
28. Larsson, J.A. Bell's inequality and detector inefficiency. *Phys. Rev. A* **1998**, *57*, 3304–3305. [CrossRef]
29. Loubenets, E.R. Separability of quantum states and the violation of Bell-type inequalities. *Phys. Rev.* **2004**, *69*, 1–8. [CrossRef]
30. Loubenets, E.R. Class of bipartite quantum states satisfying the original Bell inequality. *J. Phys. A Math. Gen.* **2006**, *38*, L653–L658. [CrossRef]
31. Loubenets, E.R. Local hidden variable modelling, classicality, quantum separability and the original Bell inequality. *J. Phys. A Math. Theor.* **2011**, *44*, 035305. [CrossRef]
32. Pitowsky, I. New Bell inequalities for the singlet state: Going beyond the Grothendieck bound. *J. Math. Phys.* **2008**, *49*, 012101. [CrossRef]
33. Tsirelson, B.S. Quantum generalizations of Bell's inequality. *Lett. Math. Phys.* **1980**, *4*, 93–100. [CrossRef]
34. Tsirelson, B.S. Quantum analogues of the Bell inequalities. The case of two spatially separated domains. *J. Soviet Math.* **1987**, *36*, 557–570. [CrossRef]
35. Werner, R.F. Quantum states with Einstein–Podolsky–Rosen correlations admitting a hidden-variable model. *Phys. Rev. A* **1989**, *40*, 4277–4281. [CrossRef]
36. Junge, M.; Palazuelos, C. Large Violation of Bell Inequalities with Low Entanglement. *Commun. Math. Phys.* **2011**, *306*, 695–746. [CrossRef]
37. Loubenets, E.R. On the probabilistic description of a multipartite correlation scenario with arbitrary numbers of settings and outcomes per site. *J. Phys. A Math. Theor.* **2008**, *41*, 445303. [CrossRef]
38. Fine, A. Hidden Variables, Joint Probability, and the Bell Inequalities. *Phys. Rev. Lett.* **1982**, *48*, 291–295. [CrossRef]
39. Loubenets, E.R. Threshold bounds for noisy bipartite states. *J. Phys. A Math. Gen.* **2006**, *39*, 5115–5123. [CrossRef]
40. For the general framework on unconditional Bell inequalities for arbitrary numbers of settings and outcomes at each site, see [41]. General bounds on the maximal quantum violation of Bell inequalities by an *N*-partite quantum state have been introduced in [42,43].
41. Loubenets, E.R. Multipartite Bell-type inequalities for arbitrary numbers of settings and outcomes per site. *J. Phys. A Math. Theor.* **2008**, *41*, 445304. [CrossRef]
42. Loubenets, E.R. Local quasi hidden variable modelling and violations of Bell-type inequalities by a multipartite quantum state. *J. Math. Phys.* **2012**, *53*, 022201. [CrossRef]
43. Loubenets, E.R. New concise upper bounds on quantum violation of general multipartite Bell inequalities. *J. Math. Phys.* **2017**, *58*, 052202. [CrossRef]
44. Horodecki, R.; Horodecki, P.; Horodecki, M. Violating Bell inequality by mixed spin-1/2 states: Necessary and sufficient condition. *Phys. Lett. A* **1995**, *200*, 340–344. [CrossRef]

45. The question of quantum upper bound for the original Bell inequality became actual in connection with studies on quantum-like modeling of psychological behavior, see related paper [46].
46. Basieva, I.; Cervantes, V.H.; Dzhafarov, E.N.; Khrennikov, A. True contextuality beats direct influences in human decision making. *arXiv* **2018**, arXiv:1807.05684.
47. Filipp, S.; Svozil, K. Boole-Bell-type inequalities in Mathematica. In *Challenging the Boundaries of Symbolic Computation*; Mitic, P., Ramsden, P., Carne, J., Eds.; Imperial College Press: London, UK, 2003; pp. 215–222.
48. Filipp, S.; Svozil, K. Tracing the bounds on Bell-type inequalities. *AIP Conf. Proc.* **2005**, *750*, 87–94.
49. Filipp, S.; Svozil, K. Generalizing Tsirelson's bound on Bell inequalities using a min-max principle. *Phys. Rev. Lett.* **2004**, *93*, 130407. [CrossRef] [PubMed]

entropy

MDPI

Article

Revisiting Entanglement within the Bohmian Approach to Quantum Mechanics

Claudia Zander [1] and Angel Ricardo Plastino [2],*

[1]   Physics Department, University of Pretoria, Pretoria 0002, South Africa; cz@up.ac.za
[2]   CeBio y Secretaria de Investigaciones, Universidad Nacional del Noroeste de la Prov. de Buenos Aires-UNNOBA y CONICET, Roque Saenz Peña 456,6000 Junín, Argentina
*   Correspondence: arplastino@unnoba.edu.ar

Received: 24 April 2018; Accepted: 12 June 2018; Published: 18 June 2018

Abstract: We revisit the concept of entanglement within the Bohmian approach to quantum mechanics. Inspired by Bohmian dynamics, we introduce two partial measures for the amount of entanglement corresponding to a pure state of a pair of quantum particles. One of these measures is associated with the statistical correlations exhibited by the joint probability density of the two Bohmian particles in configuration space. The other partial measure corresponds to the correlations associated with the phase of the joint wave function, and describes the non-separability of the Bohmian velocity field. The sum of these two components is equal to the total entanglement of the joint quantum state, as measured by the linear entropy of the single-particle reduced density matrix.

Keywords: Bohmian dynamics; entanglement indicators; linear entropy

1. Introduction

The de Broglie-Bohm approach to quantum mechanics [1–4], also referred to as the pilot-wave theory, or the quantum theory of motion, has been, since the publication of the seminal works by Bohm, a subject of constant interest in the field of the foundations of quantum mechanics. Indeed, there has been a sustained research activity on the de Broglie-Bohm formulation along the years [5–29]. In an article reviewing and celebrating the first 100 years of quantum physics, Tegmark and Wheeler included the formulation of Bohmian mechanics within the list of the most significant events in the development of this field of Science [30]. The Bohmian approach has provided stimulating new perspectives on several fundamental aspects of quantum physics, among which we can mention the quantum measurement problem [5–7], quantum chaos [8], entanglement [9,31,32], the thermal equilibrium of quantum systems [10], the concept of quantum work [11], quantum cosmology [12–15], quantum gravity [16], and quantum chemistry (in the latter case, both at the practical [17,18] and at the conceptual-philosophical [19] levels). The Bohmian point of view also constitutes the starting point of possible extensions of quantum theory, as the intriguing proposal made by Valentini illustrates, leading to specific quantitative astrophysical and cosmological predictions that might be within the reach of observational tests [20,21]. The formalism of Bohm theory has also been applied to the treatment of problems in thermal physics, such as the classical Hamilton-Jacobi formulation of Fourier heat conduction [22].

In Bohm's model of quantum mechanics the particles constituting a physical system have well defined positions in configuration space. The full description of the system is given by the particles' configuration (i.e., the particles' positions) and by a many-particle wave function. The particles' configuration evolution is determined by the wave function through a "guiding" equation, while the wave function evolves according to the standard many-particle Schrödinger equation. Even though the particles are assumed to have, at each time, well defined positions, knowledge about these positions is not accessible. All we can know about these positions is their probability density distribution, given by the squared modulus of the wave function. The position observable plays a dominant role within the Bohmian formulation. In particular, within this approach, the outcome of an experiment in which any physical observable is measured is registered by the final configuration of the particles of the experimental apparatus. The Bohmian model thus highlights a basic feature of any measurement process, which is that "... all experiments and certainly all measurements in physics are in the last analysis essentially kinematic, for they are ultimately based on observations of the position of a particle or of a pointer on a scale as a function of time" [33]. Other interesting recent approaches to the foundations of quantum mechanics, such as the entropic dynamics formulation proposed by Caticha [34], also stress the special role played by the position observable. It is worth emphasizing that Bohm's formulation of the quantum measurement process is fully consistent with Born's rule in standard quantum mechanics and, consequently, the experimental predictions of Bohm's theory coincide with those of the usual quantum mechanical formalism (see, however, Valentini's Bohm-based proposal for an extension of quantum theory [20,21]). The basic quantal non-locality is explicitly expressed in the de Broglie-Bohm formalism. In point of fact, the Bell inequalities were inspired by Bell's reaction to the work of Bohm [35].

In spite of the intense research work that has been devoted to Bohmian dynamics and its applications, relatively little attention has been paid to the quantitative analysis of entanglement within the Bohmian approach. The aim of the present contribution is to advance two quantitative indicators of the entanglement between two Bohmian particles. These quantities are explicitly formulated in terms of the Bohmian formalism. One of them corresponds to the statistical correlations exhibited by the joint probability density of the positions of the two particles. The other one is a quantitative measure of the non-separability of the Bohmian flow in the two-particle configuration space. We explore the main properties of these indicators and, as an illustrative application, we use these measures to investigate the decoherence-like process associated with two particles evolving under the effect of quantum friction.

The paper is organized as follows. In Section 2 we provide a brief review of the Bohmian theory. In Section 3 we introduce two quantitative indicators of entanglement within the Bohmian approach to quantum physics. In Section 4 we apply these measures to a system of two particles evolving under quantum friction. Finally, some conclusions are drawn in Section 5.

2. Bohmian Formulation of Quantum Dynamics

Bohmian dynamics includes, as one of its components, most of the formal apparatus of standard quantum mechanics. Indeed, a Bohmian quantum particle is endowed with a wave function $\psi(\mathbf{r}, t)$ governed by the Schrödinger equation. On top of this, the Bohmian particle has a definite position $\mathbf{r}$ that evolves in time according to the classical equation $\frac{d\mathbf{r}}{dt} = \frac{\mathbf{p}}{m} = \mathbf{v}$, where $\mathbf{p}$ and $\mathbf{v}$ respectively denote the particle's linear momentum and velocity. The position $\mathbf{r}$ of a Bohmian particle is the paradigmatic example of a hidden variable in a quantum theory. Within the Bohmian formulation it is assumed that the result of a position measurement is predetermined, even though it is not predictable. This state of affairs propagates to any other kind of physical measurement, since all of them translate, at some stage, into the position of some particles in the measuring device [34,35].

In spite of its classical flavor, there are fundamental differences between Bohmian dynamics and standard classical dynamics. In contrast to what happens in classical mechanics, the velocity $\mathbf{v}$ and the linear momentum $\mathbf{p}$ are not free variables anymore. They are instead determined, through the wave function $\psi(\mathbf{r}, t)$, by the particle's position $\mathbf{r}(t)$ at each time t. The particle moves according to a first order differential equation,

$$\frac{d\mathbf{r}}{dt} = \mathbf{v}(\mathbf{r}, t), \tag{1}$$

with the flow in configuration space given by the velocity field $\mathbf{v}(\mathbf{r}, t)$, determined by

$$\mathbf{v}(\mathbf{r}, t) = -\frac{i\hbar}{2m} \left[\frac{1}{\psi(\mathbf{r}, t)} \nabla(\psi(\mathbf{r}, t)) - \frac{1}{\psi^*(\mathbf{r}, t)} \nabla(\psi^*(\mathbf{r}, t)) \right], \tag{2}$$

where $\psi(\mathbf{r}, t)$ is a time-dependent solution of Schrödiner equation. When preparing a state of the particle, one neither has control over, nor knowledge of, the particular initial value adopted by $\mathbf{r}$. In this regard, the only accessible knowledge consists of the probability density $\rho(\mathbf{r}, t)$ corresponding to the different possible particle's positions, given by

$$\rho(\mathbf{r}, t) = |\psi(\mathbf{r}, t)|^2. \tag{3}$$

Associated with the configuration space flow (1) there is a Liouville-like continuity equation for the probability density $\rho(\mathbf{r}, t)$,

$$\frac{\partial \rho}{\partial t} + \nabla \cdot (\mathbf{v}\rho) = 0. \tag{4}$$

The form (2) of the velocity field, together with the Schrödinger equation and the continuity Equation (4), have an important consequence: if the probability density of initial positions of the Bohmian particle satisfies the relation (3), then it also satisfies this relation at all later times.

3. Entanglement Between Two Bohmian Particles: Configuration and Phase Entanglement

In this section, we are going to introduce two quantitative entanglement indicators, closely related to the main features of the Bohmian approach, for pure states of a pair of quantum particles. We are going to consider two spinless quantum particles moving in one spatial dimension (the extension of the present developments to D dimensions is straightforward) with coordinates denoted by x_1 and x_2. In terms of the standard quantum mechanical formalism, the two quantum particles are described by the pure state

$$|\psi\rangle = \int dx_1 dx_2 \psi(x_1, x_2) |x_1\rangle |x_2\rangle, \tag{5}$$

where $|x_1 x_2\rangle = |x_1\rangle |x_2\rangle$, $\psi(x_1, x_2) = \langle x_1, x_2 | \psi \rangle$ and $|x_i\rangle$ is an eigenstate of the i-particle position operator.

An important aspect of Bohm's approach is the assumption that at each point in time particles have well defined positions and, consequently, describe well defined orbits in configuration space, although the initial conditions associated with these orbits are not experimentally controllable. At the level of individual orbits of the pair of Bohmian particles that we are discussing here, quantum entanglement manifest itself by the fact that the velocity field $\mathbf{v}(x_1, x_2)$ is not separable. That is, $\mathbf{v}(x_1, x_2) \neq (v_1(x_1), v_2(x_2))$. In other words, each of the two components (v_1, v_2) of the vector field $\mathbf{v}$ depends, in general, on both particles' coordinates (x_1, x_2). This means that the behaviors of both particles are intertwined. Roughly speaking, one can say that each particle affects the behavior of the other one, even if there is no interaction potential involved, and the particles are far apart. This state of affairs is highly counter-intuitive and has been the focus of considerable attention in the literature. In fact, virtually all discussions of entanglement within the Bohmian approach have dealt, in one way or another, with this aspect of Bohmian dynamics (see, for instance [31,32] and references therein). However, and in spite of its great theoretical-philosophical interest, the study of individual Bohmian trajectories does not lend itself to a quantitative characterization of the amount of entanglement associated with a two-particle system at a given time. In point of fact, at a given instant t, two Bohmian particles following an individual trajectory are located at a specific point in configuration space with coordinates $(x_1(t), x_2(t))$, where the velocity field is given by a specific vector $\mathbf{v}(x_1, x_2)$. Now, it is not possible to assess the amount of non-separability that a vector field has at a particular location (x_1, x_2). The non-separability of a vector field is a global property that is associated with a region of configuration space. A sensible quantitative indicator of this non-separability can then be given by an average value, evaluated over such a region. In addition, Bohm's theory does involve a probability density in configuration space: the probability density $\rho(x_1, x_2) = |\psi(x_1, x_2)|^2$ of having the Bohmian particles at different locations. Consequently, it is reasonable to expect that an appropriate quantitative measure of the degree of non-separability of the Bohmian velocity field should be some sort of spatial average of non-sperability, related to the configuration space density $\rho(x_1, x_2)$ associated with the particles' positions. We thus see that it seems inevitable that the probability density in configuration space has to be involved in a quantitative treatment of entanglement within Bohmian mechanics. Now, once this density has been incorporated to the discussion, there appears another contribution to entanglement that has to be taken into account, which is the (classical-like) statistical correlations present in the configuration space probability density itself. Mathematically, these correlations are given by the non-factorizability of the density, $\rho(x_1, x_2) \neq \rho_1(x_1)\rho_2(x_2)$. On the basis of these considerations we are going to propose two indicators of the amount of entanglement of two particles at a given time, that provide quantitative measures of the above mentioned aspects of Bohmian mechanics: the non-separability of the velocity field and the classical-like correlations of the probability density in configuration space. It is worth stressing that, even though we are not going to refer explicitly to individual Bohmian orbits, our entanglement indicators are directly related to the above explained essential aspects of the Bohmian approach which originate, in turn, from the assumption that individual orbits exists. We can make here an analogy with the Gibbs approach to classical statistical mechanics. When using the canonical statistical ensemble to describe a system at thermal equilibrium one does not refer explicitly to individual orbits of a Hamiltonian system. However, it is clear that Hamiltonian dynamics still plays a fundamental role in, and is indeed at the foundations of, the canonical formulation of statistical mechanics.

The density matrix corresponding to the pure state (5) is

$$\hat{\rho} = |\psi\rangle\langle\psi| = \int dx_1 dx_2 dx_1' dx_2' \psi(x_1, x_2)\psi^*(x_1', x_2')|x_1 x_2\rangle\langle x_1' x_2'| \tag{6}$$

and has matrix elements

$$\langle x_1 x_2|\hat{\rho}|x_1' x_2'\rangle = \psi(x_1, x_2)\psi^*(x_1', x_2'). \tag{7}$$

The density matrix $\hat{\rho}$ should not be confused with the spatial density $\rho(x_1, x_2)$ mentioned before. The spatial density corresponds to the diagonal elements of the operator $\hat{\rho}$, that is $\rho(x_1, x_2) = \langle x_1 x_2|\hat{\rho}|x_1 x_2\rangle$. The marginal density matrix $\hat{\rho}_1 = \text{Tr}_2\hat{\rho}$ describing particle 1, has matrix elements

$$\begin{aligned} \langle x_1|\hat{\rho}_1|x_2\rangle &= \int \langle x_1 x_3|\hat{\rho}|x_2 x_3\rangle dx_3 \\ &= \int \psi(x_1, x_3)\psi^*(x_2, x_3) dx_3. \end{aligned} \tag{8}$$

The linear entropy of $\hat{\rho}_1$ constitutes a useful measure or indicator of the amount of entanglement of the state ψ,

$$\mathcal{E} = 1 - \text{Tr}(\hat{\rho}_1^2), \tag{9}$$

where one has,

$$\begin{aligned} \text{Tr}(\hat{\rho}_1^2) &= \int \langle x_1|\hat{\rho}_1|x_2\rangle\langle x_2|\hat{\rho}_1|x_1\rangle dx_1 dx_2 \\ &= \int \psi(x_1, x_3)\psi^*(x_2, x_3)\psi^*(x_1, x_4)\psi(x_2, x_4) dx_1 dx_2 dx_3 dx_4. \end{aligned} \tag{10}$$

Alternatively, we can express this entanglement indicator in terms of the marginal density matrix $\hat{\rho}_2 = \text{Tr}_1\hat{\rho}$ corresponding to particle 2, so one has

$$\mathcal{E} = 1 - \text{Tr}(\hat{\rho}_1^2) = 1 - \text{Tr}(\hat{\rho}_2^2). \tag{11}$$

The quantity $\mathcal{E}$ constitutes a useful, practical way to assess quantitatively the amount of entanglement exhibited by a two-particle pure state. In fact, it has been applied to the study of quantum entanglement in many different settings (see, for instance, [36]) due to its various computational advantages, both from the analytical and from the numerical points of view. However, this quantity does not have a clear interpretation in terms of the Bohmian theory. Our aim, inspired by the Bohmian approach to quantum mechanics, is to decompose the entanglement indicator $\mathcal{E}$ into two parts having a clear meaning in terms of the two basic ingredients of the dynamics of two Bohmian particles: their joint probability density in configuration space, and the joint velocity field describing the probability density flow. The above mentioned two contributions to the total entanglement, as measured by $\mathcal{E}$, constitute quantitative indicators of entanglement that we shall respectively call configuration entanglement $\mathcal{E}_c$, and phase entanglement, $\mathcal{E}_p$.

Following the Bohmian approach, we express the wave function as

$$\psi(x_1, x_2) = R^{\frac{1}{2}}(x_1, x_2) e^{i\alpha(x_1, x_2)}, \tag{12}$$

where R and α are both real functions, and $R \geq 0$. The quantity $R(x_1, x_2) = |\psi(x_1, x_2)|^2$ satisfies a continuity equation and represents the joint probability density of the two Bohmian particles in configuration space. The wave function (12) can be entangled through $R(x_1, x_2)$, through the phase $\alpha(x_1, x_2)$, or through both these quantities. Entanglement through the density R means that the probability density $R(x_1, x_2)$ cannot be factorized as $R(x_1, x_2) = R_1(x_1)R_2(x_2)$. This lack of factorizability, which we refer to as "configuration entanglement" corresponds to a correlation (in the classical sense) of the probability density $R(x_1, x_2)$. On the other hand, entanglement through the phase $\alpha(x_1, x_2)$ means that it cannot be additively decomposed as $\alpha(x_1, x_2) = \alpha_1(x_1) + \alpha_2(x_2)$. This lack of additive decomposability means that the Bohmian dynamical equations of particles 1 and 2 are not independent. To clarify this last point, lets consider the equations of motion of the two Bohmian particles,

$$\begin{aligned} \frac{dx_1}{dt} &= v_1(x_1, x_2) = \frac{\hbar}{m}\frac{\partial \alpha}{\partial x_1}, \\ \frac{dx_2}{dt} &= v_2(x_1, x_2) = \frac{\hbar}{m}\frac{\partial \alpha}{\partial x_2}. \end{aligned} \tag{13}$$

In general, if $\alpha(x_1, x_2) \neq \alpha_1(x_1) + \alpha_2(x_2)$, the two ordinary differential equations in (13) are coupled to each other. On the contrary, if $\alpha(x_1, x_2) \neq \alpha_1(x_1) + \alpha_2(x_2)$, these two equations of motion decouple, and one has,

$$\begin{aligned} \frac{dx_1}{dt} &= v_1(x_1) = \frac{\hbar}{m}\frac{d\alpha_1}{dx_1}, \\ \frac{dx_2}{dt} &= v_2(x_2) = \frac{\hbar}{m}\frac{d\alpha_2}{dx_2}. \end{aligned} \tag{14}$$

That is, in the latter case the equations of motion of the two Bohmian particles separate into two independent equations.

In the following two Subsections we are going to propose two quantitative indicators for configuration entanglement, and for phase entanglement, and determine some of their basic properties.

3.1. Configuration Entanglement

As a quantitative indicator of entanglement we propose,

$$\mathcal{E}_c = 1 - \int dx_1 dx_2 dx_3 dx_4 \sqrt{R(x_1, x_3)R(x_2, x_3)R(x_1, x_4)R(x_2, x_4)}. \tag{15}$$

The quantity $\mathcal{E}_c$ can be interpreted as an indicator of classical correlations in the probability density $R(x_1, x_2)$. Indeed, if this density is factorizable, $R(x_1, x_2) = R_1(x_1)R_2(x_2)$ we have,

$$\int dx_1 dx_2 dx_3 dx_4 \sqrt{R(x_1, x_3)R(x_2, x_3)R(x_1, x_4)R(x_2, x_4)}$$
$$= \int dx_1 dx_2 dx_3 dx_4 R_1(x_1)R_1(x_2)R_2(x_3)R_2(x_4) = 1 \tag{16}$$

due to normalization. Therefore, in the case that $R(x_1, x_2)$ is factorizable we have $\mathcal{E}_c = 0$. Now, we have

$$\int dx_1 dx_2 dx_3 dx_4 \sqrt{R(x_1, x_3)R(x_2, x_3)R(x_1, x_4)R(x_2, x_4)}$$
$$= \int dx_1 dx_2 \left\{ \int dx_3 \sqrt{R(x_1, x_3)R(x_2, x_3)} \right\} \left\{ \int dx_4 \sqrt{R(x_1, x_4)R(x_2, x_4)} \right\}$$
$$= \int dx_1 dx_2 \left\{ \int dx_3 \sqrt{R(x_1, x_3)R(x_2, x_3)} \right\}^2$$
$$\leq \int dx_1 dx_2 \left\{ \int dx_3 R(x_1, x_3)) \right\} \left\{ \int dx_4 R(x_2, x_4) \right\}$$
$$= \int dx_1 dx_2 dx_3 dx_4 R(x_1, x_3)R(x_2, x_4)$$
$$= 1. \tag{17}$$

The inequality in (17) is due to the Schwartz inequality and the final equality is due to the normalization of $R(x_1, x_2)$. It follows from (15) and (17) that the configuration entanglement $\mathcal{E}_c$ is always bounded according to

$$0 \leq \mathcal{E}_c \leq 1, \tag{18}$$

achieving its lowest bound (that is, vanishing) when the joint probability density in configuration space is factorizable.

3.2. Phase Entanglement

As a quantitative indicator of the amount of phase entanglement we propose,

$$\mathcal{E}_p = \int dx_1 dx_2 dx_3 dx_4 \{1 - \exp[i\{\alpha(x_1, x_3) - \alpha(x_2, x_3) - \alpha(x_1, x_4) + \alpha(x_2, x_4)\}]\} \times$$
$$\times R^{\frac{1}{2}}(x_1, x_3)R^{\frac{1}{2}}(x_2, x_3)R^{\frac{1}{2}}(x_1, x_4)R^{\frac{1}{2}}(x_2, x_4). \tag{19}$$

We have that

$$\mathcal{E}_p = \left\{ \int dx_1 dx_2 dx_3 dx_4 \left| \psi(x_1, x_3)\psi^*(x_2, x_3)\psi^*(x_1, x_4)\psi(x_2, x_4) \right| \right\}$$
$$- \left| \left\{ \int dx_1 dx_2 dx_3 dx_4 \psi(x_1, x_3)\psi^*(x_2, x_3)\psi^*(x_1, x_4)\psi(x_2, x_4) \right\} \right|$$
$$\geq 0. \tag{20}$$

In this last equation we used the fact that

$$\int dx_1 dx_2 dx_3 dx_4 \psi(x_1, x_3)\psi^*(x_2, x_3)\psi^*(x_1, x_4)\psi(x_2, x_4) = \mathrm{Tr}(\hat{\rho}_1^2) \tag{21}$$

is always a real positive number. It also follows from (17) and (19) (remembering that $R^{\frac{1}{2}}(x_1, x_2) = |\psi(x_1, x_2)|$) that $\mathcal{E}_p \leq 1$. Summing up, we have

$$0 \leq \mathcal{E}_p \leq 1. \tag{22}$$

In the case that the phase $\alpha(x_1, x_2)$ is additively decomposable, $\alpha(x_1, x_2) = \alpha_1(x_1) + \alpha_2(x_2)$, we have

$$
\begin{aligned}
&\alpha(x_1, x_3) - \alpha(x_2, x_3) - \alpha(x_1, x_4) + \alpha(x_2, x_4) \\
&= \alpha_1(x_1) + \alpha_2(x_3) - \alpha_1(x_2) - \alpha_2(x_3) - \alpha_1(x_1) - \alpha_2(x_4) + \alpha_1(x_2) + \alpha_2(x_4) = 0,
\end{aligned} \tag{23}
$$

implying that $\mathcal{E}_p = 0$. On the other hand, if $\mathcal{E}_p = 0$ we must have that

$$
\exp\left[i\left\{\alpha(x_1, x_3) - \alpha(x_2, x_3) - \alpha(x_1, x_4) + \alpha(x_2, x_4)\right\}\right] = \exp\left[i\delta\right] \tag{24}
$$

for some real constant δ. Therefore (assuming $\alpha(x_1, x_2)$ to be a continuous function) we have

$$
\alpha(x_1, x_3) - \alpha(x_2, x_3) - \alpha(x_1, x_4) + \alpha(x_2, x_4) = \delta. \tag{25}
$$

This relation must hold for all values of x_1, x_2, x_3, x_4. Therefore, fixing some constant values x_{20} and x_{40} for x_2 and x_4, respectively, we can write

$$
\alpha(x_1, x_3) = \alpha(x_1, x_{40}) + \alpha(x_{20}, x_3) - \alpha(x_{20}, x_{40}) + \delta. \tag{26}
$$

Defining now

$$
\begin{aligned}
\alpha_1(x_1) &= \alpha(x_1, x_{40}) - \frac{1}{2}\alpha(x_{20}, x_{40}) + \frac{\delta}{2} \\
\alpha_2(x_3) &= \alpha(x_{20}, x_3) - \frac{1}{2}\alpha(x_{20}, x_{40}) + \frac{\delta}{2}
\end{aligned} \tag{27}
$$

we have

$$
\alpha(x_1, x_3) = \alpha_1(x_1) + \alpha_2(x_3) \tag{28}
$$

and therefore the function α is additively decomposable. In summary, $\mathcal{E}_p = 0$ if and only if α is additively decomposable.

Finally, it can be verified after some algebra that the total entanglement of the two-particle pure quantum state, as measured by $\mathcal{E}$ (given by (9)), is equal to the sum of the entanglement of the configuration and the phase contributions,

$$
\mathcal{E} = \mathcal{E}_c + \mathcal{E}_p. \tag{29}
$$

Please note that the bounds (18) and (22), respectively satisfied by the configuration and the phase entanglement, are consistent with the bounds $0 \le \mathcal{E} \le 1$ satisfied by the total entanglement $\mathcal{E}$. For a factorizable quantum (pure) state of the two particles, both contributions $\mathcal{E}_c$ and $\mathcal{E}_p$ to the entanglement between the particles achieve their respective lower bounds (that is, both of them vanish). However, the configuration and the phase entanglements cannot both achieve their upper bounds (they cannot both be equal to 1) because the total entanglement satisfies $\mathcal{E}_c + \mathcal{E}_p \le 1$. This means that a state with high configuration entanglement must have low phase entanglement, and vice versa.

4. Entanglement Dynamics and Quantum Friction

We shall now apply the configuration and phase quantitative entanglement indicators to explore the entanglement dynamics of two quantum particles evolving according to a nonlinear Schrödinger equation incorporating quantum friction effects. Nonlinear Schrödinger equations have attracted considerable attention in recent years, and have been applied to the description of diverse physical phenomena. Closely related to Bohmian dynamics is the nonlinear Schrödinger equation proposed by Nassar and Miret-Artés in [6],

$$
i\hbar\frac{\partial\psi(x, t)}{\partial t} = \left[H(x, t) + i\hbar\left(W_c(x, t) + W_f(x, t)\right)\right]\psi(x, t) \tag{30}
$$

where $H = -\frac{\hbar^2}{2m}\frac{\partial^2}{\partial x^2} + V(x)$ is the standard Hamiltonian for a quantum particle of mass m moving in one dimension under the potential $V(x)$. We also have,

$$W_c(x,t) = -\kappa\left[\ln|\psi(x,t)|^2 - \langle\ln|\psi(x,t)|^2\rangle\right], \tag{31}$$

and

$$W_f(x,t) = -\frac{\nu}{2}\left[\ln\frac{\psi(x,t)}{\psi^*(x,t)} - \left\langle\ln\frac{\psi(x,t)}{\psi^*(x,t)}\right\rangle\right], \tag{32}$$

where

$$\langle\ln|\psi(x,t)|^2\rangle = \int|\psi(x,t)|^2\ln|\psi(x,t)|^2 dx, \tag{33}$$

and

$$\left\langle\ln\frac{\psi(x,t)}{\psi^*(x,t)}\right\rangle = \int|\psi(x,t)|^2\ln\left(\frac{\psi(x,t)}{\psi^*(x,t)}\right)dx. \tag{34}$$

In the above equations κ is a constant related to the resolution of position measurement and ν is a friction coefficient. The nonlinear wave Equation (30) was proposed as an effective description of the dynamics of a quantum particle undergoing a process of continuous position measurement including dissipation effects [6]. The non-linear logarithmic term W_c advanced by Nassar in [5] was motivated by Mensky's path integral formulation of continuous quantum measurements [37], whereas the term W_f considered in [6] was inspired by Kostin's work in connection with friction in quantum systems [38]. Several aspects of these kind of nonlinear evolution equations have been investigated in [39–44]. In terms of Bohmian dynamics, the friction term W_f in the non-linear Schrödinger Equation (30) leads to a new term in the equations governing the evolution of the Bohmian velocity field, that can be interpreted as describing a drag force [6].

Please note that the friction effects described by the term W_f in the nonlinear Schrödinger Equation (30) occur at the level of pure states. That is, the evolution of a quantum pure state is affected by these friction effects, but the state stays pure as it evolves. The presence of the term W_f gives rise, for instance, to the decrease (dissipation) of the energy of the time-dependent state, but not to an increase of its entropy. Equation (30) is a Schrödinger-Langevin-like equation without the stochastic force term. These kind of equations have been the focus of considerable attention and applied to diverse problems [4,6]. Nonlinear wave equations like (30) can be regarded as incorporating phenomenological descriptions of friction that describe only some aspects of the dynamics of an open system (for instance, energy dissipation). In this sense, they differ from approaches based on master equations where, in general, the entropy of the system changes as it evolves. A classical analogue may contribute to clarify this situation. A classical conservative system, such as a standard harmonic oscillator, is both conservative and deterministic. Energy is conserved during evolution, and complete knowledge of the initial conditions (represented by a point in phase space) fully determines the state of the system at a later time (represented, again, by a point in phase space). In summary: a point in phase space deterministically evolves into another point in phase space. If we incorporate friction effects described by drag forces (for instance, of the form $\mathbf{F} = -a\mathbf{v}$) the system is no longer conservative. Energy is not conserved. However, the system is still deterministic: a point in phase space still evolves into another well defined point in phase space. If one also incorporates stochastic forces (like in the Langevin equation) the system is no longer deterministic. Even if the initial conditions correspond to a point in phase space, to describe the evolution of the system one needs a time-dependent probability density in phase space. A quantum mechanical analogue of this situation is an open quantum system that evolves, for instance, according to the Linblad equation, and that has to be described by a time-dependent density matrix. On the other hand, a wave equation like (30) (or, more specifically, Equation (37) that we are going to consider later), governing the evolution of a quantum system that is at all times described by a pure state, can be regarded as a quantum mechanical analogue of a classical system that is affected by friction (drag forces) but is still deterministic. These classical systems have practical, conceptual, and historical relevance, and their properties have been the focus of investigation since long ago (a nice discussion can be found in Chapters 19 and 20 of [45]). Consequently, it is an interesting problem to explore their possible quantum mechanical counterparts.

We are now going to apply the configuration and phase entanglement indicators previously introduced, to explore the entanglement dynamics of two quantum particles governed by a particular two-dimensional version of the wave Equation (30), incorporating the friction term W_f, but not the nonlinear logarithmic term W_c. That is, in (30) we set $\kappa = 0$. We consider the evolution of a two-dimensional Gaussian wave packet describing two (entangled) particles subjected to friction and moving in one spatial dimension in a common harmonic potential well. The Gaussian ansatz is

$$\Psi(x_1, x_2, t) = e^{\lambda_0(t) + \lambda_1(t)x_1^2 + \lambda_2(t)x_2^2 + \lambda_3(t)x_1x_2 + \lambda_4(t)x_1 + \lambda_5(t)x_2}, \tag{35}$$

where the coefficients $\lambda_j(t)$ are complex functions of time,

$$\lambda_j(t) = \lambda_{jR}(t) + i\lambda_{jI}(t) \qquad \text{for} \quad j = 0, \dots, 5 \tag{36}$$

with $\lambda_{jR}, \lambda_{jI} \in \mathbb{R}$. The $\lambda_j(t)$'s are then chosen so that (35) is a solution of a non-linear Schrödinger equation incorporating a friction term. This evolution equation is given by

$$i\hbar\frac{\partial\Psi(x_1, x_2, t)}{\partial t} = \left[-\frac{\hbar^2}{2m}\left(\frac{\partial^2}{\partial x_1^2} + \frac{\partial^2}{\partial x_2^2}\right) + \frac{1}{2}m\omega^2(x_1^2 + x_2^2) + i\hbar W_f(x_1, x_2, t)\right]\Psi(x_1, x_2, t), \tag{37}$$

where

$$W_f(x_1, x_2, t) = -\frac{\nu}{2}\left[2i\alpha(x_1, x_2, t) - 2i\int dx_1 dx_2 R(x_1, x_2, t)\alpha(x_1, x_2, t)\right]. \tag{38}$$

The phase $\alpha(x_1, x_2, t)$ can be expressed in terms of the imaginary parts of the λ_j's,

$$\alpha(x_1, x_2, t) = \lambda_{0I}(t) + \lambda_{1I}(t)x_1^2 + \lambda_{2I}(t)x_2^2 + \lambda_{3I}(t)x_1x_2 + \lambda_{4I}(t)x_1 + \lambda_{5I}(t)x_2, \tag{39}$$

while $R(x_1, x_2)$ is given by the real parts of the λ_j's,

$$R(x_1, x_2, t) = e^{2(\lambda_{0R}(t) + \lambda_{1R}(t)x_1^2 + \lambda_{2R}(t)x_2^2 + \lambda_{3R}(t)x_1x_2 + \lambda_{4R}(t)x_1 + \lambda_{5R}(t)x_2)}. \tag{40}$$

Inserting the ansatz (35) into the nonlinear Schrödinger Equation (37) it is possible to prove, after some algebra, that (35) constitutes a time dependent solution of (37), provided that the real and imaginary parts of the $\lambda_j(t)$'s comply with the following set of twelve coupled ordinary differential equations,

$$\lambda'_{0R}(t) + \frac{\hbar}{m}\Big(\lambda_{1I}(t) + \lambda_{2I}(t) + \lambda_{4R}(t)\lambda_{4I}(t) + \lambda_{5R}(t)\lambda_{5I}(t)\Big) = 0$$

$$\lambda'_{0I}(t) + \frac{\hbar}{m}\Big(-\lambda_{1R}(t) - \lambda_{2R}(t) - \tfrac{1}{2}\lambda_{4R}^2(t) + \tfrac{1}{2}\lambda_{4I}^2(t) - \tfrac{1}{2}\lambda_{5R}^2(t) + \tfrac{1}{2}\lambda_{5I}^2(t)\Big) + \hbar\nu\Big(\lambda_{0I}(t) - \langle\alpha\rangle(t)\Big) = 0$$

$$\lambda'_{1R}(t) + \frac{\hbar}{m}\Big(4\lambda_{1R}(t)\lambda_{1I}(t) + \lambda_{3R}(t)\lambda_{3I}(t)\Big) = 0$$

$$\lambda'_{1I}(t) + \frac{\hbar}{m}\Big(-2\lambda_{1R}^2(t) + 2\lambda_{1I}^2(t) - \tfrac{1}{2}\lambda_{3R}^2(t) + \tfrac{1}{2}\lambda_{3I}^2(t)\Big) + \tfrac{1}{2}\frac{m\omega^2}{\hbar} + \hbar\nu\lambda_{1I}(t) = 0$$

$$\lambda'_{2R}(t) + \frac{\hbar}{m}\Big(4\lambda_{2R}(t)\lambda_{2I}(t) + \lambda_{3R}(t)\lambda_{3I}(t)\Big) = 0$$

$$\lambda'_{2I}(t) + \frac{\hbar}{m}\Big(-2\lambda_{2R}^2(t) + 2\lambda_{2I}^2(t) - \tfrac{1}{2}\lambda_{3R}^2(t) + \tfrac{1}{2}\lambda_{3I}^2(t)\Big) + \tfrac{1}{2}\frac{m\omega^2}{\hbar} + \hbar\nu\lambda_{2I}(t) = 0 \tag{41}$$

$$\lambda'_{3R}(t) + \frac{\hbar}{m}\Big(2\lambda_{1R}(t)\lambda_{3I}(t) + 2\lambda_{1I}(t)\lambda_{3R}(t) + 2\lambda_{2R}(t)\lambda_{3I}(t) + 2\lambda_{2I}(t)\lambda_{3R}(t)\Big) = 0$$

$$\lambda'_{3I}(t) + \frac{\hbar}{m}\Big(-2\lambda_{1R}(t)\lambda_{3R}(t) + 2\lambda_{1I}(t)\lambda_{3I}(t) - 2\lambda_{2R}(t)\lambda_{3R}(t) + 2\lambda_{2I}(t)\lambda_{3I}(t)\Big) + \hbar\nu\lambda_{3I}(t) = 0$$

$$\lambda'_{4R}(t) + \frac{\hbar}{m}\Big(2\lambda_{1R}(t)\lambda_{4I}(t) + 2\lambda_{1I}(t)\lambda_{4R}(t) + \lambda_{3R}(t)\lambda_{5I}(t) + \lambda_{3I}(t)\lambda_{5R}(t)\Big) = 0$$

$$\lambda'_{4I}(t) + \frac{\hbar}{m}\Big(-2\lambda_{1R}(t)\lambda_{4R}(t) + 2\lambda_{1I}(t)\lambda_{4I}(t) - \lambda_{3R}(t)\lambda_{5R}(t) + \lambda_{3I}(t)\lambda_{5I}(t)\Big) + \hbar\nu\lambda_{4I}(t) = 0$$

$$\lambda'_{5R}(t) + \frac{\hbar}{m}\Big(2\lambda_{2R}(t)\lambda_{5I}(t) + 2\lambda_{2I}(t)\lambda_{5R}(t) + \lambda_{3R}(t)\lambda_{4I}(t) + \lambda_{3I}(t)\lambda_{4R}(t)\Big) = 0$$

$$\lambda'_{5I}(t) + \frac{\hbar}{m}\Big(-2\lambda_{2R}(t)\lambda_{5R}(t) + 2\lambda_{2I}(t)\lambda_{5I}(t) - \lambda_{3R}(t)\lambda_{4R}(t) + \lambda_{3I}(t)\lambda_{4I}(t)\Big) + \hbar\nu\lambda_{5I}(t) = 0,$$

where

$$\lambda'_{jR,jI}(t) = \frac{d}{dt}\lambda_{jR,jI}(t). \tag{42}$$

The set of coupled, non-linear, first-order, ordinary differential Equation (41) need to be solved numerically. Now, $\lambda_{0R}(t)$ is determined by the normalization of the state of the system (35). So, by imposing the condition of normalization on the system, we can obtain an expression for $\lambda_{0R}(t)$ in terms of the other λ's:

$$
\begin{aligned}
1 &= \int dx_1 dx_2 \Psi^*(x_1, x_2, t)\Psi(x_1, x_2, t) \\
&= e^{2\lambda_{0R}(t)}\int dx_1 e^{2\lambda_{1R}(t)x_1^2 + 2\lambda_{4R}(t)x_1}\int dx_2 e^{2\lambda_{2R}(t)x_2^2 + 2\lambda_{3R}(t)x_1 x_2 + 2\lambda_{5R}(t)x_2}.
\end{aligned} \tag{43}
$$

The integral appearing in the right hand side of the above equation only converges if

$$\lambda_{1R}(t) < 0, \quad \lambda_{2R}(t) < 0 \quad \text{and} \quad \lambda_{3R}^2(t) < 4\lambda_{1R}(t)\lambda_{2R}(t). \tag{44}$$

Evaluating this integral and solving for λ_{0R} gives

$$
\begin{aligned}
\lambda_{0R}(t) &= \frac{1}{2}\ln\left(\frac{\sqrt{4\lambda_{1R}(t)\lambda_{2R}(t) - \lambda_{3R}^2(t)}}{\pi}\right) \\
&\quad + \frac{\lambda_{1R}(t)\lambda_{5R}^2(t) + \lambda_{2R}(t)\lambda_{4R}^2(t) - \lambda_{3R}(t)\lambda_{4R}(t)\lambda_{5R}(t)}{4\lambda_{1R}(t)\lambda_{2R}(t) - \lambda_{3R}^2(t)}.
\end{aligned} \tag{45}
$$

Since a wave function which is normalized at some time t is then automatically normalized for all time, it is sufficient to impose these conditions on the initial values of the λ_j's.

Evaluating the nested integrals (see Equations (15) and (19)) in the expression for the total entanglement (9), results in the following expression:

$$\mathcal{E}(t) = 1 - \frac{\pi^2 e^{4\lambda_{0R}(t)}e^{-\frac{4\left(\lambda_{1R}(t)\lambda_{5R}^2(t) + \lambda_{2R}(t)\lambda_{4R}^2(t) - \lambda_{3R}(t)\lambda_{4R}(t)\lambda_{5R}(t)\right)}{4\lambda_{1R}(t)\lambda_{2R}(t) - \lambda_{3R}^2(t)}}}{\sqrt{16\lambda_{1R}^2(t)\lambda_{2R}^2(t) - 4\lambda_{1R}(t)\lambda_{2R}(t)\lambda_{3R}^2(t) + \lambda_{3I}^2(t)\left(4\lambda_{1R}(t)\lambda_{2R}(t) - \lambda_{3R}^2(t)\right)}}. \tag{46}$$

Substituting for $\lambda_{0R}(t)$ from (45) and simplifying leads to a final compact expression for the total entanglement of the system:

$$\mathcal{E}(t) = 1 - \sqrt{\frac{4\lambda_{1R}(t)\lambda_{2R}(t) - \lambda_{3R}^2(t)}{4\lambda_{1R}(t)\lambda_{2R}(t) + \lambda_{3I}^2(t)}}. \tag{47}$$

Notice that the entanglement only depends on $\lambda_{1R}(t)$, $\lambda_{2R}(t)$, $\lambda_{3R}(t)$ and $\lambda_{3I}(t)$. As $\lambda_{3I}(t)$ is squared in the total entanglement expression, the sign of $\lambda_{3I}(t)$ does not affect the total entanglement. By setting $\lambda_{3I}(0) = 0$ and $\lambda_{3R}(0) = 0$, the total entanglement $\mathcal{E}(t) = 0$ and so the state is separable and remains separable. The configuration and phase entanglement of a separable state are also zero and remain zero irrespective of the choice of parameters ν and ω. The total entanglement of the initial state can be chosen to be maximal by making the numerator inside the square root in Equation (47) as small as possible or the denominator as large as possible, which is achieved by choosing $\lambda_{3R}(0)$ such that it approaches $2\sqrt{\lambda_{1R}(0)\lambda_{2R}(0)}$ or by choosing $|\lambda_{3I}(0)|$ to be very large. This makes sense as we would expect the total entanglement of our initial state to depend strongly on $\lambda_3(0)$ since this is the coefficient of the cross-term in the Gaussian state (35).

Table 1. Table of initial conditions.

	Low Entanglement	Intermediately Entangled	Generally Entangled	Highly Entangled
$\mathcal{E}(0)$	0.086	0.395	0.70717	0.923
$\lambda_{0R}(0)$	$\frac{1}{2}\ln\left(\frac{\sqrt{91}}{5\pi}\right) - \frac{5}{7}$	$\frac{1}{2}\ln\left(\frac{2\sqrt{15}}{\pi}\right) - \frac{1}{6}$	-122.79117	$\frac{1}{2}\ln\left(\frac{2\sqrt{15}}{\pi}\right) - \frac{1}{6}$
$\lambda_{0I}(0)$	1	1	4.21132	1
$\lambda_{1R}(0)$	-1	-4	-3.29371	-4
$\lambda_{1I}(0)$	1	1	-3.51362	1
$\lambda_{2R}(0)$	-1	-4	-0.613171	-4
$\lambda_{2I}(0)$	1	1	-0.86058	1
$\lambda_{3R}(0)$	$\frac{3}{5}$	2	-2.54191	2
$\lambda_{3I}(0)$	$\frac{3}{5}$	10	-3.28337	100
$\lambda_{4R}(0)$	1	1	7.56778	1
$\lambda_{4I}(0)$	1	1	3.99994	1
$\lambda_{5R}(0)$	1	1	-4.69103	1
$\lambda_{5I}(0)$	1	1	3.07785	1

We investigated numerically the evolution of the entanglement indicators for numerous initial states and for a variety of values for ν and ω. We solved numerically the equations of motion (41) for the λ_j's and evaluated, on the corresponding time-dependent Gaussian wave packet, the configuration and phase entanglement indicators, $\mathcal{E}_c$ and $\mathcal{E}_p$. The conclusion was that the particular choices of initial states given in Table 1, and of parameters ν ($\nu = 0$ or 0.1) and ω ($\omega = 0$ or 1), are representative of the behaviour of states in general. For the initial state referred to as "generally entangled" (column 3 in Table 1) the coefficients were chosen randomly, whereas for the states with low, intermediate and high entanglement listed in Table 1, the coefficients were specifically chosen. When choosing these coefficients (randomly or otherwise) we actually choose all of them except λ_{0R}, which is calculated in terms of the other coefficients to satisfy normalization. The only difference between the coefficients of the intermediately entangled and highly entangled states is that $\lambda_{3I}(0)$ is first taken to be 10 and then 100. The time evolution of the entanglement indicators for the aforementioned initial states is illustrated in Figures 1–11. Please note that for computing the numerical results displayed in the Figures we have set $\hbar$ and m to unity (i.e. we use atomic units).

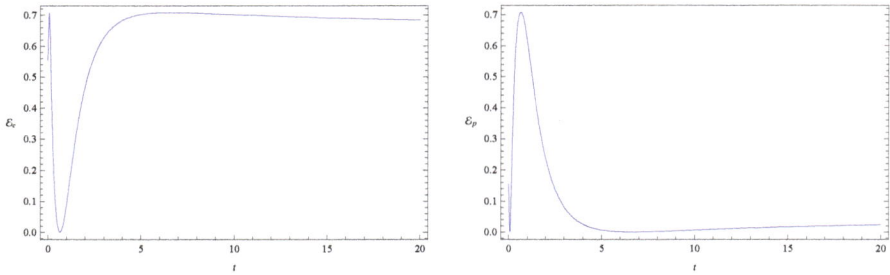

Figure 1. Plots of configuration entanglement $\mathcal{E}_c$ (left) and phase entanglement $\mathcal{E}_p$ (right) as a function of time for a free particle without friction ($\nu = 0$ and $\omega = 0$). The randomly chosen initial conditions are given in column 3 of Table 1. The total initial entanglement of the state is $\mathcal{E} = 0.70717$. The quantities $\mathcal{E}_c$ and $\mathcal{E}_p$ are dimensionless. Units of time, length and mass are chosen such that $\hbar = 1$ and $m = 1$.

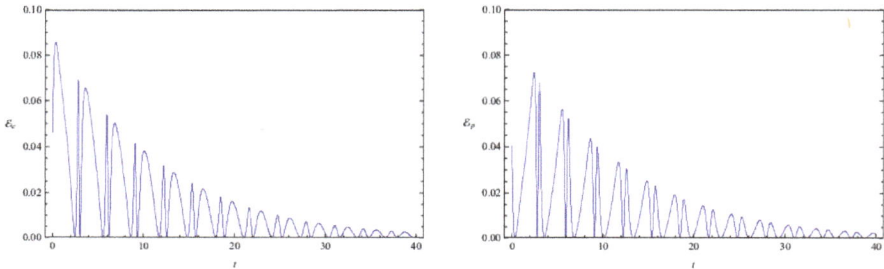

Figure 2. Configuration entanglement $\mathcal{E}_c$ (left) and phase entanglement $\mathcal{E}_p$ (right) as a function of time for an initial state of low entanglement. The initial conditions are given in column 1 of Table 1. The system parameters are $\nu = 0.1$ and $\omega = 1$, and the total initial entanglement for the state is $\mathcal{E} = 0.086$. The quantities are measured in the same units as in Figure 1.

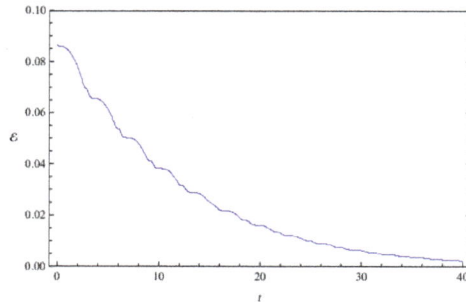

Figure 3. Total entanglement $\mathcal{E}$ as a function of time for the initial state of low entanglement considered in Figure 2. The system parameters are $\nu = 0.1$ and $\omega = 1$, and the total initial entanglement for the state is $\mathcal{E} = 0.086$. The quantities are measured in the same units as in Figure 1.

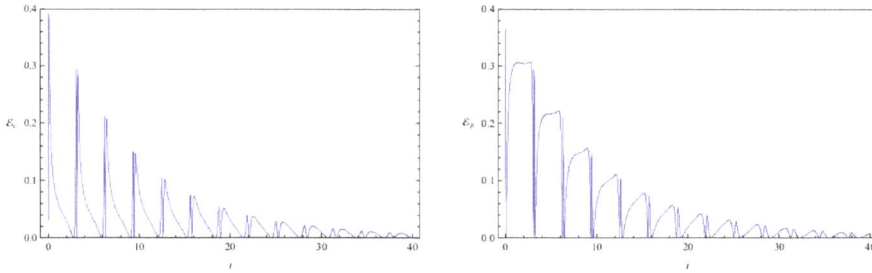

Figure 4. Configuration entanglement $\mathcal{E}_c$ (left) and phase entanglement $\mathcal{E}_p$ (right) as a function of time for an initial state of intermediate (representative) entanglement. The initial conditions are given in column 2 of Table 1. The system parameters are $\nu = 0.1$ and $\omega = 1$, and the total initial entanglement for the state is $\mathcal{E} = 0.395$. The quantities are measured in the same units as in Figure 1.

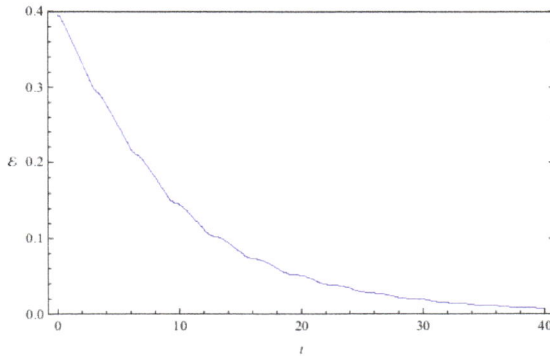

Figure 5. Total entanglement $\mathcal{E}$ as a function of time for the initial state of intermediate (representative) entanglement considered in Figure 4. The system parameters are $\nu = 0.1$ and $\omega = 1$, and the total initial entanglement for the state is $\mathcal{E} = 0.395$. The quantities are measured in the same units as in Figure 1.

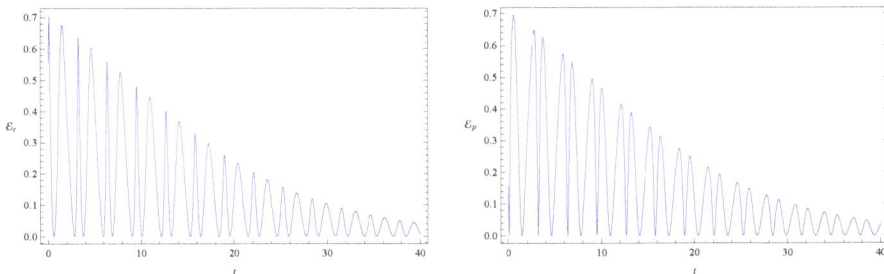

Figure 6. Plots of configuration entanglement $\mathcal{E}_c$ (left) and phase entanglement $\mathcal{E}_p$ (right) as a function of time for a randomly chosen initial state (initial conditions are given in column 3 of Table 1). The values of the physical parameters characterizing the system are $\nu = 0.1$ and $\omega = 1$, and the total initial entanglement for the state is $\mathcal{E} = 0.70717$. The quantities are measured in the same units as in Figure 1.

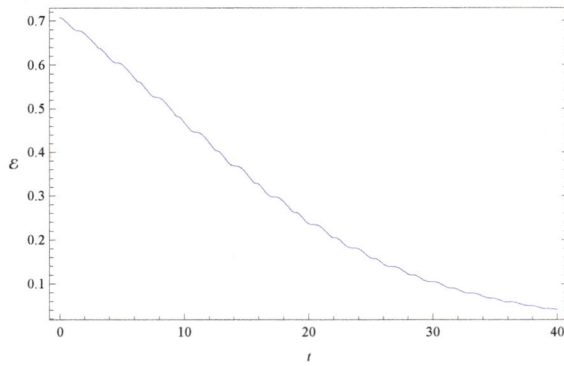

Figure 7. Total entanglement $\mathcal{E}$ as a function of time. The system parameters, initial conditions, and units used are the same as in Figure 6.

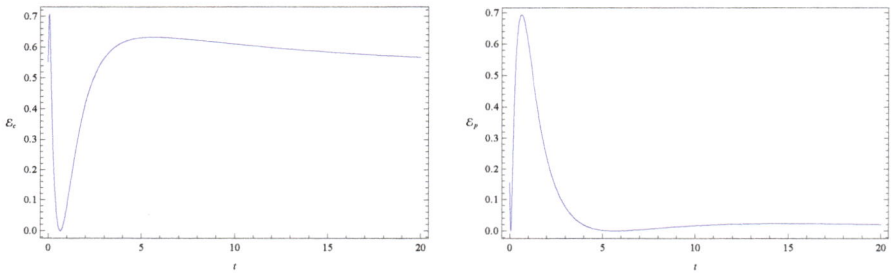

Figure 8. Plots of configuration entanglement $\mathcal{E}_c$ (left) and phase entanglement $\mathcal{E}_p$ (right) as a function of time for the same initial state as in Figures 6 and 7. The values of the physical parameters characterizing the system are $\nu = 0.1$ and $\omega = 0$, and the total initial entanglement for the state is $\mathcal{E} = 0.70717$. The quantities are measured in the same units as in Figure 1.

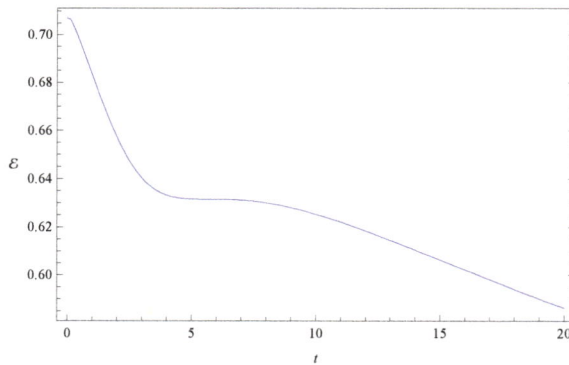

Figure 9. Total entanglement $\mathcal{E}$ as a function of time. The system parameters, initial conditions, and units used are the same as in Figure 8.

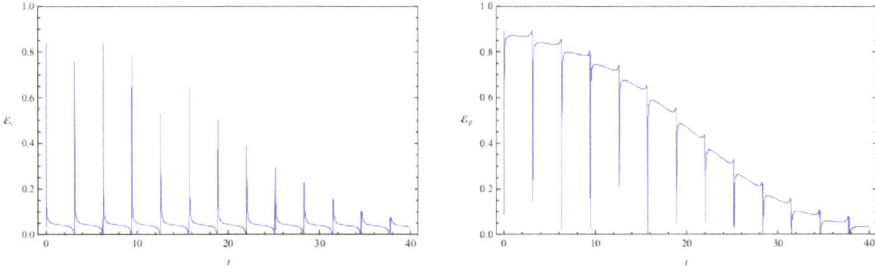

Figure 10. Plots of configuration entanglement $\mathcal{E}_c$ (left) and phase entanglement $\mathcal{E}_p$ (right) as a function of time for a highly entangled initial state (initial conditions are given in column 4 of Table 1). The system parameters are $\nu = 0.1$ and $\omega = 1$, and the total initial entanglement for the state is $\mathcal{E} = 0.923$.

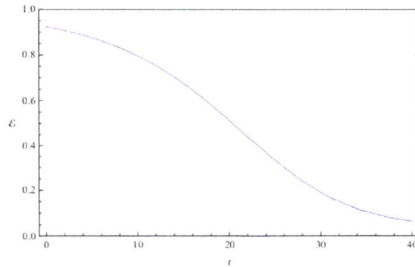

Figure 11. Total entanglement $\mathcal{E}$ as a function of time for the same highly entangled initial state considered in Figure 10. The system parameters are $\nu = 0.1$ and $\omega = 1$, and the total initial entanglement for the state is $\mathcal{E} = 0.923$.

As already mentioned, the time evolution of the entanglement indicators for the initial states listed in Table 1 is depicted in Figures 1–11. In Figure 1 we show the evolution of the configuration and phase indicators of entanglement for an entangled state of two free particles (that is, with no confining potential) moving under no friction. In this system the total entanglement $\mathcal{E}_c + \mathcal{E}_p$ is conserved, although $\mathcal{E}_c$ and $\mathcal{E}_p$ are individually time dependent. Consistently with the fact that $\mathcal{E}_c + \mathcal{E}_p$ is constant in time, we see in Figure 1 that the minima of one indicator coincides to maxima of the other one (at these points one has $(d\mathcal{E}_c/dt) = (d\mathcal{E}_p/dt) = 0$ and $(d^2\mathcal{E}_c/dt^2) = -(d^2\mathcal{E}_p/dt^2)$, implying that the extrema of one quantity coincide with the opposite extrema of the other one).

Figure 2 shows the time evolution of $\mathcal{E}_c$ and $\mathcal{E}_p$ for a pair of quantum particles evolving under the effect of friction while confined by a common harmonic potential well. The parameters characterizing the friction term and the harmonic potential are $\nu = 0.1$ and $\omega = 1$, respectively. The initial state, characterized by the coefficients appearing in the first column of Table 1, is a state of low entanglement with $\mathcal{E} = 0.086$. The time evolution of the total entanglement $\mathcal{E} = \mathcal{E}_c + \mathcal{E}_p$ is depicted in Figure 3 for the same system and initial state as in Figure 2. The time evolution of $\mathcal{E}_c$ and $\mathcal{E}_p$ is depicted in Figure 4, for a pair of particles starting with an initial state of intermediate entanglement ($\mathcal{E} = 0.395$), for the same system parameters as in Figures 2 and 3. The corresponding evolution of the total entanglement is plotted in Figure 5. The evolution of $\mathcal{E}_c$ and $\mathcal{E}_p$ for a pair of particles for the same system parameters as in Figures 4 and 5, for a randomly chosen initial state (corresponding to the third column in Table 1) is shown in Figure 6. The evolution of the total entanglement is exhibited in Figure 7. For the same initial conditions as in Figures 6 and 7, Figure 8 depicts the evolution of $\mathcal{E}_c$ and $\mathcal{E}_p$ for a pair of free particles (that is, with no external confining potential; $\omega = 0$) moving under the effect of friction ($\nu = 0.1$). The evolution of the total entanglement is shown in Figure 9. Finally, the evolution of $\mathcal{E}_c$

and $\mathcal{E}_p$ for a highly entangled initial state ($\mathcal{E} = 0.923$) with $\nu = 0.1$ and $\omega = 1$ is shown in Figure 10. The behavior of the corresponding total entanglement is depicted in Figure 11.

It can be appreciated in Figures 2, 4, and 10 that, when the particles evolve under friction in a common harmonic potential, both the configuration entanglement and the phase entanglement exhibit a strong oscillatory behaviour. The maximum values of one of the entanglement indicators tends to coincide with the minimum values of the other one. This behaviour is inherited from the corresponding behavior observed in the conservative case, which we have already discussed. This is due to the fact that, in the cases that we have studied, the time-scale of the energy dissipation is slower than the time-scale of the oscillations generated by the harmonic confining field. Consequently, during one complete harmonic period the energy stays approximately constant, and the system approximately behaves as in the conservative case.

The amplitude of the entanglement oscillations tends to decrease with time. This trend is due to the decrease in energy of the system, associated with the friction term in the nonlinear Schrödinger Equation (37). Notice that Equation (37) does not have a stochastic force term [4]. The solution of the equations of motion (41) for the λ-coefficients characterizing the evolving Gaussian wave packet, at large times, asymptotically evolves to $\lambda_{0I} = \lambda_{00} - \omega t$, $\lambda_{1R} = \lambda_{2R} = -\frac{m\omega}{2\hbar}$, λ_{0R} given by (45), and the rest of the λ's equal to zero. Here λ_{00} is a dimensionless constant. This asymptotic solution corresponds to the wave function,

$$\psi(x_1, x_2, t)_{\text{asympt}} = \sqrt{\frac{m\omega}{\hbar\pi}} \exp\left[i\left(\lambda_{00} - \omega t\right)\right] \exp\left[-\frac{m\omega}{2\hbar}\left(x_1^2 + x_2^2\right)\right], \tag{48}$$

which represents the ground state of the two particles in the harmonic potential $\frac{1}{2}m\omega^2\left(x_1^2 + x_2^2\right)$. It can be directly verified that (48) is a solution of (37). Please note that the wave function (48) describes a separable state. This is the reason, for the system here under consideration, that both entanglement indicators $\mathcal{E}_c$ and $\mathcal{E}_p$ tend to zero for large times. In summary, as the system looses energy due to friction, it relaxes towards its ground state, which has no entanglement.

The total entanglement, depicted in Figures 3, 5 and 11, also decreases in time, but in a more smooth way, with the oscillatory features highly attenuated. We see that in this system the amount of entanglement of the two-particle state (configuration entanglement, phase entanglement, and total entanglement) tends to decrease in time due to the dissipative effects. A similar decreasing trend can be observed in Figures 8 and 9, corresponding to a two-particle system experiencing friction but with no confining potential. However, in this case, where the common harmonic potential is absent, the oscillatory behaviour of $\mathcal{E}_c$ and $\mathcal{E}_p$ is less strong than the one exhibited in Figures 2, 4, and 10. The behaviour of the total entanglement in Figure 9 seems to have more structure than the corresponding behaviour in Figure 3. Some further conclusions from all the numerical investigations were that when $\omega \neq 0$, then for smaller initial total entanglement the periodicity for the total entanglement is more apparent compared to higher initial values for the total entanglement. Also, as ω increases the frequency of the entanglement oscillations (total entanglement, configuration entanglement and phase entanglement) increases, as expected for an oscillating system. As ν increases, the amplitude of the entanglement oscillations decreases more rapidly, as expected when friction plays a role.

5. Conclusions

We revisited the concept of entanglement within the Bohmian formulation of quantum mechanics. We introduced two partial measures for the amount of entanglement corresponding to a pure state of a pair of Bohmian quantum particles. These two quantities are directly related to the main ingredients of the Bohmian dynamics, and admit a clear interpretation in terms of that dynamics. One of these measures is associated with the statistical correlations exhibited by the joint probability density in configuration space corresponding to a pair of Bohmian particles. The other partial measure corresponds to the correlations associated with the phase of the joint wave function, and describes the non-separability of the Bohmian velocity field. We refer to these two measures, respectively, as the configuration entanglement indicator and the phase entanglement indicator. The sum of these two components is equal to the total entanglement of the joint quantum state, as measured by the linear entropy of the single-particle reduced density matrix. We investigated the main properties of the configuration and the phase entanglement indicators and, as an illustrative application, explored the time evolution of these quantities, corresponding to the dissipative dynamics of an initially entangled two-particle quantum system evolving under the effect of friction. The fact that the entanglement indicators advanced here are directly defined in terms of the elements of the Bohmian formalism allows for their application to the study of entanglement in extensions or modifications of Bohm's theory, such as the one recently advanced by Valentini [20,21], where some ingredients of the standard quantum formalism might be problematic. In the present work we have restricted our considerations to entanglement in pure states. It would be interesting to explore extensions to mixed states of the entanglement indicators explored here, although the Bohmian dynamics of mixed states is, in general, much less developed than that of pure states. Any further contributions along these or related lines of inquiry will be very welcome.

Author Contributions: Both authors contributed equally to this paper, and have read and approved the final manuscript.

Conflicts of Interest: The authors declare no conflict of interest.

References

1. Holland, P.R. *The Quantum Theory of Motion: An Account of the de Broglie-Bohm Causal Interpretation of Quantum Mechanics*; Cambridge University Press: Cambridge, UK, 1995.
2. Sanz, A.S.; Miret-Artés S. *A Trajectory Description of Quantum Processes. I. Fundamentals*; Springer: Berlin/ Heidelberg, Gemmery, 2012; Volume 850, pp. 1–299.
3. Sanz, A.S.; Miret-Artés S. *A Trajectory Description of Quantum Processes. II. Applications*; Springer: Berlin/ Heidelberg, Gemmery, 2014; Volume 831, pp. 1–333.
4. Nassar, A.B.; Miret-Artés S. *Bohmian Mechanics, Open Quantum Systems and Continuous Measurements*; Springer International Publishing: Basel, Switzerland, 2017; pp. 1–241.
5. Nassar, A.B. Quantum trajectories and the Bohm time constant. *Ann. Phys.* **2013**, *331*, 317–322. [CrossRef]
6. Nassar, A.B.; Miret-Artés S. Dividing line between quantum and classical trajectories in a measurement problem: Bohmian time constant. *Phys. Rev. Lett.* **2013**, *111*, 150401. [CrossRef] [PubMed]
7. Naaman-Marom, G.; Erez, N.; Vaidman, L. Position measurements in the de Broglie-Bohm interpretation of quantum mechanics. *Ann. Phys.* **2012**, *327*, 2522–2542. [CrossRef]
8. Efthymiopoulos, C.; Kalapotharakos, C.; Contopoulos, G. Origin of chaos near critical points of quantum flow. *Phys. Rev. E* **2009**, *79*, 036203. [CrossRef] [PubMed]
9. Ramsak, A. Geometrical view of quantum entanglement. *EPL* **2011**, *96*, 40004. [CrossRef]
10. Bennett, A.F. Relative dispersion and quantum thermal equilibrium in de Broglie-Bohm mechanics. *J. Phys. A Math. Theor.* **2010**, *43*, 195304. [CrossRef]
11. Sampaio, R.; Suomela, S.; Ala-Nissila, T.; Anders, J.; Philbin, T.G. Quantum work in the Bohmian framework. *Phys. Rev. A* **2018**, *97*, 012131. [CrossRef]
12. Peter, P.; Pinho, E.J.C.; Pinto-Neto, N. Noninflationary model with scale invariant cosmological perturbations. *Phys. Rev. D* **2007**, *75*, 023516. [CrossRef]
13. Pinto-Neto, N.; Falciano, F.T.; Pereira, R.; Santini, E.S. Wheeler-DeWitt quantization can solve the singularity problem. *Phys. Rev. D* **2012**, *86*, 063504. [CrossRef]
14. Pinto-Neto, N.; Santos, G.; Struyve, W. Quantum-to-classical transition of primordial cosmological perturbations in de Broglie-Bohm quantum theory. *Phys. Rev. D* **2012**, *85*, 083506. [CrossRef]
15. Letelier, P.S.; Pitelli, J.P.M. *n*-dimensional FLRW quantum cosmology. *Phys. Rev. D* **2010**, *82*, 104046. [CrossRef]
16. Resconi, G.; Licata, I.; Fiscaletti, D. Unification of quantum and gravity by non classical information entropy space. *Entropy* **2013**, *15*, 3602–3619. [CrossRef]

17. de Carvalho, F.F.; Bouduban, M.E.F.; Curchod, B.F.E.; Tavernelli, I. Nonadiabatic molecular dynamics based on trajectories. *Entropy* **2013**, *16*, 62–85. [CrossRef]
18. Benseny, A.; Albareda, G.; Sanz, A.S.; Mompart , J.; Oriols, X. Applied Bohmian mechanics. *Eur. Phys. J. D* **2014**, *68*, 286. [CrossRef]
19. Fortin, S.; Lombardi, O.; Martinez Gonzalez, J.C. The relationship between chemistry and physics from the perspective of Bohmian mechanics. *Found. Chem.* **2017**, *19*, 43–59. [CrossRef]
20. Valentini, A. Astrophysical and cosmological tests of quantum theory. *J. Phys. A Math. Theor.* **2007**, *40*, 3285. [CrossRef]
21. Valentini, A. Inflationary cosmology as a probe of primordial quantum mechanics. *Phys. Rev. D* **2010**, *82*, 063513. [CrossRef]
22. Márkus, F.; Gambár, K. Derivation of the upper limit of temperature from the field theory of thermodynamics. *Phys. Rev. E* **2004**, *70*, 055102. [CrossRef] [PubMed]
23. Valentini, A.; Westman, H. Dynamical origin of quantum probabilities. *arXiv* 2005, arXiv:quant-ph/0403034.
24. Pennini, F.; Plastino, A.R.; Plastino, A. Pilot wave approach to the NRT nonlinear Schrödinger equation. *Physica A* **2014**, *403*, 195–205. [CrossRef]
25. Plastino, A.R.; Wedemann, R.S. Nonlinear wave equations related to nonextensive thermostatistics. *Entropy* **2017**, *19*, 60. [CrossRef]
26. Nassar, A.B.; Miret-Artés, S. Bohmian trajectories of Airy packets. *Ann. Phys.* **2014**, *348*, 223–227. [CrossRef]
27. Holland, P. Computing the wavefunction from trajectories: Particle and wave pictures in quantum mechanics and their relation. *Ann. Phys.* **2005**, *315*, 505–531. [CrossRef]
28. Plastino, A.R.; Casas, M.; Plastino, A. Bohmian quantum theory of motion for particles with position-dependent effective mass. *Phys. Lett. A* **2001**, *281*, 297–304. [CrossRef]
29. Gisin, N. Why Bohmian mechanics? One- and two-time position measurements, Bell inequalities, philosophy, and physics. *Entropy* **2018**, *20*, 105. [CrossRef]
30. Tegmark, M.; Wheeler, J.A. 100 Years of quantum mysteries. *Sci. Am.* **2001**, *284*, 68–75 [CrossRef]
31. Durt, T.; Pierseaux, Y. Bohm's interpretation and maximally entangled states. *Phys. Rev. A* **2002**, *66*, 052109. [CrossRef]
32. Braverman, B.; Simon, C. Proposal to observe the nonlocality of Bohmian trajectories with entangled photons. *Phys. Rev. Lett.* **2013**, *110*, 060406. [CrossRef] [PubMed]
33. Jammer, M. *Concepts of Mass in Contemporary Physics and Philosophy*; Princeton University Press: Princeton, NJ, USA, 2009.
34. Caticha, A. Entropic dynamics, time and quantum theory. *J. Phys. A Math. Theor.* **2011**, *44*, 225303. [CrossRef]
35. Bell, J.S. *Speakable and Unspeakable in Quantum Mechanics*, 2nd ed.; Cambridge University Press: Cambridge, UK, 2004.
36. Zander, C.; Plastino, A.R.; Plastino, A.; Casas, M. Entanglement and the speed of evolution of multi-partite quantum systems. *J. Phys. A Math. Theor.* **2007**, *40*, 2861–2872. [CrossRef]
37. Mensky, M.B. *Continuous Quantum Measurement and Path Integrals*; Routledge : Abingdon, UK, 2017.
38. Kostin, M.D. On the Schrödinger-langevin equation. *J. Chem. Phys.* **1972**, *57*, 3589–3591. [CrossRef]
39. Schuch, D.; Chung, K.M.; Hartmann, H. Nonlinear Schrödinger-type field equation for the description of dissipative systems. I. Derivation of the nonlinear field equation and one-dimensional example. *J. Math. Phys.* **1983**, *24*, 1652–1660. [CrossRef]
40. Schuch, D.; Chung, K.M.; Hartmann, H. Nonlinear Schrödinger-type field equation for the description of dissipative systems. II. Frictionally damped motion in a magnetic field. *Int. J. Quantum Chem.* **1984**, *25*, 391–410. [CrossRef]
41. Schuch, D.; Chung, K.M.; Hartmann, H. Nonlinear Schrödinger-type field equation for the description of dissipative systems. III. Frictionally damped free motion as an example for an aperiodic motion. *J. Math. Phys.* **1984**, *25*, 3086–3096. [CrossRef]
42. Schuch, D. Complex nonlinear relations in classical and quantum physics. *J. Phys. Conf. Ser.* **2009**, *174*, 012042. [CrossRef]
43. Schuch, D.; Chung, K.M. From macroscopic irreversibility to microscopic reversibility via a nonlinear Schrödinger-type field equation. *Int. J. Quantum Chem.* **1986**, *29*, 1561–1573. [CrossRef]

44. Schuch, D. Relations between nonlinear Riccati equations and other equations in fundamental physics. *J. Phys. Conf. Ser.* **2014**, *538*, 012019. [CrossRef]

45. Pask, C. *Magnificent Principia: Exploring Isaac Newton's Masterpiece*; Prometheus Books: Amherst, NY, USA, 2013.

![entropy logo] *entropy*

MDPI

Article

New Forms of Quantum Value Indefiniteness Suggest That Incompatible Views on Contexts Are Epistemic

Karl Svozi

Institute for Theoretical Physics, Vienna University of Technology, Wiedner Hauptstrasse 8-10/136, 1040 Vienna, Austria; svozil@tuwien.ac.at

Received: 27 April 2018 ; Accepted: 15 May 2018; Published: 24 May 2018

Abstract: Extensions of the Kochen–Specker theorem use quantum logics whose classical interpretation suggests a true-implies-value indefiniteness property. This can be interpreted as an indication that any view of a quantum state beyond a single context is epistemic. A remark by Gleason about the ad hoc construction of probability measures in Hilbert spaces as a result of the Pythagorean property of vector components is interpreted platonically. Unless there is a total match between preparation and measurement contexts, information about the former from the latter is not ontic, but epistemic. This is corroborated by configurations of observables and contexts with a truth-implies-value indefiniteness property.

Keywords: quantum mechanics; Gleason theorem; Kochen–Specker theorem; Born rule

1. Quantum Contexts as Views on States

Contexts arise naturally in quantum mechanics: they correspond to the "greatest classical subdomains within the expanse of conceivable quantum propositions:" for all empirical matters, every observable within a particular fixed context can be assumed classical with respect to and relative to that context. Therefore, according to Gleason [1], it appears prudent to assume that classical probabilities should be applicable to such classical mini-universes; and in particular, when considering observables within a given context. Gleason formalized this in terms of frame functions and proceeded to show how the quantum probabilities, in particular, the Born rule, can be "stitched together" from these classical bits and pieces. This paper can be seen as a prolegomenon to this approach; and as a contribution to the ongoing search for its semantics.

Formally, the concept of context can be exposed in two ways: one is in terms of "largest possible" sets of orthogonal pure states; that is, in terms of (unit) vectors and their linear spans. Another one is by maximal operators and the perpendicular projection operators in their non-degenerate spectral decomposition.

Let us start by supposing that contexts can be represented by orthonormal bases of Hilbert space. Due to the spectral theorem, this immediately gives rise to an equivalent conception of context: that as a maximal observable, which is formed by some (non-degenerate) spectral sum of the mutually orthogonal perpendicular projection operators corresponding to the basis states. This is just the expression of the dual role of perpendicular projection operators in quantum mechanics: they represent both pure states, as well as observable bits; that is, elementary yes-no propositions.

For the sake of an elementary example, suppose one is dealing with (lossless) electron spin state (or photon polarization) measurements. As there are two outcomes, the associated Hilbert space is two-dimensional. The two outcomes can be identified with two arbitrary orthogonal normalized vectors therein, forming an orthonormal basis. Suppose, for the sake of further simplicity, that we parametrize this basis to be the standard Cartesian basis in two-dimensional Hilbert space, its two vectors being (Equation (1.8), [2]) $|0\rangle = \left(1,0\right)^{\mathsf{T}}$ and $|1\rangle = \left(0,1\right)^{\mathsf{T}}$, where the superscript symbol

"⊤" indicates transposition. Their dyadic products $\mathbf{E}_0 = |0\rangle\langle 0| = \left(1,0\right)^{\mathsf{T}} \otimes \left(1,0\right) = \begin{pmatrix} 1 & 0 \\ 0 & 0 \end{pmatrix}$,

$\mathbf{E}_1 = \begin{pmatrix} 0 & 0 \\ 0 & 1 \end{pmatrix}$ form the corresponding (mutually) orthogonal perpendicular projection operators. These contexts can be either represented in terms of vectors, like $\mathcal{C} = \{|0\rangle, |1\rangle\}$, or in terms of perpendicular projection operators, like $\mathcal{C} = \{\mathbf{E}_0, \mathbf{E}_1\}$.

Any two distinct numbers $\lambda_0 \neq \lambda_1$ define a maximal operator through the "weighted" spectral sum:

$$\mathbf{A} = \lambda_0 \mathbf{E}_0 + \lambda_1 \mathbf{E}_1 = \lambda_0 |0\rangle\langle 0| + \lambda_1 |1\rangle\langle 1| = \begin{pmatrix} \lambda_0 & 0 \\ 0 & \lambda_1 \end{pmatrix}. \tag{1}$$

The term "maximal" refers to the fact that $\mathbf{A}$ "spans" a "classical sub-universe" of mutually commuting operators through variations of $f(\mathbf{A}) = f(\lambda_0)\mathbf{E}_0 + f(\lambda_1)\mathbf{E}_1$, where $f : \mathbb{R} \mapsto \mathbb{R}$ represents some real valued polynomial or function of a single real argument (§ 84, Theorems 1 and 2, p. 171, [3]). In particular, this includes the context $\mathcal{C} = \{\mathbf{E}_0, \mathbf{E}_1\}$ through the two binary functions $f_i(\lambda_j) = \delta_{ij}$, with $i, j \in \{0, 1\}$.

2. Probabilities on Contexts in Quantum Mechanics

Let us concentrate on probabilities next. As already mentioned, Gleason [1] observed that classical observables should obey classical probabilities (this should be the same for Bayesian and frequentist approaches). Can we, therefore, hope for the existence of some "Realding", that is some global ontology, some enlarged panorama of "real physical properties", behind these stitched probabilities? As it turns out, relative to reasonable assumptions and the absence of exotic options, this is futile.

Formally, this issue can be rephrased by recalling that the main formal entities of quantum mechanics are all based on Hilbert space; that is, on vectors, as well as their relative position and permutations. A pure state represented as a vector $|\psi\rangle$ can be conveniently parameterized or encoded by coordinates referring to the respective bases. Because of their convenience, one chooses orthonormal bases, that is contexts, for such a parametrization. Why is convenience important? Because, as has been noted earlier, in finite dimensions D, any such context $\mathcal{C} \equiv \{|\mathbf{e}_1\rangle, |\mathbf{e}_2\rangle, \ldots, |\mathbf{e}_D\rangle\}$ can also be interpreted as a maximal set of co-measurable propositions $\mathcal{C} \equiv \{\mathbf{E}_1, \mathbf{E}_2, \ldots, \mathbf{E}_D\}$ with $\mathbf{E}_i = |\mathbf{e}_i\rangle\langle\mathbf{e}_i|$, $1 \leq i \leq D$, as the latter refers to a complete system of orthogonal perpendicular projections, which are a resolution of the identity operator $\mathbb{I}_D = \sum_{i=1}^{D} \mathbf{E}_i$. For any such context, classical Kolmogorov probability theory requires the probabilities P to satisfy the following axioms:

A1   probabilities are real-valued and non-negative: $P(\mathbf{E}_i) \in \mathbb{R}$, and $P(\mathbf{E}_i) \geq 0$ for all $\mathbf{E}_i \in \mathcal{C}$, or, equivalently, $1 \leq i \leq D$;

A2   probabilities of mutually-exclusive observables within contexts are additive: $P\left(\sum_{i=1}^{k \leq D} \mathbf{E}_i\right) = \sum_{i=1}^{k \leq D} P(\mathbf{E}_i)$;

A3   probabilities within one context add up to one: $P(\mathbb{I}_D) = P\left(\sum_{i=1}^{D} \mathbf{E}_i\right) = 1$.

How can probabilities $P_\psi(\mathbf{E})$ of propositions formalized by perpendicular projection operators (or, more generally, observables whose spectral sums contain such propositions) on given states $|\psi\rangle$ be formed that adhere to these axioms? As already Gleason pointed out in the second paragraph of (Section 1, p. 885, [1]), there is an ad hoc way to obtain a probability measure on Hilbert spaces: a vector $|\psi\rangle$ can be "viewed" through a "probing context" $\mathcal{C}$ as follows:

(i)   For each closed subspace spanned by the vectors $|\mathbf{e}_i\rangle$ in the context $\mathcal{C}$, take the projection $\mathbf{E}_i |\psi\rangle$ of $|\psi\rangle$ onto $|\mathbf{e}_i\rangle$.

(ii)   Take the absolute square of the length (norm) of this projection and identify it with the probability $P_\psi\left(\mathbf{E}_i\right)$ of finding the quantum system that is in state $|\psi\rangle$ to be in state $|e_i\rangle$; that is (the symbol "†" stands for the Hermitian adjoint):

$$P_\psi\left(\mathbf{E}_i\right) = \left(\mathbf{E}_i|\psi\rangle\right)^\dagger \mathbf{E}_i|\psi\rangle = \langle\psi|\mathbf{E}_i^\dagger \mathbf{E}_i|\psi\rangle$$
$$= \langle\psi|e_i\rangle \underbrace{\langle e_i|e_i\rangle}_{=1} \langle e_i|\psi\rangle = \langle\psi|e_i\rangle\langle e_i|\psi\rangle = \|\langle e_i|\psi\rangle\|^2. \tag{2}$$

Because of the mutual orthogonality of the elements in the context $\mathcal{C}$, the Pythagorean theorem enforces the third axiom **A3** as long as all vectors involved are normalized; that is, has length (norm) one. This situation is depicted in Figure 1.

The situation is symmetric in a sense that reflects the duality between observable and state observed: Suppose now that the state $|\psi\rangle$ is "completed" by other vectors to form an entire context $\mathcal{C}'$. Then, one could consider this context $\mathcal{C}'$, including $|\psi\rangle$ to be "probe" vectors, now identified as states, in the original context $\mathcal{C}$. Very similarly, probability measures adhering to Axioms **A1**–**A3** can be constructed by, say, for instance, $P_{\mathbf{E}_\psi}\left(\mathbf{E}_i\right)$

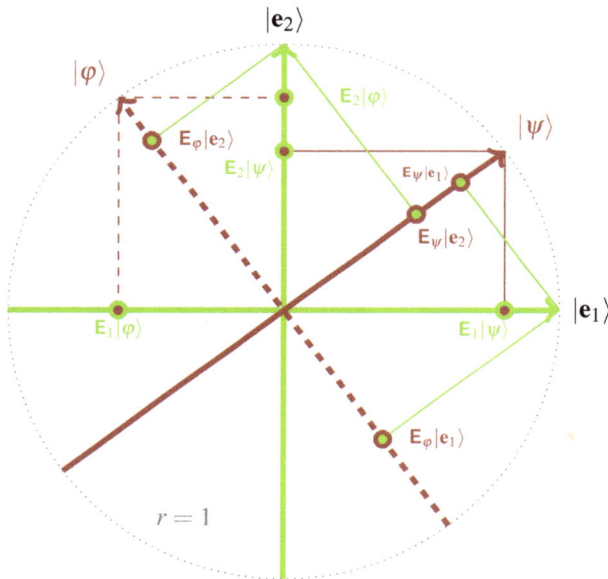

Figure 1. An orthonormal basis forming a context $\mathcal{C} = \{|e_1\rangle, |e_2\rangle\}$ represents a frame of reference from which a "view" on a state $|\psi\rangle$ can be obtained. Formally, if the vectors $|\psi\rangle$ and $|\varphi\rangle$ are normalized, such that $\langle\psi|\psi\rangle = \langle\varphi|\varphi\rangle = 1$, then the absolute square of the length (norm) of the projections $\mathbf{E}_1|\psi\rangle = |e_1\rangle\langle e_1|\psi\rangle$ and $\mathbf{E}_2|\psi\rangle = |e_2\rangle\langle e_2|\psi\rangle$, as well as $\mathbf{E}_1|\varphi\rangle = |e_1\rangle\langle e_1|\varphi\rangle$ and $\mathbf{E}_2|\varphi\rangle = |e_2\rangle\langle e_2|\varphi\rangle$ adds up to one. Conversely, a second context $\mathcal{C}' = \{|\psi\rangle, |\varphi\rangle\}$ grants a frame of reference from which a "view" on the first context $\mathcal{C}$ can be obtained.

It is important to keep in mind that, although Gleason's ansatz is about a single context $\mathcal{C}$, it is valid for all contexts; indeed, formally, for a continuum of contexts represented by the continuum of possible orthonormal bases of D-dimensional Hilbert space. Every such context entails a particular view on the state $|\psi\rangle$; and there is a continuum of such views on the state $|\psi\rangle$.

Furthermore, there is a symmetry between the two contexts $\mathcal{C}$ and $\mathcal{C}'$ involved. We may call $\mathcal{C}'$ the "preparation context" and $\mathcal{C}$ the "measurement context," but these denominations are purely conventional. In this sense, it is a matter of convention if we consider "$\mathcal{C}$ probing $\mathcal{C}'$" or "$\mathcal{C}'$ probing $\mathcal{C}$."

There is one "privileged view" on the preparation context $\mathcal{C}'$, that is the view obtained if both the preparation and measurement contexts coincide: $\mathcal{C} = \mathcal{C}'$. Under such circumstances, the observables are value definite: their values coincide with those of the preparation.

3. Contexts in Partition Logics and Their Probabilities

This section is a reminder rather than an exposition [4–10] of partition logics. Suffice it to say that partition logics are probably the most elementary generalization of Boolean algebras: they are the Boolean subalgebras associated with sets of partitions of a given set, which are "pasted" or "stitched" together at their common elements; similar to contexts (blocks, subalgebras) in quantum logic. The main difference is that the latter is a continuous logic based on geometrical entities (vectors), whereas partition logics are discrete, finite algebraic structures based on sets of partitions of a given set. Nevertheless, for empirical purposes, it is always possible to come up with a partition logic mimicking the respective quantum logic [11]. Partition logics have two known model realization: automaton logics [12–14] and generalized urn models [15–17].

Just like classical probabilities on Boolean logics, the probabilities on Boolean structures are formed by a convex summation of all two-valued measures [9,10,18], corresponding to ball types. Such probabilities will henceforth be called (quasi)classical.

4. Probabilities on Pastings or Stitchings of Contexts

From dimension $D \geq 3$ onwards, contexts can be non-trivially connected or intertwined [1] in up to $D - 2$ common elements. Such intertwining chains of contexts give rise to various apparently "non-classical" logics; and a wealth (some might say a plethora) of publications dealing with ever-increasing "strange" or "magic" properties of observables hitherto unheard of in classical physics. The following logics have a realization in (mostly three-dimensional if not stated otherwise) Hilbert space. For concrete parametrizations, the reader is either referred to the literature or to a recent survey (Chapter 12, [10]).

On such pastings of contexts, (quasi)classical probabilities and their bounds, termed conditions of possible experience by Boole (p. 229, [19]), can be obtained in three steps [8–10,18]:

(i)   Enumerate all truth assignments (or two-$\{0, 1\}$-valued measures or states) v_i.
(ii) The (quasi)classical probabilities are obtained by the formation of the convex sum $\sum_i \lambda_i v_i$ over all such states obtained in (i), with $0 \leq \lambda_i \leq 1$ and $\sum_i \lambda_i = 1$.
(ii) The Bell-type bounds on probabilities and expectations are obtained by bundling these truth assignments into vectors, one per two-valued measure, with the coordinates representing the respective values of those states on the atoms (propositions, observables) of the logic; and by subsequently solving the hull problem for a convex polytope whose vertices are identified with the vectors formed by all truth assignments [20–23].

In what follows, some such quantum logics will be enumerated whose quantum probabilities co-exist and sometimes violate their (quasi)classical probabilities, if they exist. Such violations can be expected to occur quite regularly, as (although in both cases, the probability Axioms **A1**–**A3** are satisfied for mutually-compatible observables) the quantum probabilities are formed very differently from the (quasi)classical ones; that is, not by convex sums as in the (quasi)classical case, but by scalar products among vectors.

4.1. Triangular and Square Logics in Four Dimensions

For geometric and algebraic reasons, there is no cyclic pasting of three or four contexts in three dimensions, but in four dimensions, this is possible; as depicted in Figure 2. The (quasi)classical probabilities are enumerated in Appendices A and B.

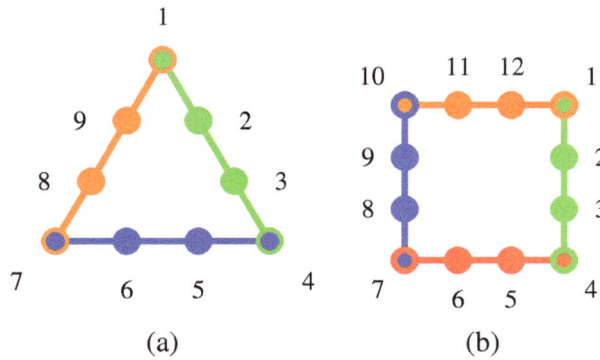

(a) (b)

Figure 2. Informally, Greechie (or, in different wording, orthogonality) diagrams [24] represent contexts by smooth curves such as straight lines or circles. The atoms are represented by circles. Two intertwining contexts are represented by "broken" (not smooth), but connected lines. (**a**) Greechie orthogonality diagram of triangle logic in four dimensions, realized by (from the top) $1: \frac{1}{2}(1,1,1,1)^{\mathsf{T}}$, $2: \frac{1}{\sqrt{2}}(1,0,-1,0)^{\mathsf{T}}$, $3: \frac{1}{\sqrt{2}}(0,1,0,-1)^{\mathsf{T}}$, $4: \frac{1}{2}(-1,1,-1,1)^{\mathsf{T}}$, $5: \frac{1}{\sqrt{2}}(0,1,1,0)^{\mathsf{T}}$, $6: \frac{1}{\sqrt{2}}(1,0,0,1)^{\mathsf{T}}$, $7: \frac{1}{2}(1,1,-1,-1)^{\mathsf{T}}$, $8: \frac{1}{\sqrt{2}}(0,0,1,-1)^{\mathsf{T}}$ and $9: \frac{1}{\sqrt{2}}(1,-1,0,0)^{\mathsf{T}}$. (**b**) Greechie orthogonality diagram of triangle logic in four dimensions, realized by (from the top right) $1: (1,0,0,0)^{\mathsf{T}}$, $2: \frac{1}{\sqrt{2}}(0,1,0,1)^{\mathsf{T}}$, $3: \frac{1}{\sqrt{2}}(0,1,0,-1)^{\mathsf{T}}$, $4: (0,0,1,0)^{\mathsf{T}}$, $5: \frac{1}{\sqrt{2}}(1,1,0,0)^{\mathsf{T}}$, $6: \frac{1}{\sqrt{2}}(1,-1,0,0)^{\mathsf{T}}$, $7: (0,0,0,1)^{\mathsf{T}}$ and $8: \frac{1}{\sqrt{2}}(1,0,1,0)^{\mathsf{T}}$, $9: \frac{1}{\sqrt{2}}(1,0,-1,0)^{\mathsf{T}}$, $10: (0,1,0,0)^{\mathsf{T}}$, $11: \frac{1}{\sqrt{2}}(0,0,1,1)^{\mathsf{T}}$, $12: \frac{1}{\sqrt{2}}(0,0,1,-1)^{\mathsf{T}}$. (Not all orthogonality relations are represented.) The associated (quasi)classical probabilities are obtained from a convex summation over all truth assignments, and listed in Appendices A and B.

Summation of the (quasi)classical probabilities on the intertwining atoms of the triangle logic yields $p_1 + p_4 + p_7 = \lambda_1 + \lambda_2 + \lambda_7 + \lambda_{12} + \lambda_{13} + \lambda_{14} \leq 1$. However, the axioms of probability theory are too restrictive to allow for quantum violations of these probabilities: after all, these adjacent vertices are mutually orthogonal and thus are in the same context (augmented with the fourth atom of that context). Other inequalities, such as $p_1 + p_2 = \lambda_1 + \lambda_2 \leq p_5 + p_6 = (\lambda_1 + \lambda_3 + \lambda_4 + \lambda_8 + \lambda_9) + (\lambda_2 + \lambda_5 + \lambda_6 + \lambda_{10} + \lambda_{11})$, compare vertices with the adjacent "inner" atoms; but again, due to the probability Axiom **A3**, the quantum probabilities must obey these inequalities, as well.

Komei Fukuda's cddlib package [25] can be employed for a calculation of the hull problem, yielding all Bell-type inequalities associated with the convex polytope, the vertices of which are associated with the 14 or 34 truth assignments (two-valued measures) on the respective triangle and square logics. It turns out that all of them are expressions of Axioms **A1**–**A3**, which are mandatory also for the quantum probabilities within contexts.

4.2. Pentagon (Pentagram) Logic

The pentagon (graph theoretically equivalent to a pentagram) logic is a cyclic stitching or pasting of five contexts [26–32], as depicted in Figure 3. The (quasi)classical probabilities (p. 289, Figure 11.8, [9]) can be obtained by taking the convex sum of all 11 two-valued measures [26], as listed in Appendix C.

Because of the convex sum of all λ's adds up to one, the sum of the (quasi)classical probabilities enumerated in Equation (A3), taken merely on the five intertwining observables, yields:

$$
\begin{aligned}
& p_1 + p_3 + p_5 + p_7 + p_9 \\
& = \lambda_1 + \lambda_4 + \lambda_7 + \lambda_9 + \lambda_{10} + 2\left(\lambda_2 + \lambda_3 + \lambda_5 + \lambda_6 + \lambda_8\right) \\
& \leq 2 \sum_{i=1}^{11} \lambda_i = 2.
\end{aligned}
\tag{3}
$$

This inequality is in violation of quantum predictions [30,32] of $\sqrt{5} > 2$. Note that, in order to obtain the probabilities on the five intertwining observables (vertices), all of them need to be determined. However, only adjacent pairs share a common context. Therefore, at least three incompatible measurement types are necessary.

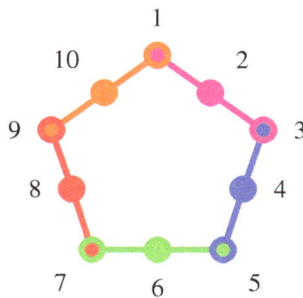

Figure 3. Greechie orthogonality diagram of the pentagon (pentagram) logic. The associated (quasi)classical probabilities are obtained from a convex summation over all truth assignments, and listed in Appendix C.

4.3. Specker Bug Logic with the True-Implies-False Property

A pasting of two pentagon logics, the "Specker bug" logic, has been introduced (Figure 1, p. 182, [33]) and used (Γ_1, p. 68, [34]) by Kochen and Specker and discussed by many researchers [35–37]; see also (Figure B.l, p. 64, [38]), (pp. 588–589, [39]), (Section IV, Figure 2, [40]) and (p. 39, Figure 2.4.6, [41]). It is a pasting [27,42] of seven contexts in such a tight way (cf. Figure 4a) that preparation of a (quasi)classical system in state **a** entails the non-occurrence of observable **b**. As has been observed by Stairs (pp. 588–589, [39]) and Clifton (Sections II and III, Figure 1, [40,43,44]), this is no longer the case for quantum states and quantum observables. Therefore, if one prepares a system in a state $|\mathbf{a}\rangle$ and measures $\mathbf{E}_b = |\mathbf{b}\rangle\langle\mathbf{b}|$, associated with state $|\mathbf{b}\rangle$, then the mere occurrence of $|\mathbf{b}\rangle$ implies the non-classicality of the quantized system.

Again, the (quasi)classical probabilities (p. 286, Figure 11.5(iii), [9]) enumerated in Appendix D can be obtained by taking the convex sum of all 14 two-valued measures (p. 579, Table 7, [8]). Pták and Pulmannová (p. 39, Figure 2.4.6, [41]), as well as Pitowsky in (p. 402, Figure 2, [36]) and (pp. 224, 225, Figure 10.2, [37]) noted that, for (quasi)classical probabilities, including ones on partition logics, the sum of the probabilities on $|\mathbf{a}\rangle$ and $|\mathbf{b}\rangle$ must not exceed $\frac{3}{2}$. Therefore, both cannot be true at the same time, because this would result in their sum being two. This might be called a true-implies-false property [45] (also known as the one-zero rule [46]) on the atoms **a** and **b**.

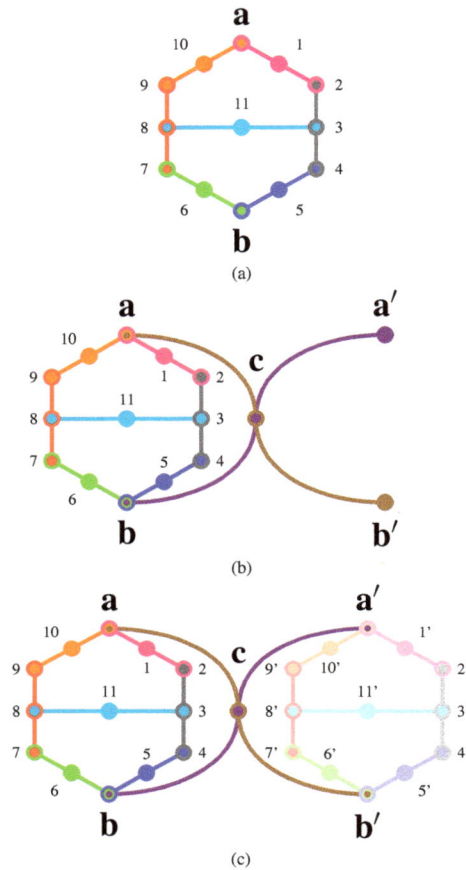

Figure 4. Greechie orthogonality diagram of (**a**) the Specker bug logic (Figure 1, p. 182, [33]). A proof that, if the system is prepared in state **a**, then classical (non-contextual) truth assignments require **b** not to occur proceeds as follows: In such a truth assignment, as per Axiom **A3**, there is only one true atom per context; all the others have to be false. In a proof by contradiction, suppose that both **a** and **b** are true. Then, all atoms connected to them (2,4,7,9) must be false. This in turn requires that the observables (3,8) connecting them must both be true. Alas, those two observables (3,8) are connected by a "middle" context {3, 11, 8} . But the occurrence of two true observables within the same context is forbidden by Axiom **A3**. The only consistent alternative is to disallow **b** to be true if **a** is assumed to be true; or conversely, to disallow **a** to be true if **b** is assumed to be true. (**b**) Greechie orthogonality diagram of a Specker bug logic extended by two contexts, which has the true-implies-true property on **a**', given **a** to be true (Γ_1, p. 68, [34]). (**c**) Greechie orthogonality diagram of a combo of two Specker bug logics (Γ_3, p. 70, [34]). If **a** is assumed to be true, then the remaining atoms in the context {**a**, **c**, **b**'} connecting **a** with **b**' and, in particular, **c** have to be false. Furthermore, if **a** is true, then **b** is false. Therefore, **a**' needs to be true if **b** and **c** both are false, because they form the context {**b**, **c**, **a**'}. This argument is valid even in the absence of a second Specker bug logic. Introduction of a second Specker bug logic ensures the converse: whenever **a**' is true, **a** must be true, as well. Therefore, **a** and **a**' (and by symmetry, also **b** and **b**') cannot be separated by any truth assignment.

Actually, this classical bound can be tightened by explicity summing the (quasi)classical probabilities of **a** and **b** enumerated in Equation (A5). Because of the convex sum of all λ's adds up to one, this yields:

$$p_{\mathbf{a}} + p_{\mathbf{b}} = \lambda_1 + \lambda_2 + \lambda_3 + \lambda_6 + \lambda_{13} + \lambda_{14} \leq \sum_{i=1}^{14} \lambda_i = 1. \tag{4}$$

This inequality is in violation of quantum predictions for a system prepared in state $|\mathbf{a}\rangle$; in this case [47], $\frac{10}{9} > 1$.

Indeed, Cabello [47] (see also his dissertation (pp. 55–56, [48])) pointed out that in three dimensions, $|\mathbf{a}\rangle$ and $|\mathbf{b}\rangle$ must be at least an angle $\angle(\mathbf{a}, \mathbf{b}) \geq \operatorname{arcsec}(3) = \arccos\left(\frac{1}{3}\right) = \frac{\pi}{2} - \operatorname{arccot}\left(2\sqrt{2}\right) = \arctan\left(2\sqrt{2}\right)$ apart. Therefore, the probability of finding a state prepared along $|\mathbf{a}\rangle \equiv \left(1, 0, 0\right)^{\mathsf{T}}$ in a state $|\mathbf{b}\rangle \equiv \left(\cos\angle(\mathbf{a}, \mathbf{b}), \sin\angle(\mathbf{a}, \mathbf{b}), 0\right)^{\mathsf{T}}$ cannot exceed $|\langle\mathbf{b}|\mathbf{a}\rangle|^2 = 1/9$. Thus, in at most one-ninth of all cases will quantum mechanical probabilities violate the classical ones, as the classical prediction demands zero probability to measure **b**, given **a** (this prediction is relative to the assumption of non-contextuality, such that the truth assignment is independent of the particular context). For a concrete "optimal" realization (p. 206, Figure 1, [49]) (see also (Figure 4, p. 5387, [50])), take $|\mathbf{a}\rangle = \frac{1}{\sqrt{3}}\left(1, \sqrt{2}, 0\right)^{\mathsf{T}}$ and $|\mathbf{b}\rangle = \frac{1}{\sqrt{3}}\left(-1, \sqrt{2}, 0\right)^{\mathsf{T}}$, which yield $|\langle\mathbf{b}|\mathbf{a}\rangle| = \frac{1}{3}$.

Another true-implies-false configuration depicted in Figure 5a has an immediate quantum realization (Table 1, p. 102201-7, [51]) for $|\langle\mathbf{a}|\mathbf{b}\rangle|^2 = \frac{1}{2}$ and can be constructively (i.e., algorithmically computable) extended to arbitrary angles between non-collinear and non-orthogonal vectors.

4.4. Combo of Specker Bug Logic with the True-Implies-True, as Well as Inseparability Properties

This non-classical behavior can be "boosted" by an extension of the Specker bug logic (Γ_1, p. 68, [34]), including two additional contexts $\{\mathbf{a}, \mathbf{c}, \mathbf{b}'\}$, as well as $\{\mathbf{b}, \mathbf{c}, \mathbf{a}'\}$, as depicted in Figure 4b. It implements a true-implies-true property [45] (also known as the one-one rule [46]) for **a** and **a**'. Cabello's bound on the angle $\angle(\mathbf{a}, \mathbf{b})$ between **a** and **b** mentioned earlier results in bounds between **a** and **a**', as well as **b** and **b**': since **a** and **b**', as well as **b** and **a**' are orthogonal, that is, $\angle(\mathbf{a}, \mathbf{b}') = \angle(\mathbf{b}, \mathbf{a}') = \frac{\pi}{2}$, it follows for planar configurations that $\angle(\mathbf{a}, \mathbf{a}') = \angle(\mathbf{b}, \mathbf{a}') - \angle(\mathbf{a}, \mathbf{b}) \leq \frac{\pi}{2} - \arccos\left(\frac{1}{3}\right) = \operatorname{arccot}\left(2\sqrt{2}\right) = \operatorname{arccsc}(3) = \arcsin\left(\frac{1}{3}\right)$. For symmetry reasons, the same estimate holds for planar configurations between **b** and **b**'. For non-planar configurations, the angles must be even less than for planar ones.

True-implies-true properties have also been studied by Stairs (pp. 588–589, note added in proof, [39]); Clifton (Sections II and III, Figure 1, [40,43,44]) presents a similar argument, based on another true-implies-true logic inspired by Bell (Figure C.l, p. 67, [38]) (cf. also Pitowsky (p. 394, [52])), on the Specker bug logic (Section IV, Figure 2, [40]). More recently, Hardy [53–55], as well as Cabello and García-Alcaine and others [32,56–60] have discussed such scenarios.

Another true-implies-true configuration depicted in Figure 5b has an immediate quantum realization (Table 1, p. 102201-7, [51]) for $|\langle\mathbf{a}|\mathbf{b}\rangle|^2 = \frac{1}{2}$ and can be extended to arbitrary angles between non-collinear and non-orthogonal vectors.

A combo of Specker bug logics renders a non-separable set of two-valued states (Γ_3, p. 70, [34]): in the logic depicted in Figure 4c, **a** and **a**', as well as **b** and **b**' cannot be "separated" from one another by any non-contextual (quasi)classical truth assignment enumerated in Appendix D. Kochen and Specker (Theorem 0, p. 67, [34]) pointed out that this type of inseparability is a necessary and sufficient condition for a logic to be not embeddable in any classical Boolean algebra. Therefore, whereas both the Specker bug logic, as well as its extension true-implies-true logic can be represented by a partition logic, the combo Specker bug logic cannot.

(a)

(b)

(c)

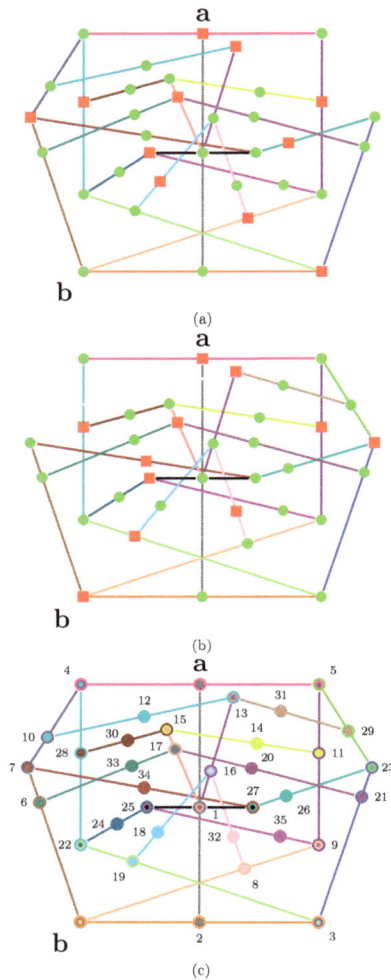

Figure 5. Greechie orthogonality diagram of a logic (Figure 2, p. 102201-8, [51]) realizable in $\mathbb{R}^3$ (**a**) with the true-implies-false property, (**b**) with the true-implies-true property and (**c**) with the true-implies-value indefiniteness (neither true nor false) property on the atoms **a** and **b**, respectively. (a,b) contain the single (out of 13) value assignment that is possible and for which **a** is true. All eight value assignments of the logic depicted in (c) require **a** to be false.

4.5. Logics Inducing Partial Value (In)Definiteness

Probably the strongest forms of value indefiniteness [61,62] are theorems [51,63,64] stating that relative to reasonable (admissibility, non-contextuality) assumptions, if a quantized system is prepared in some pure state $|\mathbf{a}\rangle$, then any observable that is not identical or orthogonal to $|\mathbf{a}\rangle$ is undefined. That is, there exist finite systems of quantum contexts whose pastings are demanding that any pure state $|\mathbf{b}\rangle$ not belonging to some context with $|\mathbf{a}\rangle$ can neither be true, nor false; else a complete contradiction would follow from the assumption of classically pre-existent truth values on some pasting of contexts such as the Specker bug logic.

What does "strong" mean here? Suppose one prepares the system in a particular context $\mathcal{C}$ such that a single vector $|\mathbf{a}\rangle \in \mathcal{C}$ is true; that is, $|\mathbf{a}\rangle$ has probability measure of one when measured along $\mathcal{C}$. Then, if one measures a complementary variable $|\mathbf{b}\rangle$ and $|\mathbf{b}\rangle$ is sufficiently separated from $|\mathbf{a}\rangle$ (more precisely, at least an angle $\arccos\left(\frac{1}{3}\right)$ apart for the Specker bug logic), then intertwined quantum propositional structures (such as the Specker bug logic) exist, which, interpreted (quasi)classically, demand that $|\mathbf{b}\rangle$ can never occur (cannot be true); and yet, quantum systems allow $|\mathbf{b}\rangle$ to occur. Likewise, other intertwined contexts that correspond to true-implies-true configurations of quantum observables (termed Hardy-like [53–55] by Cabello [60]) (quasi)classically imply that some endpoint $|\mathbf{b}'\rangle$ must always occur, given $|\mathbf{a}\rangle$ is true. Yet, quantum mechanically, since $|\mathbf{a}\rangle$ and $|\mathbf{b}'\rangle$ are not collinear, quantum mechanics predicts that occasionally, $|\mathbf{b}'\rangle$ does not occur. In the "strongest" form [51,63,64] of classical "do's and don'ts", there are no possibilities whatsoever for an observable proposition to be either true or false. That is, even if the Specker bug simultaneously allows some $|\mathbf{a}\rangle$ to be true and $|\mathbf{b}\rangle$ to be false (although disallowing the latter to be true), there is another, supposedly more sophisticated finite configuration of intertwined quantum contexts, that can be constructively enumerated and that disallows $|\mathbf{b}\rangle$ even to be false (it cannot be true either).

For the sake of an explicit example, take the logic (Figure 2, p. 102201-8, [51]) depicted in Figure 5c. It is the composite of two logics depicted in Figure 5a,b, which perform very differently at **b** given **a** to be true: whereas (a) implements a true-implies-false property, (b) has a true-implies-true property for the atoms **a** and **b**, respectively. Both (a) and (b) are proper subsets (lacking two contexts) of the logic in Figure 5c; and apart from their difference in four contexts, are identical.

More precisely, as explicated in Appendix E, both of these logics (a) and (b) allow 13 truth assignments (two-valued states), but only a single one allows **a** to be true on either of them (this uniqueness is not essential to the argument). The logic in (c) allows for eight truth assignments, but all of them assign falsity to **a**. By combining the logics (a) and (b), one obtains (c) which, if **a** is assumed to be true, implies that **b** can neither be true (this would contradict the true-implies-false property of (a)) nor can it be false, because this would contradict the true-implies-true property of (b). Hence, we are left with the only consistent alternative (relative to the assumptions): that a system prepared in state **a** must be value indefinite for observable **b**. Thereby, as the truth assignment on **b** is not defined, it must be partial on the entire logic depicted in Figure 5c.

The scheme of the proof is as follows:

(i) Find a logic (collection of intertwined contexts of observables) exhibiting a true-implies-false property on the two atoms **a** and **b**.

(ii) Find another logic exhibiting a true-implies-true property on the same two atoms **a** and **b**.

(iii) Then, join (paste) these logics into a larger logic, which, given **a**, neither allows **b** to be true nor false. Consequently, **b** must be value indefinite.

The most suggestive candidate for such a pasting is, however, unavailable: it is the combination of a Specker bug logic and another, extended Specker bug logic, as depicted in Figure 6. Such a logic cannot be realized in three dimensions, as the angles cannot be chosen consistently; that is, obeying the Cabello bounds on the relative angles, respectively.

The latter result about the partiality of the truth assignment has already been discussed by Pitowsky [61], and later by Hrushovski and Pitowsky [62]. It should also be mentioned that the logic (c) has been realized with a particular configuration in three-dimensional real Hilbert space (Tables I and II, p. 102201-7, [51]), which are an angle $\angle(\mathbf{a}, \mathbf{b}) = \arccos\left(\frac{1}{\sqrt{2}}\right)$ apart, but as has been mentioned earlier, this kind of value indefiniteness on any particular state **b**, given that the system has been prepared in state **a**, can be constructively obtained by an extension of the above configuration whenever **a** and **b** are neither collinear (in this case, **b** would be true) nor orthogonal (in this case, **b** would be false). Therefore, basically, all states not identical (or orthogonal) to the state prepared must be value indefinite.

Entropy **2018**, *20*, 406

All three logics in Figure 6a–c have another non-classical feature: they are non-unital [49], meaning that the truth assignments on some of their atoms can only acquire the value as false, regardless of the preparation. That is, in this "state-independent" form, whenever a proposition corresponding to such an atom is measured to be true, this can be interpreted as the indication of non-classicality (note that one can always rotate the entire set of rays so that this particular atom coincides with some observable measured.).

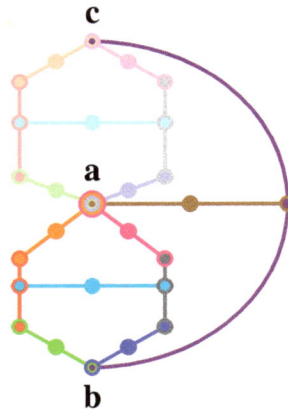

Figure 6. Greechie orthogonality diagram of a logic that is value indefinite on **b** (as well as on **c** for symmetry reasons), given **a** is true; alas, such a logic has no realization in three-dimensional Hilbert space, as the angles $\angle(\mathbf{a}, \mathbf{b})$ between **a** and **b** should simultaneously obey $1.2 \approx \text{arcsec}(3) \leq \angle(\mathbf{a}, \mathbf{b}) \leq \text{arccsc}(3) \approx 0.3$.

5. Propositional Logic Does Not Uniquely Determine Probabilities

By now, it should be clear that the propositional structure does in general not uniquely determine its probabilities. The Specker bug in Figure 4a serves as a good example of that: it supports (quasi)classical probabilities, explicitly enumerated in(p. 286, Figure 11.5(iii), [9]) and (p. 91, Figure 12.10, [10]), which are formed by convex combinations of all two-valued states on them.

Other propositional structures such as the pentagon logic support "exotic" probability measures [26], which do not vanish at their interlink observables and are equally weighted with value $\frac{1}{2}$ there. This measure is neither realized in the (quasi)classical partition logic setup explicitly discussed in (p. 289, Figure 11.8, [9]) and (p. 88, Figure 12.8, [10]), nor in quantum mechanics. It remains to be seen if a more general theory of probability measures based on Axioms **A1**–**A3** can be found.

6. Some Platonist Afterthoughts

The author's not-so-humble reading of all these aforementioned "mind-boggling" non-classical quantum predictions is a rather sober one: in view of the numerous indications that classical value definiteness cannot be extended to more than a single context, the most plausible supposition is that, besides exotic possibilities [65,66], ontologically, there is only one such "Realding" (indeed, a rather obvious candidate suggesting itself as ontology): a single vector, or rather a single context. Quantized systems can be completely and exhaustively characterized by a unique context and a "true" proposition within this context.

Suppose for a moment that this hypothesis is correct and that there is no ontology, no "Realding," beyond a single context. There is one preferred view, namely the context identical to the context in which the system has been prepared, and all but one epistemic view.

Yet, a confusing experience is the apparent ease with which an experimenter appears to measure, without any difficulty, a context or (maximal) observable not (or only partly through intertwines) matching the preparation context. In such a situation, one may assume that the measurement grants an "imperfect" view on the preparation context. In this process, information, in particular the relative locatedness of the measurement context with respect to the preparation context, is augmented by properties of the measurement device, thereby effectively generating entanglement [67,68] via context translation [69]. Frames of reference that do not coincide with the "Realding" or preparation context necessarily include stochastic elements that are not caused or determined by any property of the formerly individual "Realding." One may conclude [70] with Bohr's 1972 Como lecture (p. 580, [71]) that "any observation of atomic phenomena will involve an interaction with the agency of observation not to be neglected. Accordingly, an independent reality in the ordinary physical sense can neither be ascribed to the phenomena nor to the agencies of observation." That is, any interaction between the previously separated individual object and the measurement device results in a joint physical state that is no longer determined by the states of the (previously) individual constituents [68,72]. Instead, the joint state exhibits what Schrödinger later called entanglement [67]. Entanglement is characterized by a value definite relational [73] or collective (re-)encoding of information with respect to the constituent parts, thereby (since the unitary quantum evolution is injective) resulting in the value indefiniteness of the previously individual and separate parts. As a result, knowledge about observations obtained by different contexts than the preparation context are necessarily (at least partially in the sense of the augmented information from the measurement device) epistemic.

Another possible source of perplexity might be the various types of algebraic or logical structures involved. Classically, empirical logics are Boolean algebras. Then, in a first step towards non-classicality, there are partition logics that are not Boolean any longer (they feature complementarity through non-distributivity), but nevertheless still allow for a certain type of (quasi)classicality; that is, a separating and unital set of two-valued states. Then, further on this road, there are (finite) quantum logics that do not allow any definite state at all.

One might be puzzled by the fact that there exist "intermediate" logics, such as the Specker bug or the pentagon (pentagram) logic discussed in Sections 4.2 and 4.3 that still allow (even classical) simultaneous value indefiniteness, although they contain observables that are mutually complementary (non-collinear and non-orthogonal). However, this apparent paradox should rather be interpreted epistemically, as means (configuration) relative [74]: in the case of the pentagon, we have decided to concentrate on 10 observables in a cyclic pasting of five contexts, but we have thereby implicitly chosen to "look the other way" and disregard the abundance of other observables that impose much more stringent conditions on the value definiteness of the observables in the pentagon logic than the pentagon logic itself.

Therefore, properties such as the true-implies-false, the true-implies-true properties, as well as inseparability and even value indefiniteness are means relative and valid only if one restricts or broadens one's attention to sometimes very specific, limited sublogics of the realm of all conceivable quantum logics, which are structures formed by perpendicular projection operators in Hilbert spaces of dimension larger than two.

Pointedly stated, sets of intertwining contexts connecting two (or more) relevant complementary observables **a** and **b** should be considered as totally arbitrary when it comes to the inclusion or exclusion of particular contexts interconnecting them: there is neither a necessity nor even a compelling reason to take into account one such structure and disregard another, or favor one over the other. Indeed, in an extreme, sui generis form of the argument, suppose a single quantum is prepared in some state **a**. Then, every single outcome of a measurement of every complementary (non-collinear and non-orthogonal relative to the state prepared) quantum observable may be considered as "proof" or "certification of non-classicality" (or, in different terminology, "contextuality"). Those observable can be identified with the "endpoint" **b** of either some true-implies-false, or alternatively true-implies-true configuration (say the one sketched in Figure 6a,b), depending on whether the classical false or true predictions need to contradict the particular outcome, respectively. For quantum logics with a unital set of two-valued states, such as the logics depicted by Tkadlec (p. 207, Figure 2, [49]) or the ones in Figure 6a,b, one could even get rid of the state preparation if **b** occurs and is identified with an observable that, according to the classical predictions associated with that logic, cannot occur. There is no principle that could prevent us from arguing that way if we insist on the simultaneous existence of multiple contexts encountered in quantum mechanics. Indeed, are not intertwining contexts scholastic [75] sophisms in desperate need of deconstruction?

An interesting historical question arises: Kochen and Specker, in a succession of papers on partial algebras [33,34,76], have insisted that logical operations should only be defined within contexts and must not be applied to propositions outside of it. Yet, they have considered extended counterfactual structures of pasted context, ending up in a holistic argument involving complementary observables. Of course, an immediate reply might be that without intertwined contexts, there cannot be any non-trivial (non-classical, non-Boolean) configuration of observables that is of any interest.

For the reasons mentioned earlier, the emphasis should not be on "completing" quantum mechanics by some sort of hidden parameter theory, such as, for instance, Valentini [77] envisioning a theory that is to quantum mechanics as statistical physics is to thermodynamics, but just the opposite: the challenge is to acknowledge the scarcity of resources, the "Realding" or physical state as a mere vector, despite the continuum of possible views on it, resulting in an illusory over-abundance and over-determination.

In this line of thought, the question of what might be the reason behind the futility to co-define non-commuting quantum observables (from two or more different contexts) simultaneously should be answered in terms of a serious lack of a proper perspective of what one is dealing with: metaphorically speaking, it is almost as if one pretends to take a 360° panorama of what lies in the outside world, while actually merely taking photos from some sort of echo chamber, or house of mirrors, partly reflecting what is in it, and partly reproducing the observer (photographer) in almost endless reflections. Stitching together photos from these reflections yields a panorama of one and the same object in seemingly endless varieties. In this way, one might end up with a horribly distorted image of this situation; and with the inside turned outside.

This is not dissimilar to what Plato outlined in the Republic's cave metaphor (Book 7, 515c, p. 221, [78]): "what people in this situation would take for truth would be nothing more than the shadows of the manufactured objects." In the quantum transcription of this metaphor, the vectors are the objects, and the shadows taken for truth are the views on these objects, mediated or translated [69] by arbitrary mismatching contexts.

Acknowledgments: Federico Holik has constantly inspired me to (re-)think probabilities and invited me to Argentina. I kindly thank Adán Cabello and José R. Portillo for numerous explanations and discussions during an ongoing collaboration. I am also deeply indebted to Alastair A. Abbott and Cristian S. Calude for their contributions and collaboration on the localization of value indefiniteness. Josef Tkadlec has kindly provided a Pascal program to compute two-valued states on logics, given the contexts (blocks). Christoph Lemell and the Nonlinear Dynamics group at the Vienna University of Technology have provided the computational framework for the hull calculations performed.

Conflicts of Interest: The author declares no conflict of interest.

Appendix A. Two-Valued States, (Quasi)Classical Probabilities on the Triangular Logic in Four Dimensions

The two-valued states (also known as truth tables) have been enumerated by Josef Tkadlec's Pascal program 2states [79].Implicitly, the convex sums over the respective probabilities encode the truth tables, as, on any particular atom, the i'th truth table entry is one if λ_i appears in the listing of the classical probability p_i. Otherwise, the i'th truth table entry is zero.

The bounds for classical probabilities have been obtained by Komei Fukuda's cddlib package [25].

There are nine propositions forming three contexts $\{1, 2, 3, 4\}$, $\{4, 5, 6, 7\}$ and $\{7, 8, 9, 1\}$ allowing 14 (separating, unital) two-valued states whose convex sum yields the following (quasi)classical probabilities:

$$
\begin{aligned}
p_1 &= \lambda_1 + \lambda_2, \\
p_2 &= \lambda_3 + \lambda_4 + \lambda_5 + \lambda_6 + \lambda_7, \\
p_3 &= \lambda_8 + \lambda_9 + \lambda_{10} + \lambda_{11} + \lambda_{12}, \\
p_4 &= \lambda_{13} + \lambda_{14}, \\
p_5 &= \lambda_1 + \lambda_3 + \lambda_4 + \lambda_8 + \lambda_9, \\
p_6 &= \lambda_2 + \lambda_5 + \lambda_6 + \lambda_{10} + \lambda_{11}, \\
p_7 &= \lambda_7 + \lambda_{12}, \\
p_8 &= \lambda_3 + \lambda_5 + \lambda_8 + \lambda_{10} + \lambda_{13}, \\
p_9 &= \lambda_4 + \lambda_6 + \lambda_9 + \lambda_{11} + \lambda_{14}.
\end{aligned}
\tag{A1}
$$

Appendix B. Truth Assignments, (Quasi)Classical Probabilities on the Square Logic in Four Dimensions

There are 12 propositions forming four contexts $\{1, 2, 3, 4\}$, $\{4, 5, 6, 7\}$, $\{7, 8, 9, 10\}$ and $\{10, 11, 12, 1\}$ allowing 34 (separating, unital) two-valued states whose convex sum yields the following (quasi)classical probabilities:

$$
\begin{aligned}
p_1 =\ & \lambda_1 + \lambda_2 + \lambda_3 + \lambda_4 + \lambda_5, \\
p_2 =\ & \lambda_6 + \lambda_7 + \lambda_8 + \lambda_9 + \lambda_{10} + \lambda_{11} \\
& + \lambda_{12} + \lambda_{13} + \lambda_{14} + \lambda_{15} + \lambda_{16} + \lambda_{17}, \\
p_3 =\ & \lambda_{18} + \lambda_{19} + \lambda_{20} + \lambda_{21} + \lambda_{22} + \lambda_{23} \\
& + \lambda_{24} + \lambda_{25} + \lambda_{26} + \lambda_{24} + \lambda_{28} + \lambda_{29}, \\
p_4 =\ & \lambda_{30} + \lambda_{31} + \lambda_{32} + \lambda_{33} + \lambda_{34}, \\
p_5 =\ & \lambda_1 + \lambda_2 + \lambda_6 + \lambda_7 + \lambda_8 + \lambda_9 \\
& + \lambda_{10} + \lambda_{18} + \lambda_{19} + \lambda_{20} + \lambda_{21} + \lambda_{22}, \\
p_6 =\ & \lambda_3 + \lambda_4 + \lambda_{11} + \lambda_{12} + \lambda_{13} + \lambda_{14} \\
& + \lambda_{15} + \lambda_{23} + \lambda_{24} + \lambda_{25} + \lambda_{26} + \lambda_{27}, \\
p_7 =\ & \lambda_5 + \lambda_{16} + \lambda_{17} + \lambda_{28} + \lambda_{29}, \\
p_8 =\ & \lambda_1 + \lambda_3 + \lambda_6 + \lambda_7 + \lambda_{11} + \lambda_{12} \\
& + \lambda_{18} + \lambda_{19} + \lambda_{23} + \lambda_{24} + \lambda_{30} + \lambda_{31}, \\
p_9 =\ & \lambda_2 + \lambda_4 + \lambda_8 + \lambda_9 + \lambda_{13} + \lambda_{14} \\
& + \lambda_{20} + \lambda_{21} + \lambda_{25} + \lambda_{26} + \lambda_{32} + \lambda_{33}, \\
p_{10} =\ & \lambda_{10} + \lambda_{15} + \lambda_{22} + \lambda_{27} + \lambda_{34}, \\
p_{11} =\ & \lambda_6 + \lambda_8 + \lambda_{11} + \lambda_{13} + \lambda_{16} + \lambda_{18} \\
& + \lambda_{20} + \lambda_{23} + \lambda_{25} + \lambda_{28} + \lambda_{30} + \lambda_{32}, \\
p_{12} =\ & \lambda_7 + \lambda_9 + \lambda_{12} + \lambda_{14} + \lambda_{17} + \lambda_{19} \\
& + \lambda_{21} + \lambda_{24} + \lambda_{26} + \lambda_{29} + \lambda_{31} + \lambda_{33}.
\end{aligned}
\tag{A2}
$$

Appendix C. Two-Valued States, (Quasi)Classical Probabilities on the Pentagon (Pentagram) Logic in Three Dimensions

There are five contexts $\{1,2,3\}$, $\{3,4,5\}$, $\{5,6,7\}$, $\{7,8,9\}$ and $\{9,10,1\}$ allowing 11 (separating, unital) two-valued states [26] whose convex sum yields the following (quasi)classical probabilities:

$$
\begin{aligned}
p_1 &= \lambda_1 + \lambda_2 + \lambda_3, \\
p_2 &= \lambda_4 + \lambda_5 + \lambda_7 + \lambda_9 + \lambda_{11}, \\
p_3 &= \lambda_6 + \lambda_8 + \lambda_{10}, \\
p_4 &= \lambda_1 + \lambda_2 + \lambda_4 + \lambda_7 + \lambda_{11}, \\
p_5 &= \lambda_3 + \lambda_5 + \lambda_9, \\
p_6 &= \lambda_1 + \lambda_4 + \lambda_6 + \lambda_{10} + \lambda_{11}, \\
p_7 &= \lambda_2 + \lambda_7 + \lambda_8, \\
p_8 &= \lambda_1 + \lambda_3 + \lambda_9 + \lambda_{10} + \lambda_{11}, \\
p_9 &= \lambda_4 + \lambda_5 + \lambda_6, \\
p_{10} &= \lambda_7 + \lambda_8 + \lambda_9 + \lambda_{10} + \lambda_{11}.
\end{aligned}
\tag{A3}
$$

Appendix D. Truth Assignments, (Quasi)Classical Probabilities on the Specker Bug Combo Logic

The logic depicted in Figure 4c contains 27 propositions forming 16 contexts $\{\mathbf{a}, 1, 2\}$, $\{2, 3, 4\}$, $\{4, 5, \mathbf{b}\}$, $\{\mathbf{b}, 6, 7\}$, $\{7, 8, 9\}$, $\{9, 10, \mathbf{a}\}$, $\{3, 8, 11\}$, $\{\mathbf{a}, \mathbf{c}, \mathbf{b}'\}$, $\{\mathbf{b}, \mathbf{c}, \mathbf{a}'\}$, $\{\mathbf{a}', 1', 2'\}$, $\{2', 3', 4'\}$, $\{4', 5', \mathbf{b}'\}$, $\{\mathbf{b}', 6', 7'\}$, $\{7', 8', 9'\}$, $\{9', 10', \mathbf{a}'\}$ and $\{3', 8', 11'\}$, allowing 82 non-separating on $\mathbf{a}/\mathbf{a}'$ and $\mathbf{b}/\mathbf{b}'$, unital two-valued states (not enumerated here because of volume). Nine and nine of these permit $\mathbf{a}$, as well as $\mathbf{a}'$ and $\mathbf{b}$, as well as $\mathbf{b}'$ to be true, respectively.

The logic depicted in Figure 4b contains 16 propositions forming nine contexts $\{\mathbf{a}, 1, 2\}$, $\{2, 3, 4\}$, $\{4, 5, \mathbf{b}\}$, $\{\mathbf{b}, 6, 7\}$, $\{7, 8, 9\}$, $\{9, 10, \mathbf{a}\}$, $\{3, 8, 11\}$, $\{\mathbf{a}, \mathbf{c}, \mathbf{b}'\}$ and $\{\mathbf{b}, \mathbf{c}, \mathbf{a}'\}$, allowing 22 (separating and unital) two-valued states, which, through their convex summation, yield the (quasi-)classical probabilities:

$$p_{\mathbf{a}} = \lambda_1 + \lambda_2 + \lambda_3,$$
$$p_{\mathbf{b}} = \lambda_8 + \lambda_{21} + \lambda_{22},$$
$$p_{\mathbf{a}'} = \lambda_1 + \lambda_2 + \lambda_3 + \lambda_5 + \lambda_7 + \lambda_{10}$$
$$\qquad + \lambda_{12} + \lambda_{14} + \lambda_{16} + \lambda_{18} + \lambda_{20},$$
$$p_{\mathbf{b}'} = \lambda_5 + \lambda_7 + \lambda_8 + \lambda_{10} + \lambda_{12} + \lambda_{14}$$
$$\qquad + \lambda_{16} + \lambda_{18} + \lambda_{20} + \lambda_{21} + \lambda_{22},$$
$$p_{\mathbf{c}} = \lambda_4 + \lambda_6 + \lambda_9 + \lambda_{11} + \lambda_{13} + \lambda_{15} + \lambda_{17} + \lambda_{19},$$
$$p_1 = \lambda_4 + \lambda_5 + \lambda_6 + \lambda_7 + \lambda_8 + \lambda_9$$
$$\qquad + \lambda_{10} + \lambda_{11} + \lambda_{12} + \lambda_{13} + \lambda_{14},$$
$$p_2 = \lambda_{15} + \lambda_{16} + \lambda_{17} + \lambda_{18} + \lambda_{19} + \lambda_{20} + \lambda_{21} + \lambda_{22},$$
$$p_3 = \lambda_1 + \lambda_4 + \lambda_5 + \lambda_6 + \lambda_7 + \lambda_8,$$
$$p_4 = \lambda_2 + \lambda_3 + \lambda_9 + \lambda_{10} + \lambda_{11} + \lambda_{12} + \lambda_{13} + \lambda_{14}, \qquad \text{(A4)}$$
$$p_5 = \lambda_1 + \lambda_4 + \lambda_5 + \lambda_6 + \lambda_7 + \lambda_{15}$$
$$\qquad + \lambda_{16} + \lambda_{17} + \lambda_{18} + \lambda_{19} + \lambda_{20},$$
$$p_6 = \lambda_2 + \lambda_4 + \lambda_5 + \lambda_9 + \lambda_{10} + \lambda_{11}$$
$$\qquad + \lambda_{12} + \lambda_{15} + \lambda_{16} + \lambda_{17} + \lambda_{18},$$
$$p_7 = \lambda_1 + \lambda_3 + \lambda_6 + \lambda_7 + \lambda_{13} + \lambda_{14} + \lambda_{19} + \lambda_{20},$$
$$p_8 = \lambda_2 + \lambda_9 + \lambda_{10} + \lambda_{15} + \lambda_{16} + \lambda_{21},$$
$$p_9 = \lambda_4 + \lambda_5 + \lambda_8 + \lambda_{11} + \lambda_{12} + \lambda_{17} + \lambda_{18} + \lambda_{22},$$
$$p_{10} = \lambda_6 + \lambda_7 + \lambda_9 + \lambda_{10} + \lambda_{13} + \lambda_{14}$$
$$\qquad + \lambda_{15} + \lambda_{16} + \lambda_{19} + \lambda_{20} + \lambda_{21},$$
$$p_{11} = \lambda_3 + \lambda_{11} + \lambda_{12} + \lambda_{13} + \lambda_{14} + \lambda_{17}$$
$$\qquad + \lambda_{18} + \lambda_{19} + \lambda_{20} + \lambda_{22}.$$

Note that, for all configurations, $p_{\mathbf{a}} = \lambda_1 + \lambda_2 + \lambda_3 \leq p_{\mathbf{a}'}$, implying that, whenever **a** is true, **a**′ must be true, as well.

The Specker bug logic depicted in Figure 4a contains 13 propositions forming seven contexts $\{\mathbf{a}, 1, 2\}$, $\{2, 3, 4\}$, $\{4, 5, \mathbf{b}\}$, $\{\mathbf{b}, 6, 7\}$, $\{7, 8, 9\}$, $\{9, 10, \mathbf{a}\}$ and $\{3, 8, 11\}$, allowing 14 (separating and unital) two-valued states:

$$p_{\mathbf{a}} = \lambda_1 + \lambda_2 + \lambda_3,$$
$$p_{\mathbf{b}} = \lambda_6 + \lambda_{13} + \lambda_{14},$$
$$p_1 = \lambda_4 + \lambda_5 + \lambda_6 + \lambda_7 + \lambda_8 + \lambda_9,$$
$$p_2 = \lambda_{10} + \lambda_{11} + \lambda_{12} + \lambda_{13} + \lambda_{14},$$
$$p_3 = \lambda_1 + \lambda_4 + \lambda_5 + \lambda_6,$$
$$p_4 = \lambda_2 + \lambda_3 + \lambda_7 + \lambda_8 + \lambda_9,$$
$$p_5 = \lambda_1 + \lambda_4 + \lambda_5 + \lambda_{10} + \lambda_{11} + \lambda_{12}, \qquad \text{(A5)}$$
$$p_6 = \lambda_2 + \lambda_4 + \lambda_7 + \lambda_8 + \lambda_{10} + \lambda_{11},$$
$$p_7 = \lambda_1 + \lambda_3 + \lambda_5 + \lambda_9 + \lambda_{12},$$
$$p_8 = \lambda_2 + \lambda_7 + \lambda_{10} + \lambda_{13},$$
$$p_9 = \lambda_4 + \lambda_6 + \lambda_8 + \lambda_{11} + \lambda_{14},$$
$$p_{10} = \lambda_5 + \lambda_7 + \lambda_9 + \lambda_{10} + \lambda_{12} + \lambda_{13},$$
$$p_{11} = \lambda_3 + \lambda_8 + \lambda_9 + \lambda_{11} + \lambda_{12} + \lambda_{14}.$$

Note that, for all configurations, whenever **a** is true, **b** is false, and vice versa.

Appendix E. Truth Assignments, (Quasi)Classical Probabilities on Truth-Implies-Value Indefiniteness Logic in Three Dimensions

Figure 6c depicts 37 propositions $\{\mathbf{a}, \mathbf{b}, 1, 2, 3, \ldots, 35\}$ in 26 contexts $\{\mathbf{a}, 1, 2\}$, $\{\mathbf{b}, 2, 3\}$, $\{4, \mathbf{a}, 5\}$, $\{\mathbf{b}, 6, 7\}$, $[\{7, 10, 4\}]_{(a),(c)}$, $[\{10, 12, 13\}]_{(a),(c)}$, $[\{5, 29, 23\}]_{(b),(c)}$, $[\{13, 31, 29\}]_{(b),(c)}$, $\{3, 21, 23\}$, $\{4, 28, 22\}$, $\{22, 19, 3\}$, $\{\mathbf{b}, 8, 9\}$, $\{9, 11, 5\}$, $\{28, 30, 15\}$, $\{15, 14, 11\}$, $\{6, 33, 17\}$, $\{17, 20, 21\}$, $\{7, 34, 27\}$, $\{27, 26, 23\}$, $\{22, 24, 25\}$, $\{25, 35, 9\}$, $\{15, 17, 1\}$, $\{13, 16, 1\}$, $\{16, 18, 19\}$, $\{16, 32, 8\}$ and $\{25, 1, 27\}$, allowing eight (non-separating, non-unital on $\mathbf{a}$, 2, 13, 15, 16, 17, 25, 27) two-valued states whose convex sum yields the following weights:

$$
\begin{aligned}
p_{\mathbf{a}} &= p_2 = p_{13} = p_{15} = p_{16} = p_{17} = p_{25} = p_{27} = 0, \\
p_{\mathbf{b}} &= \lambda_1 + \lambda_2 + \lambda_3 + \lambda_4, \\
p_1 &= \lambda_1 + \lambda_2 + \lambda_3 + \lambda_4 + \lambda_5 + \lambda_6 + \lambda_7 + \lambda_8 = 1, \\
p_3 &= + \lambda_5 + \lambda_6 + \lambda_7 + \lambda_8, \\
p_4 &= \lambda_1 + \lambda_2 + \lambda_5 + \lambda_6, \\
p_5 &= \lambda_3 + \lambda_4 + \lambda_7 + \lambda_8, \\
p_6 &= \lambda_5 + \lambda_6 + \lambda_7, \\
p_7 &= \lambda_8, \; p_9 = \lambda_6, \; p_{22} = \lambda_4, \; p_{23} = \lambda_2, \\
p_8 &= \lambda_5 + \lambda_7 + \lambda_8, \\
p_{10} &= \lambda_3 + \lambda_4 + \lambda_7, \\
p_{11} &= \lambda_1 + \lambda_2 + \lambda_5, \\
p_{12} &= \lambda_1 + \lambda_2 + \lambda_5 + \lambda_6 + \lambda_8, \\
p_{14} &= \lambda_3 + \lambda_4 + \lambda_6 + \lambda_7 + \lambda_8, \\
p_{18} &= \lambda_4 + \lambda_5 + \lambda_6 + \lambda_7 + \lambda_8, \\
p_{19} &= \lambda_1 + \lambda_2 + \lambda_3, \\
p_{20} &= + \lambda_2 + \lambda_5 + \lambda_6 + \lambda_7 + \lambda_8, \\
p_{21} &= \lambda_1 + \lambda_3 + \lambda_4, \\
p_{24} &= \lambda_1 + \lambda_2 + \lambda_3 + \lambda_5 + \lambda_6 + \lambda_7 + \lambda_8, \\
p_{26} &= \lambda_1 + \lambda_3 + \lambda_4 + \lambda_5 + \lambda_6 + \lambda_7 + \lambda_8, \\
p_{28} &= \lambda_3 + \lambda_7 + \lambda_8, \\
p_{29} &= \lambda_1 + \lambda_5 + \lambda_6, \\
p_{30} &= \lambda_1 + \lambda_2 + \lambda_4 + \lambda_5 + \lambda_6, \\
p_{31} &= \lambda_2 + \lambda_3 + \lambda_4 + \lambda_7 + \lambda_8, \\
p_{32} &= \lambda_1 + \lambda_2 + \lambda_3 + \lambda_4 + \lambda_6, \\
p_{33} &= \lambda_1 + \lambda_2 + \lambda_3 + \lambda_4 + \lambda_8, \\
p_{34} &= \lambda_1 + \lambda_2 + \lambda_3 + \lambda_4 + \lambda_5 + \lambda_6 + \lambda_7, \\
p_{35} &= \lambda_1 + \lambda_2 + \lambda_3 + \lambda_4 + \lambda_5 + \lambda_7 + \lambda_8.
\end{aligned}
\tag{A6}
$$

The logics in Figure 6a,b contain 35 observables in 24 contexts, which are the same as before in Figure 6c, lacking two contexts $[\{5, 29, 23\}]_{(b),(c)}$ and $[\{13, 31, 29\}]_{(b),(c)}$, as well as $[\{7, 10, 4\}]_{(a),(c)}$ and $[\{10, 12, 13\}]_{(a),(c)}$, respectively.

The logic in Figure 6a allows 13 (non-unital on 16) two-valued states whose convex sum yields the following weights:

$$p_{\mathbf{a}} = \lambda_1,$$
$$p_{\mathbf{b}} = \lambda_2 + \lambda_3 + \lambda_4 + \lambda_5 + \lambda_6 + \lambda_7,$$
$$p_{16} = 0,$$
$$p_1 = \lambda_2 + \lambda_3 + \lambda_4 + \lambda_5 + \lambda_6 + \lambda_7 + \lambda_8 + \lambda_9 + \lambda_{10} + \lambda_{11},$$
$$p_2 = \lambda_{12} + \lambda_{13},$$
$$p_3 = \lambda_1 + \lambda_8 + \lambda_9 + \lambda_{10} + \lambda_{11},$$
$$p_4 = \lambda_2 + \lambda_3 + \lambda_8 + \lambda_9 + \lambda_{12},$$
$$p_5 = \lambda_4 + \lambda_5 + \lambda_6 + \lambda_7 + \lambda_{10} + \lambda_{11} + \lambda_{13},$$
$$p_6 = \lambda_8 + \lambda_9 + \lambda_{10} + \lambda_{12},$$
$$p_7 = \lambda_1 + \lambda_{11} + \lambda_{13},$$
$$p_8 = \lambda_1 + \lambda_8 + \lambda_{10} + \lambda_{11} + \lambda_{13},$$
$$p_9 = \lambda_9 + \lambda_{12},$$
$$p_{10} = \lambda_4 + \lambda_5 + \lambda_6 + \lambda_7 + \lambda_{10},$$
$$p_{11} = \lambda_1 + \lambda_2 + \lambda_3 + \lambda_8,$$
$$p_{12} = \lambda_2 + \lambda_3 + \lambda_8 + \lambda_9 + \lambda_{11},$$
$$p_{13} = \lambda_1 + \lambda_{12} + \lambda_{13},$$
$$p_{14} = \lambda_4 + \lambda_5 + \lambda_6 + \lambda_7 + \lambda_9 + \lambda_{10} + \lambda_{11} + \lambda_{13},$$
$$p_{15} = \lambda_{12}, \tag{A7}$$
$$p_{17} = \lambda_1 + \lambda_{13},$$
$$p_{18} = \lambda_1 + \lambda_5 + \lambda_7 + \lambda_8 + \lambda_9 + \lambda_{10} + \lambda_{11},$$
$$p_{19} = \lambda_2 + \lambda_3 + \lambda_4 + \lambda_6 + \lambda_{12} + \lambda_{13},$$
$$p_{20} = \lambda_3 + \lambda_6 + \lambda_7 + \lambda_8 + \lambda_9 + \lambda_{10} + \lambda_{11},$$
$$p_{21} = \lambda_2 + \lambda_4 + \lambda_5 + \lambda_{12},$$
$$p_{22} = \lambda_5 + \lambda_7,$$
$$p_{23} = \lambda_3 + \lambda_6 + \lambda_7 + \lambda_{13},$$
$$p_{24} = \lambda_2 + \lambda_3 + \lambda_4 + \lambda_6 + \lambda_8 + \lambda_9 + \lambda_{10} + \lambda_{11} + \lambda_{12},$$
$$p_{25} = \lambda_1 + \lambda_{13},$$
$$p_{26} = \lambda_1 + \lambda_2 + \lambda_4 + \lambda_5 + \lambda_8 + \lambda_9 + \lambda_{10} + \lambda_{11},$$
$$p_{27} = \lambda_{12},$$
$$p_{28} = \lambda_1 + \lambda_4 + \lambda_6 + \lambda_{10} + \lambda_{11} + \lambda_{13},$$
$$p_{30} = \lambda_2 + \lambda_3 + \lambda_5 + \lambda_7 + \lambda_8 + \lambda_9,$$
$$p_{32} = \lambda_2 + \lambda_3 + \lambda_4 + \lambda_5 + \lambda_6 + \lambda_7 + \lambda_9 + \lambda_{12},$$
$$p_{33} = \lambda_2 + \lambda_3 + \lambda_4 + \lambda_5 + \lambda_6 + \lambda_7 + \lambda_{11},$$
$$p_{34} = \lambda_2 + \lambda_3 + \lambda_4 + \lambda_5 + \lambda_6 + \lambda_7 + \lambda_8 + \lambda_9 + \lambda_{10},$$
$$p_{35} = \lambda_2 + \lambda_3 + \lambda_4 + \lambda_5 + \lambda_6 + \lambda_7 + \lambda_8 + \lambda_{10} + \lambda_{11}.$$

Therefore, whenever **a** is true, that is, $p_{\mathbf{a}} = \lambda_1 = 1$, **b** has to be false, because $p_{\mathbf{b}} = \lambda_2 + \lambda_3 + \lambda_4 + \lambda_5 + \lambda_6 + \lambda_7 = 0$.

Conversely, the logic in Figure 6b allows 13 (non-separating on 15/27 and non-unital on 16) two-valued states whose convex sum yields the following weights:

$$p_{\mathbf{a}} = \lambda_1,$$
$$p_{\mathbf{b}} = \lambda_1 + \lambda_2 + \lambda_3 + \lambda_4 + \lambda_5,$$
$$p_{16} = 0,$$
$$p_1 = \lambda_2 + \lambda_3 + \lambda_4 + \lambda_5 + \lambda_6 + \lambda_7 + \lambda_8 + \lambda_9 + \lambda_{10} + \lambda_{11},$$
$$p_2 = \lambda_{12} + \lambda_{13},$$
$$p_3 = \lambda_6 + \lambda_7 + \lambda_8 + \lambda_9 + \lambda_{10} + \lambda_{11},$$
$$p_4 = \lambda_2 + \lambda_3 + \lambda_6 + \lambda_7 + \lambda_8 + \lambda_9 + \lambda_{12},$$
$$p_5 = \lambda_4 + \lambda_5 + \lambda_{10} + \lambda_{11} + \lambda_{13},$$
$$p_6 = \lambda_6 + \lambda_7 + \lambda_{10} + \lambda_{13},$$
$$p_7 = \lambda_8 + \lambda_9 + \lambda_{11} + \lambda_{12},$$
$$p_8 = \lambda_6 + \lambda_8 + \lambda_{10} + \lambda_{11} + \lambda_{12} + \lambda_{13},$$
$$p_9 = \lambda_7 + \lambda_9,$$
$$p_{11} = \lambda_1 + \lambda_2 + \lambda_3 + \lambda_6 + \lambda_8 + \lambda_{12},$$
$$p_{13} = \lambda_1 + \lambda_{12} + \lambda_{13},$$
$$p_{14} = \lambda_4 + \lambda_5 + \lambda_7 + \lambda_9 + \lambda_{10} + \lambda_{11},$$
$$p_{15} = p_{27} = \lambda_{13},$$
$$p_{17} = \lambda_1 + \lambda_{12}, \tag{A8}$$
$$p_{18} = \lambda_5 + \lambda_6 + \lambda_7 + \lambda_8 + \lambda_9 + \lambda_{10} + \lambda_{11} + \lambda_{13},$$
$$p_{19} = \lambda_1 + \lambda_2 + \lambda_3 + \lambda_4 + \lambda_{12},$$
$$p_{20} = \lambda_3 + \lambda_6 + \lambda_7 + \lambda_8 + \lambda_9 + \lambda_{10} + \lambda_{11},$$
$$p_{21} = \lambda_2 + \lambda_4 + \lambda_5 + \lambda_{13},$$
$$p_{22} = \lambda_5 + \lambda_{13},$$
$$p_{23} = \lambda_1 + \lambda_3 + \lambda_{12},$$
$$p_{24} = \lambda_2 + \lambda_3 + \lambda_4 + \lambda_6 + \lambda_7 + \lambda_8 + \lambda_9 + \lambda_{10} + \lambda_{11},$$
$$p_{25} = \lambda_1 + \lambda_{12},$$
$$p_{26} = \lambda_2 + \lambda_4 + \lambda_5 + \lambda_6 + \lambda_7 + \lambda_8 + \lambda_9 + \lambda_{10} + \lambda_{11},$$
$$p_{28} = \lambda_1 + \lambda_4 + \lambda_{10} + \lambda_{11},$$
$$p_{29} = \lambda_2 + \lambda_6 + \lambda_7 + \lambda_8 + \lambda_9,$$
$$p_{30} = \lambda_2 + \lambda_3 + \lambda_5 + \lambda_6 + \lambda_7 + \lambda_8 + \lambda_9 + \lambda_{12},$$
$$p_{31} = \lambda_3 + \lambda_4 + \lambda_5 + \lambda_{10} + \lambda_{11},$$
$$p_{32} = \lambda_1 + \lambda_2 + \lambda_3 + \lambda_4 + \lambda_5 + \lambda_7 + \lambda_9,$$
$$p_{33} = \lambda_2 + \lambda_3 + \lambda_4 + \lambda_5 + \lambda_8 + \lambda_9 + \lambda_{11},$$
$$p_{34} = \lambda_1 + \lambda_2 + \lambda_3 + \lambda_4 + \lambda_5 + \lambda_6 + \lambda_7 + \lambda_{10},$$
$$p_{35} = \lambda_2 + \lambda_3 + \lambda_4 + \lambda_5 + \lambda_6 + \lambda_8 + \lambda_{10} + \lambda_{11} + \lambda_{13}.$$

Therefore, whenever **a** is true, that is, $p_{\mathbf{a}} = \lambda_1 = 1$, **b** has to be true, because $p_{\mathbf{b}} = \lambda_1 + \lambda_2 + \lambda_3 + \lambda_4 + \lambda_5 = \lambda_1 = 1$.

References

1. Gleason, A.M. Measures on the closed subspaces of a Hilbert space. *J. Math. Mech.* **1957**, *6*, 885–893. [CrossRef]
2. Mermin, D.N. *Quantum Computer Science*; Cambridge University Press: Cambridge, UK, 2007.

3. Halmos, P.R. *Finite-Dimensional Vector Spaces*; Undergraduate Texts in Mathematics; Springer: New York, NY, USA, 1958.
4. Svozil, K. *Randomness & Undecidability in Physics*; World Scientific: Singapore, 1993.
5. Dvurečenskij, A.; Pulmannová, S.; Svozil, K. Partition Logics, Orthoalgebras and Automata. *Helv. Phys. Acta* **1995**, *68*, 407–428.
6. Svozil, K. *Quantum Logic*; Springer: Singapore, 1998.
7. Svozil, K. Logical equivalence between generalized urn models and finite automata. *Int. J. Theor. Phys.* **2005**, *44*, 745–754. [CrossRef]
8. Svozil, K. Contexts in quantum, classical and partition logic. In *Handbook of Quantum Logic and Quantum Structures*; Engesser, K., Gabbay, D.M., Lehmann, D., Eds.; Elsevier: Amsterdam, The Netherland, 2009; pp. 551–586.
9. Svozil, K. Generalized event structures and probabilities. In *Information and Complexity*; Burgin, M., Calude, C.S., Eds.; World Scientific Series in Information Studies; World Scientific: Singapore, 2016; Volume 6, Chapter 11, pp. 276–300.
10. Svozil, K. *Physical [A]Causality. Determinism, Randomness and Uncaused Events*; Springer: Cham, Switzerland; Berlin/Heidelberg, Germany; New York, NY, USA, 2018.
11. Chevalier, G. Commutators and decompositions of orthomodular lattices. *Order* **1989**, *6*, 181–194. [CrossRef]
12. Moore, E.F. Gedanken-Experiments on Sequential Machines. In *Automata Studies*; Shannon, C.E., McCarthy, J., Eds.; Princeton University Press: Princeton, NJ, USA, 1956; pp. 129–153.
13. Schaller, M.; Svozil, K. Automaton partition logic versus quantum logic. *Int. J. Theor. Phys.* **1995**, *34*, 1741–1750. [CrossRef]
14. Schaller, M.; Svozil, K. Automaton logic. *Int. J. Theor. Phys.* **1996**, *35*. [CrossRef]
15. Wright, R. Generalized urn models. *Found. Phys.* **1990**, *20*, 881–903. [CrossRef]
16. Svozil, K. Staging quantum cryptography with chocolate balls. *Am. J. Phys.* **2006**, *74*, 800–803. [CrossRef]
17. Svozil, K. Non-contextual chocolate ball versus value indefinite quantum cryptography. *Theor. Comput. Sci.* **2014**, *560*, 82–90. [CrossRef]
18. Svozil, K. On generalized probabilities: Correlation polytopes for automaton logic and generalized urn models, extensions of quantum mechanics and parameter cheats. *arXiv* **2000**, arXiv:quant-ph/0012066.
19. Boole, G. On the Theory of Probabilities. *Philos. Trans. R. Soc. Lond.* **1862**, *152*, 225–252. [CrossRef]
20. Froissart, M. Constructive generalization of Bell's inequalities. *Il Nuovo Cimento B* **1981**, *64*, 241–251. [CrossRef]
21. Cirel'son, B.S. Some results and problems on quantum Bell-type inequalities. *Hadron. J. Suppl.* **1993**, *8*, 329–345.
22. Pitowsky, I. The range of quantum probabilities. *J. Math. Phys.* **1986**, *27*, 1556–1565. [CrossRef]
23. Pitowsky, I. George Boole's 'Conditions of Possible Experience' and the Quantum Puzzle. *Br. J. Philos. Sci.* **1994**, *45*, 95–125. [CrossRef]
24. Richard, J. Orthomodular lattices admitting no states. *J. Comb. Theory Ser. A* **1971**, *10*, 119–132.
25. Fukuda, K. cdd and cddplus Homepage, cddlib Package cddlib-094h, 2000. Available online: http://www.inf.ethz.ch/personal/fukudak/cdd_home/ (accessed on 1 July 2017).
26. Wright, R. The state of the pentagon. A nonclassical example. In *Mathematical Foundations of Quantum Theory*; Marlow, A.R., Ed.; Academic Press: New York, NY, USA, 1978; pp. 255–274.
27. Kalmbach, G. *Orthomodular Lattices (London Mathematical Society Monographs)*; Academic Press: London, UK; New York, NY, USA, 1983; Volume 18.
28. Beltrametti, E.G.; Mączyński, M.J. On the range of non-classical probability. *Rep. Math. Phys.* **1995**, *36*. [CrossRef]
29. Klyachko, A.A.; Can, M.A.; Binicioğlu, S.; Shumovsky, A.S. Simple Test for Hidden Variables in Spin-1 Systems. *Phys. Rev. Lett.* **2008**, *101*, 020403. [CrossRef] [PubMed]
30. Bub, J.; Stairs, A. Contextuality and Nonlocality in 'No Signaling' Theories. *Found. Phys.* **2009**, *39*. [CrossRef]
31. Bub, J.; Stairs, A. Contextuality in Quantum Mechanics: Testing the Klyachko Inequality. *arXiv* **2010**, arXiv:1006.0500.
32. Badziąg, P.; Bengtsson, I.; Cabello, A.; Granström, H.; Larsson, J.A. Pentagrams and Paradoxes. *Found. Phys.* **2011**, *41*. [CrossRef]

33. Kochen, S.; Specker, E.P. Logical Structures arising in quantum theory. In *The Theory of Models, Proceedings of the 1963 International Symposium at Berkeley*; North Holland: Amsterdam, The Netherland; New York, NY, USA; Oxford, UK, 1965; pp. 177–189.

34. Kochen, S.; Specker, E.P. The Problem of Hidden Variables in Quantum Mechanics. *J. Math. Mech.* **1967**, *17*, 59–87. [CrossRef]

35. Redhead, M. *Incompleteness, Nonlocality, and Realism: A Prolegomenon to the Philosophy of Quantum Mechanics*; Clarendon Press: Oxford, UK, 1990.

36. Pitowsky, I. Betting on the outcomes of measurements: A Bayesian theory of quantum probability. *Stud. Hist. Philos. Sci. Part B Stud. Hist. Philos. Mod. Phys.* **2003**, *34*, 395–414. [CrossRef]

37. Pitowsky, I. Quantum Mechanics as a Theory of Probability. In *Physical Theory and Its Interpretation*; Demopoulos, W., Pitowsky, I., Eds.; The Western Ontario Series in Philosophy of Science; Springer: Dordrecht, The Netherlands, 2006; Volume 72, pp. 213–240.

38. Belinfante, F.J. *A Survey of Hidden-Variables Theories*; International Series of Monographs in Natural Philosophy; Pergamon Press, Elsevier: Oxford, UK; New York, NY, USA, 1973; Volume 55.

39. Stairs, A. Quantum logic, realism, and value definiteness. *Philos. Sci.* **1983**, *50*, 578–602. [CrossRef]

40. Getting contextual and nonlocal elements-of-reality the easy way. *Am. J. Phys.* **1993**, *61*, 443–447. [CrossRef]

41. Pták, P.; Pulmannová, S. *Orthomodular Structures as Quantum Logics. Intrinsic Properties, State Space and Probabilistic Topics*; Fundamental Theories of Physics; Kluwer Academic Publishers, Springer: Dordrecht, The Netherlands, 1991; Volume 44.

42. Navara, M.; Rogalewicz, V. The pasting constructions for orthomodular posets. *Math. Nachr.* **1991**, *154*, 157–168. [CrossRef]

43. Johansen, H.B. Comment on Getting contextual and nonlocal elements-of-reality the easy way. *Am. J. Phys.* **1994**, *62*, 471. [CrossRef]

44. Vermaas, P.E. Comment on Getting contextual and nonlocal elements-of-reality the easy way. *Am. J. Phys.* **1994**, *62*, 658. [CrossRef]

45. Cabello, A.; Portillo, J.R.; Solís, A.; Svozil, K. Minimal true-implies-false and true-implies-true sets of propositions in noncontextual hidden variable theories. *arXiv* **2013**, arXiv:1805.00796.

46. Svozil, K. Quantum Scholasticism: On Quantum Contexts, Counterfactuals, and the Absurdities of Quantum Omniscience. *Inf. Sci.* **2009**, *179*, 535–541. [CrossRef]

47. Cabello, A. A simple proof of the Kochen-Specker theorem. *Eur. J. Phys.* **1994**, *15*, 179–183. [CrossRef]

48. Cabello, A. Pruebas Algebraicas de Imposibilidad de Variables Ocultas en Mecánica Cuántica. Ph.D. Thesis, Universidad Complutense de Madrid, Madrid, Spain, 1996.

49. Tkadlec, J. Greechie diagrams of small quantum logics with small state spaces. *Int. J. Theor. Phys.* **1998**, *37*, 203–209. [CrossRef]

50. Svozil, K.; Tkadlec, J. Greechie diagrams, nonexistence of measures in quantum logics and Kochen–Specker type constructions. *J. Math. Phys.* **1996**, *37*, 5380–5401. [CrossRef]

51. Abbott, A.A.; Calude, C.S.; Svozil, K. A variant of the Kochen-Specker theorem localising value indefiniteness. *J. Math. Phys.* **2015**, *56*, 102201. [CrossRef]

52. Pitowsky, I. Substitution and Truth in Quantum Logic. *Philos. Sci.* **1982**, *49*. [CrossRef]

53. Hardy, L. Quantum mechanics, local realistic theories, and Lorentz-invariant realistic theories. *Phys. Rev. Lett.* **1992**, *68*, 2981–2984. [CrossRef] [PubMed]

54. Hardy, L. Nonlocality for two particles without inequalities for almost all entangled states. *Phys. Rev. Lett.* **1993**, *71*, 1665–1668. [CrossRef] [PubMed]

55. Boschi, D.; Branca, S.; De Martini, F.; Hardy, L. Ladder Proof of Nonlocality without Inequalities: Theoretical and Experimental Results. *Phys. Rev. Lett.* **1997**, *79*, 2755–2758. [CrossRef]

56. Cabello, A.; García-Alcaine, G. A hidden-variables versus quantum mechanics experiment. *J. Phys. A Math. Gen. Phys.* **1995**, *28*. [CrossRef]

57. Cabello, A.; Estebaranz, J.M.; García-Alcaine, G. Bell-Kochen-Specker theorem: A proof with 18 vectors. *Phys. Lett. A* **1996**, *212*, 183–187. [CrossRef]

58. Cabello, A. No-hidden-variables proof for two spin- particles preselected and postselected in unentangled states. *Phys. Rev. A* **1997**, *55*, 4109–4111. [CrossRef]

59. Chen, J.L.; Cabello, A.; Xu, Z.P.; Su, H.Y.; Wu, C.; Kwek, L.C. Hardy's paradox for high-dimensional systems. *Phys. Rev. A* **2013**, *88*, 062116. [CrossRef]

60. Cabello, A.; Badziag, P.; Terra Cunha, M.; Bourennane, M. Simple Hardy-Like Proof of Quantum Contextuality. *Phys. Rev. Lett.* **2013**, *111*, 180404. [CrossRef] [PubMed]
61. Pitowsky, I. Infinite and finite Gleason's theorems and the logic of indeterminacy. *J. Math. Phys.* **1998**, *39*, 218–228. [CrossRef]
62. Hrushovski, E.; Pitowsky, I. Generalizations of Kochen and Specker's theorem and the effectiveness of Gleason's theorem. *Stud. Hist. Philos. Sci. Part B Stud. Hist. Philos. Mod. Phys.* **2004**, *35*, 177–194. [CrossRef]
63. Abbott, A.A.; Calude, C.S.; Conder, J.; Svozil, K. Strong Kochen-Specker theorem and incomputability of quantum randomness. *Phys. Rev. A* **2012**, *86*, 062109. [CrossRef]
64. Abbott, A.A.; Calude, C.S.; Svozil, K. Value-indefinite observables are almost everywhere. *Phys. Rev. A* **2014**, *89*, 032109. [CrossRef]
65. Pitowsky, I. Deterministic model of spin and statistics. *Phys. Rev. D* **1983**, *27*, 2316–2326. [CrossRef]
66. Meyer, D.A. Finite precision measurement nullifies the Kochen-Specker theorem. *Phys. Rev. Lett.* **1999**, *83*, 3751–3754. [CrossRef]
67. Schrödinger, E. Die gegenwärtige Situation in der Quantenmechanik. *Naturwissenschaften* **1935**, *23*, 807–812. [CrossRef]
68. London, F.; Bauer, E. The Theory of Observation in Quantum Mechanics. In *Quantum Theory and Measurement*; Princeton University Press: Princeton, NJ, USA, 1983; pp. 217–259.
69. Svozil, K. Quantum information via state partitions and the context translation principle. *J. Mod. Opt.* **2004**, *51*, 811–819. [CrossRef]
70. Howard, D. Who Invented the "Copenhagen Interpretation"? A Study in Mythology. *Philos. Sci.* **2004**, *71*, 669–682. [CrossRef]
71. Bohr, N. The quantum postulate and the recent development of atomistic theory. *Nature* **1928**, *121*, 580–590. [CrossRef]
72. von Neumann, J. *Mathematische Grundlagen der Quantenmechanik*, 2nd ed.; Springer: Berlin/Heidelberg, Germany, 1996.
73. Zeilinger, A. A Foundational Principle for Quantum Mechanics. *Found. Phys.* **1999**, *29*, 631–643. [CrossRef]
74. Myrvold, W.C. Statistical mechanics and thermodynamics: A Maxwellian view. *Stud. Hist. Philos. Sci. Part B Stud. Hist. Philos. Mod. Phys.* **2011**, *42*, 237–243. [CrossRef]
75. Specker, E. Die Logik nicht gleichzeitig entscheidbarer Aussagen. *Dialectica* **1960**, *14*, 239–246. [CrossRef]
76. Kochen, S.; Specker, E.P. The calculus of partial propositional functions. In Proceedings of the 1964 International Congress for Logic, Methodology and Philosophy of Science, Jerusalem, Israel, 26 August–2 September 1964; North Holland: Amsterdam, The Netherlands, 1965; pp. 45–57.
77. Valentini, A. *The de Broglie-Bohm Pilot-Wave Theory*; Lecture Series on Foundations of Physics: Scientific Realism; University of Vienna: Vienna, Austria, 2018.
78. Plato. *The Republic*; Cambridge Texts in the History of Political Thought; Ferrari, G.R.F., Ed.; Cambridge University Press: Cambridge, UK, 2000.
79. Tkadlec, J. (Czech Technical University in Prague). Personal communication, 23 August 2017.

entropy

MDPI

Article

Adiabatic Quantum Computation Applied to Deep Learning Networks

Jeremy Liu [1,2,*], **Federico M. Spedalieri** [2,3], **Ke-Thia Yao** [2], **Thomas E. Potok** [4], **Catherine Schuman** [4], **Steven Young** [4], **Robert Patton** [4], **Garrett S. Rose** [5] **and Gangotree Chamka** [5]

[1] Department of Computer Science, University of Southern California, Los Angeles, CA 90089, USA
[2] Information Sciences Institute, University of Southern California, Marina del Rey, CA 90292, USA;
     fspedali@isi.edu (F.M.S.); kyao@isi.edu (K.-T.Y.)
[3] Department of Electrical Engineering, University of Southern California, Los Angeles, CA 90089, USA
[4] Computational Data Analytics Group, Oak Ridge National Laboratory, Oak Ridge, TN 37830, USA;
     potokte@ornl.gov (T.E.P.); schumancd@ornl.gov (C.S.); youngsr@ornl.gov (S.Y.); pattonrm@ornl.gov (R.P.)
[5] Department of Electrical Engineering & Computer Science, University of Tennessee, Knoxville, TN 37996,
     USA; garose@utk.edu (G.S.R.); gchakma@vols.utk.edu (G.C.)
* Correspondence: jeremyjl@usc.edu

Received: 6 April 2018; Accepted: 16 May 2018; Published: 18 May 2018

Abstract: Training deep learning networks is a difficult task due to computational complexity, and this is traditionally handled by simplifying network topology to enable parallel computation on graphical processing units (GPUs). However, the emergence of quantum devices allows reconsideration of complex topologies. We illustrate a particular network topology that can be trained to classify MNIST data (an image dataset of handwritten digits) and neutrino detection data using a restricted form of adiabatic quantum computation known as quantum annealing performed by a D-Wave processor. We provide a brief description of the hardware and how it solves Ising models, how we translate our data into the corresponding Ising models, and how we use available expanded topology options to explore potential performance improvements. Although we focus on the application of quantum annealing in this article, the work discussed here is just one of three approaches we explored as part of a larger project that considers alternative means for training deep learning networks. The other approaches involve using a high performance computing (HPC) environment to automatically find network topologies with good performance and using neuromorphic computing to find a low-power solution for training deep learning networks. Our results show that our quantum approach can find good network parameters in a reasonable time despite increased network topology complexity; that HPC can find good parameters for traditional, simplified network topologies; and that neuromorphic computers can use low power memristive hardware to represent complex topologies and parameters derived from other architecture choices.

Keywords: deep learning; quantum computing; neuromorphic computing; high performance computing

1. Introduction

A neural network is a machine learning concept originally inspired by studies of the visual cortex of the brain. In biology, neural networks are the neurons of the brain connected to each other via synapses; accordingly, in machine learning, they are graphical models where variables are connected to each other with certain weights. Both are highly useful in analyzing image data, but practical considerations regarding network topology limit the potential of simulating neural networks on computers. Simulated networks tend to divide neurons into different layers and prohibit intralayer connections. Many-layered networks are called deep learning networks, and the restriction of intralayer connections allows rapid training on graphical processing units (GPUs).

We explain some current limitations of deep learning networks and offer approaches to help mitigate them. For this article we focus on a quantum adiabatic computing approach, which is one of a trio in a larger project to survey machine learning in non-traditional computing environments, though we also describe the other approaches at a high level to offer comparison and context for experiment designs. The second approach uses a high performance computing environment to automatically discover good network topologies, albeit they remain restricted from using intralayer connections. The third approach uses neuromorphic computing as a low-power alternative for representing neural networks. Rather than explicitly choosing one solution or another, these approaches are meant to augment each other. Describing these different approaches necessitates a brief description of various machine learning models and networks including Boltzmann machines (BMs), convolutional neural networks (CNNs), and spiking neural networks (SNNs). Results obtained from CNNs and SNNs, while important to our project, are not the focus of this article and are presented in the appendix.

1.1. Boltzmann Machines

A Boltzmann machine is an energy-based generative model of data. BMs contain binary units, and each possible configuration of units is assigned a certain energy based on edge weights. The goal of training is to find edge weights that result in low energy configurations for patterns more likely to occur in data. Since BMs can be represented as Ising models, and because the D-Wave processor is designed to natively solve Ising models, BMs are particularly attractive for our purposes. We tend to view BMs as probabilistic neural networks with symmetrically connected units [1]. BMs are well suited to solving constraint satisfaction tasks with many weak constraints, including digit and object recognition, compression/coding, and natural language processing.

A common algorithm for training BMs exposes a BM to input data and updates the weights in order to maximize the likelihood that the underlying model of the BM reproduces the data set. This method requires computing certain quantities which, due to the specific form of the BM, turn out to be the values of certain correlation functions in thermal equilibrium. However, training is a slow and arduous task if we allow models with unrestricted topology. Connectivity loops slow down the convergence of many algorithms used to estimate thermal equilibrium properties. Simulated annealing is a generic and widely used algorithm to reach this thermal equilibrium, but this remains a slow and expensive process for large networks. This forces us to either use tiny networks or to give up complex topologies, with the latter option leading to the popular choice of using restricted Boltzmann machines (RBMs) [2].

Units in RBMs are categorized as "visible" or "hidden." During training, the visible units of a RBM represent the input dataset whereas the hidden units represent latent factors that control the data distribution. After undergoing the above training process, an RBM will produce a distribution of visible unit states that should closely match the input dataset. Additionally, only bipartite connectivity between the two types is allowed, which makes parallel computation feasible. Figure 1 shows an example of this bipartite connectivity. Approximation algorithms make training tractable in practice, and RBMs can be stacked together to form deep belief networks (DBNs) [3].

1.2. Convolutional Neural Networks

Of the many designs for deep learning networks, CNNs have become the most widely used for analyzing image data [4]. As with other deep learning networks, CNNs contain many layers of neural units with many connections between different layers but no connections between units of a particular layer. They also use standard stochastic gradient descent and back-propagation combined with labeled data to train. What separates a CNN from other networks are its unique connectivity arrangement and different types of layers. See Figure 2 for a high-level diagram of the CNN architecture.

One type of layer in CNNs is the convolutional layer. Unlike in other neural networks, a convolutional layer uses a kernel, or small set of shared weights, to produce a feature map of the input to the layer, and many convolutional layers operate in succession. Other networks would typically have every input unit connected to every processing unit in a layer whereas a CNN is satisfied with using convolution to produce sparse connections between layers—see Figure 1 for the dense connectivity of a BM and compare it against the sparse CNN connectivity shown in Figure 3. A kernel captures a certain feature from the input, and convolving a kernel with the data finds this feature across the whole input. For example, a kernel that detects diagonal lines can be convolved with an image to produce a feature map that can be interpreted as identifying all areas of an image that contain diagonal lines.

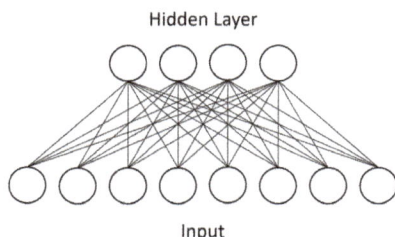

Figure 1. A Boltzmann machine is divided into a visible layer, representing the data input, and a hidden layer, which represents latent factors controlling the data distribution. This diagram shows the restricted Boltzmann machine, or RBM, in which intralayer connections are prohibited. Each connection between units is a separate weight parameter which is discovered through training.

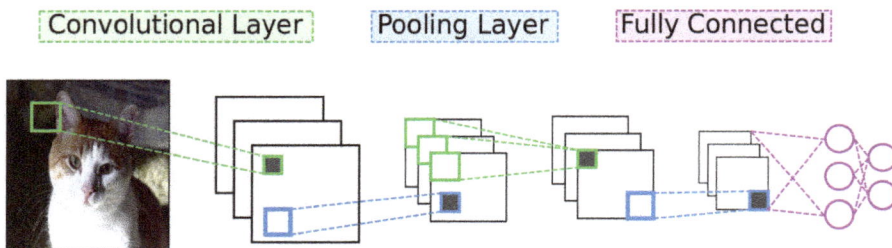

Figure 2. A convolutional neural network is composed of a series of alternating convolutional and pooling layers. Each convolutional layer extracts features from its preceding layer to form feature maps. These feature maps are then down-sampled by a pooling layer to exploit data locality. A perceptron, a simple type of classification network, is placed as the last layer of the CNN.

The second type of layer is the pooling layer. Pooling layers use the many feature maps produced by convolutional layers as input and subsample them to produce smaller feature maps to help take advantage of data locality within images. CNNs use alternating layers of convolutional and pooling layers to extract and abstract image features. Pooling operations makes feature detection in CNNs resilient to position shifts in images [5].

Convolutional Layer

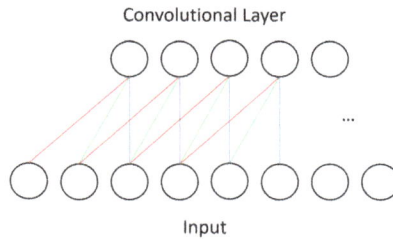

Input

Figure 3. The connectivity in a CNN is sparse relative to the previously shown BM model. Additionally, the set of weights is shared between units, unlike in BMs. In this illustration we symbolize this with the red, green, and blue connections to show that each unit in the convolutional layer applies the same operation to different segments of the input.

1.3. Spiking Neural Networks

SNNs differ from both BMs and CNNs by incorporating the extra dimension of time into how information is processed. BMs and CNNs do not have a sense of time built into their architectures—neural unit activity is iteratively calculated on a layer-by-layer basis. SNNs instead use integrate-and-fire neurons, units that collect activation potential over time and fire or "spike" upon reaching a threshold, after which they cannot fire during what is known as a refractory period. Additionally, synapses in a SNN can include programmable delay components, where larger delay values on the synapse correspond to longer propagation time of signals along that synapse. Additionally, there is not necessarily a division of units into well-organized layers in a SNN, and input is fed to the network over time.

SNNs have great potential in moving away from the traditional implementation of machine learning algorithms on the CPU/memory von Neumann architecture. For example, the CPU/memory model, while useful on many diverse applications, has the drawback of high power requirements. Nature's biological neural networks have extremely low power requirements by comparison. There are many different ways to implement neuromorphic systems, but one of the more promising device types to include in neuromorphic systems is memristors. Development of memristive technology opens the potential of running spiking neural networks using low power consumption on neuromorphic architectures.

A key challenge associated with SNNs in general and SNNs for neuromorphic systems in particular is determining the correct training or learning algorithm with which to build the SNN. Though there have been efforts to map existing architectures like CNNs to equivalent spiking neuromorphic systems [6,7], there is also potential to develop independent deep learning architectures that exploit the temporal processing power of SNNs.

1.4. Challenges

Complex networks pose enormous problems for deep learning, three of which we identify. How we tackle each of these challenges is the basis of our project, where we seek relief from these issues through quantum adiabatic computing, high performance computing, and neuromorphic computing.

The first of these challenges comes from complex network topology in neural networks. By complex network topology we mean bidirectional connections and looping connectivity between neural units, which slow training to a crawl. The training algorithms we know for such complex networks have greater than polynomial runtime, making them effectively intractable and untenable for practical purposes. Deep networks deployed on real-world problems, like the previously discussed CNN architecture, instead impose limitations on network topology. Removing intralayer connections or enforcing strict rules for network topology allows faster and tractable training algorithms to run. However, doing so takes away some of the representational power of the network [8], and these restricted or limited networks do not reflect models found in nature. While tractable models perform remarkably well on specialized classification tasks, we speculate that other more complex and generalized tasks may benefit from the additional representational power offered by complex networks. We believe quantum adiabatic computing offers part of a potential solution through its ability to sample from complex probability distributions such as those generated by neural networks containing intralayer connections.

The second challenge is automatically discovering optimal or near-optimal network hyperparameters and topologies. Hyperparameters in deep learning refer to the model parameters, i.e., the activation function used, the number of hidden units in a layer, the kernel size of a convolutional layer, and the learning rate of the solver. Currently the best deep learning models are discovered by creating, training, testing, and tuning many models on some well-known reference dataset and reporting the best model in the literature. However, if the dataset has not been examined before, it is difficult to know how to tune networks for optimal performance. GPU-based high performance computing provides an opportunity to automate much of this process—to train, test, and evolve thousands of deep learning networks to find optimally-performing network hyperparameters and network topologies.

The last challenge is power consumption, which we can help address through neuromorphic computing. Machine learning's computational needs have so far been met with power-hungry CPUs and more recently GPUs. The switch from CPUs to GPUs has significantly sped up computation and lowered computation costs, but GPU efficiency in training networks still pales in comparison to the efficiency of biological brains. For an image recognition task, it might take many server farms and a hydroelectric dam to compete with a mundane human brain running on a bit of glucose. Neuromorphic computing offers a potential solution by developing specialized low-power hardware that can implement SNNs approximating trained networks derived from more orthodox architectures.

This article focuses on deep learning's challenges related to quantum adiabatic computing. Though high performance and neuromorphic computing are an integral part of our project, we move discussions of these topics to the appendix to better fit our focus for this journal, though mentions of both appear as necessary through the rest of the article. Our experiments use the MNIST dataset [9], an image dataset of handwritten digits, and a neutrino particle detection dataset produced by Fermi National Accelerator Laboratory. Next, we will review works related to quantum computing; then we provide our experimental approach, results, and future research.

2. Related Work

We look at the current state-of-art quantum computing as it relates to the previously discussed challenges in deep learning. Work related to high performance computing and neuromorphic computing are presented in Appendix A. Though the papers and articles referenced in the appendix are not strictly related to quantum adiabatic computing, they provide context for the larger ORNL project and present existing or proposed systems that can be compared against our own quantum computing efforts.

Feynman first discussed quantum computing within the context of simulation, noting that simulating a quantum system using a classical computer seems to be intractable [10]. Interest in quantum computing surged with the introduction by Shor of a polynomial-time algorithm for factoring integers [11], giving an exponential speedup over the best known classical algorithm and threatening to break most modern encryption systems. As with Turing's work, these theories for quantum computing were developed before quantum hardware was available. Different models of quantum computing have since been developed in order to explore the power of quantum information processing. In the quantum circuit model (on which Shor's original algorithm relies), a sequence of unitary transformations are applied to a set of quantum bits (qubits), in a way analogous to the logical gates that are applied to classical bits in classical computing. In the measurement-based quantum computing model [12], a special quantum state is prepared beforehand, and a computation is performed by adaptively applying quantum gates to each qubit and measuring them. In the adiabatic quantum computing (AQC) model [13], a quantum state encoding the solution of a problem is prepared using the adiabatic theorem of quantum mechanics. All three models have the same computational power but also offer different trade-offs. Quantum information is extremely fragile, and any source of noise (like thermal fluctuations, unwanted interactions with an uncontrolled environment, etc.) can destroy the quantum features that are expected to provide a computational speedup. The AQC model has been considered as the most robust implementation, and hence the development of actual devices based on AQC has led that of the other two approaches, both of which are more susceptible to noise and require very large overhead to overcome the effects of that noise. However, currently available devices such as the D-Wave processor are still limited in many aspects, the most important being the fact that they operate at a finite temperature and that the effects of this noise in the performance of the device is still an active area of research. We typically refer to these devices operating at a finite temperature as quantum annealers.

Quantum annealers are in principle designed to solve a particular optimization problem, typically finding the ground state of an Ising Hamiltonian. Unfortunately, thermal fluctuations due to interactions with a finite temperature reservoir, in addition to unwanted quantum interactions with other systems in the environment, tend to kick the system out of its ground state and into an excited state. These unavoidable features make the quantum annealer behave more like a sampler than an exact optimizer in practice. However, this seemingly counterproductive property may be turned into an advantage since the ability to draw samples from complicated probability distributions is essential to probabilistic deep learning approaches such as the Boltzmann machine, which relies heavily upon sampling complex distributions in both training and output. Quantum annealers could then help us overcome the problem of complex topologies mentioned before. BMs in their unrestricted form are impractical to train on classical machines, a fact that led to the development of RBMs that eliminate intralayer edges and introduce bipartite connectivity [2]. Bipartite graphs allow the use of an algorithm known as contrastive divergence that approximates samples from a RBM in linear time, which is a critical tool for the practical usage of BMs because sampling is the core engine for training BMs. Quantum annealing hardware allows us to partially pull back from this bipartite limitation. Quantum annealers provide a novel way to sample richer topologies, and several approaches exploit this feature for different choices of graphs and topologies on D-Wave hardware [14–16].

3. Approach and Data

Quantum adiabatic computation, high performance computing, and neuromorphic computing differ significantly from each other in how they process data. As such, the amount of data each can support dictated our choice of deep learning problem that could be adapted to each of these three heterogeneous paradigms. At the time that the results were collected, D-Wave supported 1000 qubits (now 2000 qubits), which limited the size of problems we could solve. With this in mind we chose to examine two datasets we refer to as MNIST and neutrino data.

The Modified National Institute of Standards and Technology (MNIST) data set is a well-known collection of hand-written digits extensively studied in the deep learning community. The dataset is composed of images, each of which contains a handwritten digit and an associated label identifying the digit. The digit images are only $28 \times 28 = 784$ pixels, which fits within the 1000 qubit D-Wave hardware and onto HPC and neuromorphic architectures. Our later experiments used neutrino particle detection data down-sampled and adjusted to 32×32 pixels.

The neutrino scattering dataset was collected at Fermi National Accelerator Laboratory as part of the MINERvA experiment that is focused on vertex reconstruction [17]. In the Main Injector Experiment for v-A (MINERvA) experiment, many scintillator strips were arranged in planes orthogonal to the neutrino beam within the detector aligned across three different orientations or "views". We utilized both the energy lattice and the time lattice information in the dataset. In particular, we represented the energy lattice as an image, where the intensity of each pixel in the image corresponds to the average energy over time in the detection event. The images show the trajectory of particles over time from the view of one particular plane. We also used the time lattice in one of our experiments. For the time lattice, each data point in a detection event corresponds to the time at which every level exceeds a certain threshold. Associated with each detection event is a number corresponding to a specific detection plate within the chamber; this number indicates which plate a neutrino strikes. This number can then be utilized to determine in which detector region or segment the vertex of the event was located.

In BM experiments we used down-sampled and collated image data from one single plane. We did not use the original data because the quantum annealer has limited space for storing problems and because BMs are not well-suited to handling temporal data. However, the SNN experiments did take advantage of temporal data because SNNs are designed to handle such data. We offer more explanation on SNNs in Section 1.3.

Consideration of which deep learning networks to study on these platforms came next. Initially, CNNs seemed like an appropriate option, especially for HPC, but we ran into problems when considering the quantum environment. CNNs had consistently provided superior performance on standard datasets and had proven quite popular in the deep learning community. On a quantum platform, however, it became unclear how to effectively implement a CNN. Neither the circuit nor adiabatic optimization models offered good fits for CNNs, which operate using many successive layers of units. On the other hand, BMs and their probabilistic units were more like the sort of optimization problem that D-Wave hardware solves. Additionally, the quantum architecture allowed for intralayer connections between units that would normally be intractable for conventional machines to compute. Meanwhile, neuromorphic hardware running SNNs provided native time-based analysis models. BMs running on D-Wave and SNNs running on neuromorphic hardware were potentially offering distinct capabilities we believed could augment or strengthen CNN models trained in an HPC environment.

With this in mind we hope the following sections will illustrate the benefits of these different platforms. First we describe how we used a quantum annealer to train a BM containing intralayer connections and utilized the hardware to approximate samples from more complex probability distributions. Then, we show how we used a high performance computing cluster to automatically discover near-optimal topologies and parameters with evolutionary algorithms. Finally, we discuss how we natively implemented trained models produced by the previous two platforms on memristive hardware running spiking neural networks.

Because the adiabatic quantum computation portion of this project is of particular interest for this article, we next provide a more detailed description of the process and of the annealing hardware. Descriptions of the corresponding HPC and neuromorphic portions are left to Appendices B.1 and B.2.

3.1. Adiabatic Quantum Computation

Adiabatic quantum computation (AQC) is an implementation of the ideas of quantum computing that relies on the adiabatic theorem of quantum mechanics. This result states that if a system is in the ground state of a particular Hamiltonian and the parameters of this Hamiltonian are changed slowly enough, the system will remain in the ground state of the time-dependent Hamiltonian. This idea was used by Farhi et al. [13] to propose an alternative to the quantum circuit model of quantum computing. The main idea is to start with a Hamiltonian whose ground state is easy to construct, and slowly change it into one whose ground state encodes the answer to a particular problem.

One application of AQC is to solve combinatorial optimization problems, a particular example of which is finding the ground state of an Ising model. This model describes a system of interacting magnetic moments subject to local biases. This problem was shown by Barahona [18] to be NP-hard,

so many other optimization problems of practical interest can be recast in this form. If we consider
a set of spin variables $S_i = \pm 1$, the energy of the system is given by a quadratic expression of the form

$$E_{Ising}(s) = \sum_i h_i\, s_i + \sum_{i,j} J_{ij} s_i s_j \tag{1}$$

Solving this problem means finding a spin configuration that minimizes this energy function.
In a quantum approach, we consider a quantum system of interacting spins described by the
Ising Hamiltonian

$$H_{Ising} = \sum_i h_i\, \sigma_i^z + \sum_{i,j} J_{ij} \sigma_i^z \otimes \sigma_j^z \tag{2}$$

where h_i represent local magnetic fields and J_{ij} are couplings between spin pairs. This Hamiltonian is
diagonal in the σ^z basis, and its ground state can be used to construct the corresponding configuration
that minimizes the Ising energy above.

To solve this problem in the context of AQC we can choose an initial Hamiltonian of the form

$$H_0 = -\sum_i \sigma_i^x \tag{3}$$

that represents the effects of a transverse field applied to all spins. The ground state of H_0 consists in
all spins being in the $|+\rangle = (|0\rangle + |1\rangle)/\sqrt{2}$ state. If we consider the spins as little magnetic moments,
this corresponds to all spins pointing in the x direction. Quantum mechanically this state is separable,
easy to construct (just apply a strong magnetic field in the x direction), and when expressed in the
computational basis it is an equal superposition of all possible states.

The computation is performed by slowly changing the relative weights of H_0 and H_{Ising} during
the interval $[0, T]$

$$H(t) = (1 - (t/T))H_0 + (t/T)H_{Ising}. \tag{4}$$

This process is known as *quantum annealing*. The change must be slow compared to the time scale
associated with the minimum energy gap of the time-dependent Hamiltonian, where we define the
gap as the energy difference between the energies of the first excited state and the ground state [19–21].
If the change is too fast the system can transition to an excited state, and the state at the end of the
annealing will not be the ground state of the Ising Hamiltonian. On the other hand, if the change is too
slow the computation will take a long time. The main challenges in adiabatic quantum computing are
to understand the connection between this energy gap (i.e., the runtime) and the size of the problem,
and to find Hamiltonians that solve a given problem while possessing a larger gap [22]. However,
other issues are also important for practical implementations, in particular how unavoidable noise
affects the system due to the system's interaction with the environment.

3.2. The Superconducting Quantum Adiabatic Processor

The architecture and physical details of the quantum adiabatic processor we studied are described
in detail in [23]. In essence, it is designed to represent the Ising Hamiltonian as an array of
superconducting flux qubits with programmable interactions. The qubits are implemented using
superconducting quantum interference devices (SQUIDs) composed of a Niobium loop elongated
in one direction. Several loops and Josephson junctions are added to the design to both allow for
the required controls to implement quantum annealing and to compensate for the slight differences
between the physical properties of any two SQUIDs due to fabrication variations. The processor has a
unit-cell structure composed of 8 qubits with four arranged horizontally and four vertically such that
each vertical qubit intersects every horizontal one. At these intersections another SQUID is placed to
control the magnetic coupling between the corresponding horizontal and vertical qubits. These are
the only couplings allowed (i.e., horizontal qubits are not coupled to other horizontal qubits). This
architecture results in a coupling graph that is fully bipartite at the unit cell level. The processor is

then built by adjoining more unit cells in a square lattice such that the horizontal qubits in one cell are coupled to the horizontal qubits in the neighboring cells to the right and the left, and the vertical qubits are coupled to the vertical qubits on top and on the bottom. A visualization of this setup, also known as a Chimera graph, is shown in Figure 4.

Programmable interactions and biases are used to implement the Ising Hamiltonian in Equation (2). The parameters h_i represent local magnetic fields while the parameters J_{ij} are the couplings between two spins. Their values are restricted to the range $[-2, 2]$ for the local local fields, and $[-1, 1]$ for the couplings. It is understood that the couplings J_{ij} are only nonzero when there is a physical coupler associated with that particular pair of qubits on the chip. A transverse field term can also be implemented on each qubit, resulting in a driver Hamiltonian of the form shown in Equation (3). The adiabatic quantum computation is implemented by combining the two Hamiltonians above and changing their relative weight adiabatically, such that the system remains always in the ground state. In other words, the processor implements the Hamiltonian

$$H(t) = A(t)H_x + B(t)H_{Ising} \tag{5}$$

where the functions A and B satisfy $A(0) >> B(0)$ and $A(T) << B(T)$, for some final annealing time T. At $t = 0$, the system is in the ground state of the transverse field Hamiltonian H_x, corresponding to all the qubits being in the same eigenstate of σ^x, or in other words, a superposition of all possible states in the computational basis. For the closed system case (where there are no interactions with the environment), if the quantum annealing is done slowly enough, the adiabatic theorem of quantum mechanics guarantees that the state of the system at time T is with high probability the ground state of H_{Ising}. How slow is "slowly enough" depends on the details of the Hamiltonian, in particular the inverse of the energy gap between the ground state and the first excited state, and this feature is the main factor in determining a lower bound on the run time of the device. However, real devices are not ideal closed systems, so unwanted interactions with the environment will try to kick the system out of its ground state.

The current generations of D-Wave machines are designed for experimental use and are not optimized for turnaround time, unlike relatively mature CPU or GPU platforms. Rather than directly competing against existing classical solutions to machine learning, we focus on showing it is viable to use a quantum annealer to help train a neural network with complex topologies using architectures and approximations that differ from what has been used before [14–16]. For this reason, instead of using clock timings, we measure error metrics against the number of training epochs. As quantum annealing technology becomes more developed, machine learning algorithms may see benefits from using this new type of hardware. Regardless, clock timings are still important to consider. We next describe the computational workflow for each problem using D-Wave machines and communication latency between a client machine and a D-Wave machine; later we describe the timings over various operations on the hardware.

Each problem is sent across a network using D-Wave's Solver API (Matlab or Python) to the worker queue. Workers can concurrently process multiple requests and submit post-processed requests to the quantum processing unit (QPU) queue. Each request is then run sequentially on the QPU. Finally, the workers return the results back to the client. In one study D-Wave reported the mean turnaround time for each request was approximately 340 ms. Timings can vary depending on network latency-request latency can be reduced by placing the client physically next to the annealer, for example.

Communication latency aside, we also look at how long it takes to define and solve a problem on D-Wave. Loading and defining a problem on D-Wave hardware takes around $t_d = 10$ ms. Drawing a sample from the defined distribution via annealing takes around $t_a = 20$ μs. Reading out the unit states from a sample takes around $t_r = 120$ μs. We repeat the sampling and read-out stages $k = 100$ times for each MNIST image or neutrino detection instance in our experiments. So for each data point within our datasets, it takes $T = t_d + k(t_a + t_r)$ time to process. Currently the problem definition time t_d and read-out time t_r dominate wall-clock timing, but we again stress that we are looking to future developments and advancements in quantum annealing hardware that will reduce such overhead. We find the low annealing time particularly appealing because it scales well in algorithmic terms. That is, we can add additional hardware qubits or connectivity to produce more complex networks but the sampling time (annealing time t_a for our experiments) will not increase, which is not the case for simulating equivalent networks in software.

The number of physical couplers restricts the set of problems that can be natively implemented on the processor, and it represents one of the main limitations of the devices. Minor graph embeddings can overcome this limitation but at the expense of utilizing more than one qubit per graph node [24]. As we will show in the next section, our approach turns this problem on its head. Instead of trying to fit a problem into a particular topology, we start with our hardware topology using RBMs that have no intralayer couplings and study the advantages gained from adding additional couplers.

3.3. Implementing a Boltzmann Machine on D-Wave

We used D-Wave's adiabatic quantum computer located at the University of Southern California Lockheed Martin Quantum Computing Center. We implemented a Boltzmann machine to represent the MNIST digit recognition problem and neutrino particle detection problem. Deep learning using BMs has been proposed before, but as discussed in Section 2, learning is intractable for fully connected topologies because we need to compute expected values over an exponentially large state space [1,25]. RBMs address this by restricting network topology to bipartite connectivity to introduce conditional independence among "visible" units (representing the dataset and RBM output) given the "hidden" units (representing latent factors that control the data distribution), and vice versa, though they lose some representational power in the process. The quantum annealing hardware gave us an opportunity to first implement an RBM to establish baseline performance and then ease some topology restraints to investigate how more complex topologies could improve our results.

Our RBM used 784 visible units to represent each pixel in a 28×28 MNIST digit image and 80 hidden units on a D-Wave adiabatic quantum computer. We added an additional 10 visible units as a digit classification layer where the unit with highest probability was chosen as the label. Similarly we used $32 \times 32 = 1024$ units to represent the neutrino data, 80 hidden units, and 11 classification units to represent the 11 collision sites in the neutrino detection chamber, where the classification unit with the highest probability was chosen as the BM's guess for which plate the particle struck. The BMs were trained over 25 epochs on a training set and then evaluated against a validation set.

Next, as mentioned above, we loosened some of the topology restrictions of RBMs. RBMs enforce bipartite connectivity (see Figure 1), meaning hidden units are not connected to one another. We partially removed this restriction and allowed some of our hidden units to communicate with each other. We called this semi-restricted BM a "limited" Boltzmann machine (LBM). LBMs can be viewed as a superset of RBMs, the only difference being a set of extra available connections between hidden units. The previously described superconducting quantum adiabatic processor has physical constraints that limit connectivity to a chimera topology, so LBMs remain a subset of BMs.

Because D-Wave hardware faces a physical constraint on the number of possible units and connections, we would have had to employ the minor embedding approach mentioned above if we wanted to represent all of a BMs units on hardware. This would result in a large overhead in the number of qubits required, restricting our approach to small BMs. However, we can still try to exploit the quantum features of the D-Wave by restricting the topology of our model and only embedding part of it in the device. In our implementation we chose to represent only the hidden units, used the annealer as a sampler for the interconnected hidden units to estimate required quantities needed to update the weights, and left representation of the visible units to a classical machine. We were primarily interested in the interaction between hidden/latent units because they can represent abstract features extracted from the data. Figure 4 visualizes the extra connectivity we added to the LBM model and Figure 5 shows how we represented LBMs on the D-Wave's chimera topology.

Using D-Wave hardware to adjust LBM parameters may help tackle the intractability issue because the quantum annealer does not rely on conditional independence between units within a layer. We give a short explanation of the training process for BMs to illustrate.

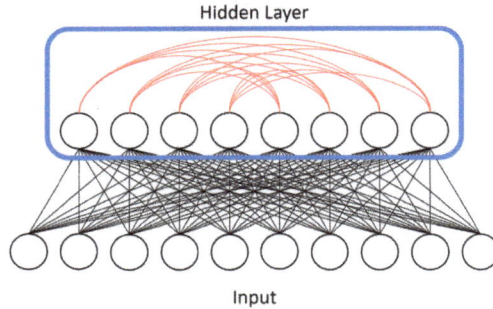

Figure 4. Our LBM model added connectivity between units in the hidden layer, shown in red. RBMs prohibit such intralayer connections because they add too much computational complexity for classical machines. We represented the hidden layer (outlined in blue) on the D-Wave device. The connections between hidden units were 4-by-4 bipartite due to the device's physical topology constraints.

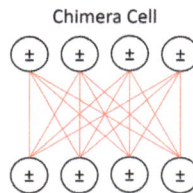

Figure 5. The hidden layer from Figure 4 is represented in one of D-Wave's chimera cells here, with the cell's bipartite connectivity made more obvious. The input/visible units of the LBM are left on a classical machine. Their contributions to the activity of the hidden units is reduced to an activity bias (represented with ± symbols) on those units. Figure 6 shows the overall chimera topology of the D-Wave device.

The configuration x of binary states s of units has an energy E defined by

$$E(x) = -\sum_i s_i b_i - \sum_{i<j} s_i s_j w_{ij} \tag{6}$$

where b is the bias of a unit and w_{ij} is the mutual weight between two units i and j. The partition function is $\sum_u e^{-E(u)}$, and the probability the BM produces a particular configuration x is

$$P(x) = e^{-E(x)} / \sum_u e^{-E(u)}. \tag{7}$$

$P(x)$ is difficult to compute in a full BM because it requires a sum over an exponentially large state space. If we want to determine the probability of some hidden unit i is on (equal to 1) without any guarantee of conditional independence, we would have to calculate $P(h_i = 1) = P(h_i = 1|v, h_{-i})$ where v is the state configuration of visible units and h is state configuration of the hidden units. However, if we use RBMs to restrict ourselves to bipartite connectivity between v and h, this probability factorizes and we can write $P(h_i = 1) = \prod_{j=1}^n P(h_i = 1|v_j)$. Our first RBM baseline experiment used this standard procedure with 1-step Gibbs sampling. In our LBM experiment, we did not need to rely on conditional independence or Gibbs sampling because we used quantum annealing instead to approximate samples from the more complicated probability distribution.

The training procedure for BMs compares the distribution of the data against the expected distribution according to the model and uses the difference to adjust the weight matrix w. Sampling from the model is difficult so we approximate using Markov Chain Monte Carlo (MCMC) sampling. The first "positive" phase of training locks the states of visible units to a configuration determined by the data—for example, a 28×28 pixel image from the MNIST dataset. The hidden unit distribution according to the data is found in this phase. The second "negative" phase unlocks the visible units and the system is allowed to settle. Sampling during this phase is difficult so we approximate samples using contrastive divergence with one step of MCMC and find the unit distributions according to the BM model. The weight matrix is then updated with the following equation:

$$\Delta w_{ij} = \epsilon(\langle v_i h_j \rangle_{data} - \langle v_i h_j \rangle_{reconstruction}) \tag{8}$$

where ϵ is the learning rate, $\langle v_i h_j \rangle_{data}$ is the product of visible and hidden unit state probabilities in the positive phase, and $\langle v_i h_j \rangle_{reconstruction}$ is the product of visible and hidden unit probabilities in the negative phase.

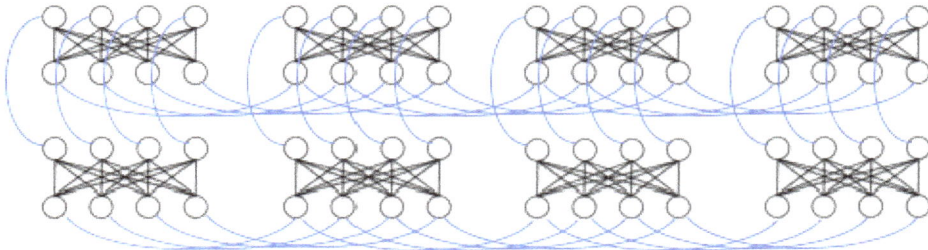

Figure 6. Chimera graphs are composed of 8-qubit cells featuring bipartite connectivity. Each cell's partition is connected to another partition in the adjacent cells.

For the MNIST problem we used 6000 images from the MNIST digit dataset to train the RBM and LBM. Each 28×28 image was represented with a 784-length vector with 10 units using 1-hot encoding to represent the class of digit. In training the labels were hidden and the BM attempted to reconstruct them to guess what the image label was. The classification unit with the highest probability of being "on" was chosen as the BM's label guess. The neutrino experiment used the same setup except the images were 32×32 pixels and thus there were 1024 visible units. The weight matrices were randomly initialized from a standard normal distribution and updated using the rule in Equation (8).

We wanted to further explore how connections between hidden units, referred to as couplers, contributed to problem solving in an LBM topology. To do so we limited the visible-to-hidden connectivity in the next experiment such that each hidden unit was only allowed to see a 4×4 box of pixels in the input images. These boxes did not overlap with each other. Reconstructing the input image became a much harder problem and the hope was that the addition of couplers would allow hidden units to trade information about input pixels in boxes they normally could not communicate with and improve results. This setup was somewhat inspired by CNN convolutional layers but we decided to make the "convolution" non-overlapping to use fewer qubits. In the future we will expand to use more qubits.

We believed this setup would make couplers relatively more important to the LBM because we reduced the ratio of visible-hidden connections to couplers. An input image with $32 \times 32 = 1024 = 2^{10}$ data points and 64 hidden units has $2^{10} \times 2^6 = 2^{16}$ visible-to-hidden connections for 168 couplers. However, hidden units with only 4×4 boxes of pixel visibility would instead have $2^4 \times 2^6 = 2^{10}$ visible-to-hidden connections for 168 couplers.

4. Results

We trained our RBM and LBM using the same parameters over 25 epochs (complete runs over all the training data). We followed common guidelines for choosing and adjusting hyperparameters [26]. We selected the learning rate ϵ to be 0.1 for weights between visible-to-hidden weights and 0.1 for hidden-to-hidden units for our experiments, excepting our first one shown in Figure 7. Setting ϵ too low means a BM learns slowly and may get trapped in local minima whereas setting it too high can cause the network to travel wildly in parameter space and be unable to learn coherently.

Before implementing the RBM running on MNIST data we wanted to get initial results indicating there was some merit to the LBM topology. Using simulated data, we mapped a BM to a quantum annealing simulator and trained two configurations, one where intralayer connections were disabled and one that had random intralayer connections. Ten epochs of training an RBM and LBM in Figure 7 show that LBM has some advantage.

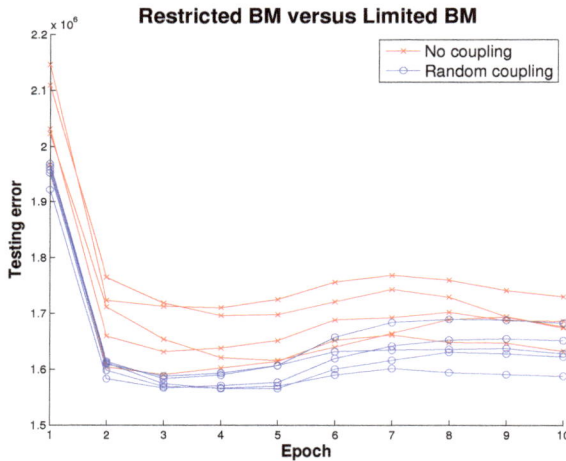

Figure 7. An initial experiment to demonstrate LBM utility. Reconstruction error (sum of squared error) of BMs trained on simulated data using no intralayer connections and using random intralayer connections with a small (0.0001) hidden-to-hidden weight learning rate. Here we show five RBMs (**red**) and five LBMs (**blue**), and the results suggest even just the presence of relatively static intralayer connections gives LBMs a performance advantage over RBMs. We obtained these results from the quantum annealing simulator provided by D-Wave.

As discussed, our first experiment was to establish performance baselines in RBMs so we could later compare LBMs against them. Figure 8 displays reconstruction error (sum of squared error between the actual data and BM reconstruction data mentioned in Section 3.3) and classification rate. This figure is included to confirm that the RBM did indeed learn to model the MNIST digit data distribution. Figure 9 contains a comparison of RBM performance and LBM performance on the MNIST digit recognition problem.

The RBM and LBM were both implemented on D-Wave and on MNIST images using the same number of hidden and visible units. For this test we trained over 10 epochs. The RBM configuration, as discussed, had no intra-layer connections, whereas the LBM configuration had limited connections between the hidden nodes.

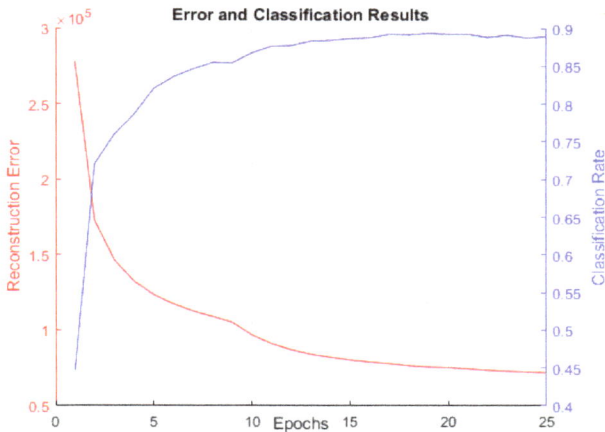

Figure 8. Reconstruction error and classification rate over 25 training epochs using 6000 MNIST images for training and 6000 for testing. Reconstruction error decreases as classification rate rises, confirming that the RBM learns the MNIST data distribution.

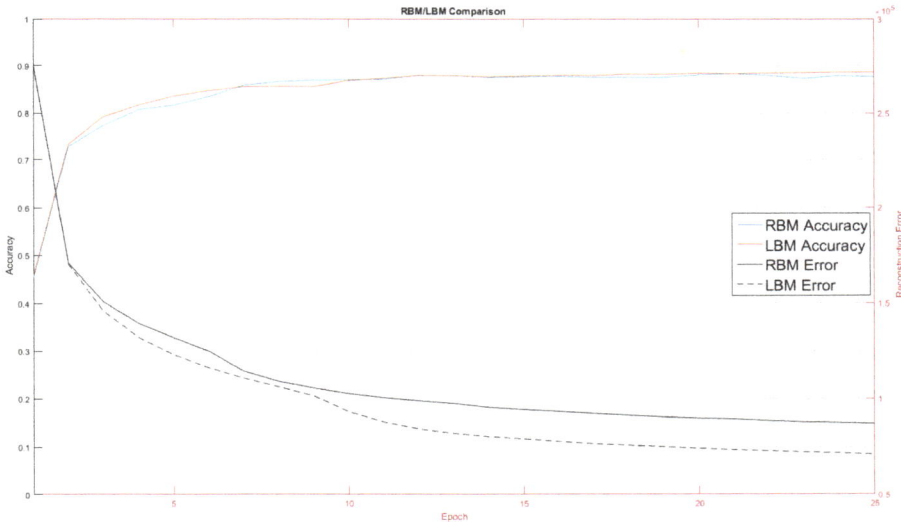

Figure 9. RBM and LBM performance on the MNIST digit classification task. The LBM tends to label the digits slightly better and produces lower reconstruction error than the RBM.

One quirk we found was LBM configuration initially performed worse than the RBM configuration. This was unexpected and we adopted a hybrid learning approach where the intralayer connections were reassigned from a random normal distribution for the first three training epochs. Afterwards the intralayer couplers were allowed to evolve according to the standard training rule. Our choice of a 3-epoch delay for intralayer training was rather arbitrary; further exploration into the mechanics involved will be explored in future work where we will pre-train models as RBMs on classical machines and then later hand over training to a quantum annealer.

The LBM achieved a classification rate of 88.53 percent, seen in Figure 7, and was comparable to other RBM results on MNIST [27].

Our LBM setup mapped only the hidden units to the D-Wave hardware whereas most other works map a whole BM. The latter approach requires down-sampling and graph embedding. We hope our approach scales better with problem size because we represent the visible input units on classical machines and still use contrastive divergence as a training method.

Our experiments on neutrino data and limited visible-to-hidden connectivity were run on both simulation software and D-Wave hardware. We used both because hardware has physical limits regarding parameter ranges and experiences parameter warping, so the inclusion of software results provides additional support if both environments produce comparable results. Parameters on the hardware for Ising models have around 4–5 bit precision and can only take on values within a small range, typically $[-2, 2]$ for h or $[-1, 1]$ for J. Software simulators do not have this limited precision and their parameters are not limited to any particular range.

We show the simulator results in Figure 10. Results from the simulator suggest the addition of couplers in this new setup improved performance, which led to our move to experiment on the quantum annealing hardware. Our experiments in Figure 11 were similar to the previous ones, albeit we first trained an RBM on a classical machine. We then took this lightly trained RBM model and moved it to the D-Wave hardware, used its semi-trained parameters to initialize the weights of the D-Wave RBM and LBM, enabled 168 couplers, then continued training for an additional 20 epochs. We again performed the RBM experiment five times and the LBM experiment five times.

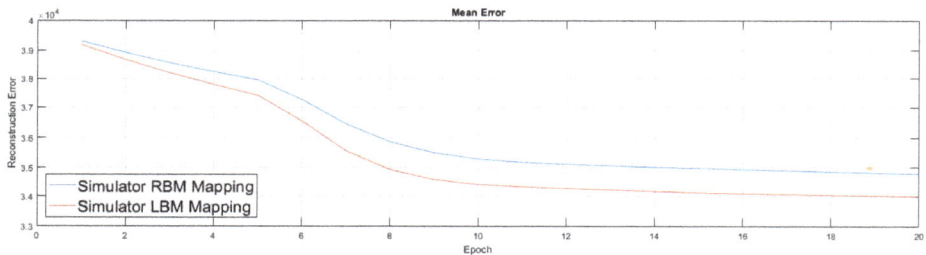

Figure 10. Comparison of RBM against LBM trained on neutrino data using a software simulator. Weights are randomly initialized from a normal distribution. The change in learning rate at epoch 5 is due to a change in the momentum parameter in the algorithm that is designed to speed the rate of training. The graph shows the mean performance of five different RBMs and five different LBMs and suggests the mean reconstruction error of RBM and LBM are significantly different.

In the LBM experiment we did not remap qubits in any scheme more complicated that a linear fashion. That is, we designated each qubit to oversee a 4×4 box in a horizontal order and simply assigned each qubit to unit cells according to this order. In future work we will argue this is suboptimal and that we can improve our results even more by considering smarter remappings of qubits to take advantage of locality within image data. For now we leave the comparison as RBM results versus LBM results without any special qubit remapping.

One aspect of superconducting technology worth mentioning is power consumption. The energy consumption of a system such as the D-Wave hardware is dominated by the cooling of the processor. When programming the device, the control signals inject some energy into the system that can increase the temperature by a few million Kelvin. This energy needs to extracted, resulting in a few pico Watts of power being dissipated in this step. However, the actual computation requires a negligible amount of energy. The cooling requirement has remained flat for four generations of the D-Wave device and is not expected to change in the foreseeable future. While the energy consumption of quantum annealers is typically not a highlighted advantage over classical systems, power efficiency may eventually become an important reason for preferring quantum computing systems in the future.

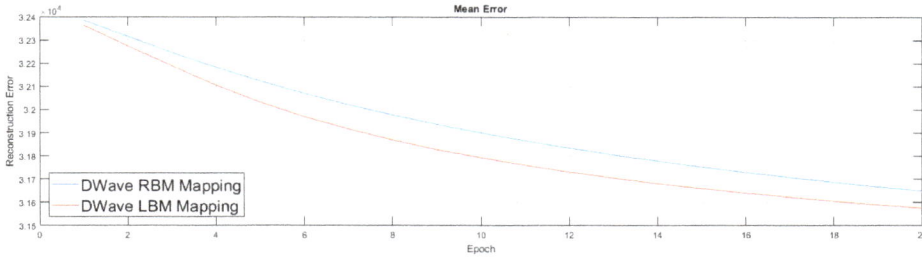

Figure 11. Another comparison of RBM against LBM run on neutrino data using D-Wave hardware. Both the RBM and LBM are initialized from the same pre-trained model. The pre-trained model is an RBM run for three epochs on a classical machine. The graph shows the mean performance of five different RBMs and five different LBMs, suggesting the performance difference between RBM and LBM persists on hardware.

5. Alternative Approaches

We have mentioned HPC and neuromorphic technology as two other platforms that can be utilized to benefit deep learning networks. Each has certain qualities that are not found in our adiabatic quantum computation approach due to fundamental differences between the platforms. Quantum annealers can handle complex topology but are limited in number; HPC exploits massive parallelization for computation speed but still uses classical machines; neuromorphic hardware is low power but tricky to train. We envision an integrated future where we can call upon the strengths of each platform to augment machine learning efforts. In this section we describe results from our HPC and neuromorphic efforts and how they can also contribute to training deep learning networks.

5.1. HPC

In previously reported work [28] we demonstrated that improved network hyperparameters can be found by using an evolutionary algorithm [29] and the Titan supercomputer, a collection of 300,000 cores and 18,000 Nvidia Tesla K20x GPUs. These results demonstrated that near optimal hyperparameters for CNN architectures can be found for the MNIST handwritten digit dataset by combining evolutionary algorithms and high performance computing. The kernel size and the number of hidden units per layer were the hyperparameters that were optimized. This work utilized 500 nodes of Titan for 3 h in order to evaluate 16,000 hyperparameter sets.

An improved version of the aforementioned evolutionary algorithm has been developed such that not only can hyperparameters of a fixed topology be optimized, but the topology of the network itself can be optimized [30]. This improved algorithm can evolve the number of layers and the type of each layer in addition to each individual layer's hyperparameters. This work has been applied to the MINERvA vertex reconstruction problem, which we have referred to as the neutrino particle detection problem in this paper, and has yielded improved results over standard networks. This approach is able to achieve an accuracy of 82.11% after evaluating nearly 500,000 networks on Titan in under 24 h utilizing 18,000 nodes of Titan, which represents a significant improvement over the baseline network that achieved 77.88%. Manually designing a network to attain such an improvement could take weeks or months due to the limited ability of a human to design, evaluate, and interrogate the performance of their networks in order to propose improved designs.

These HPC results are relevant to our quantum annealing approach because efforts to apply AQC to deep learning networks can benefit from this ability to pick good hyperparameters. When we designed our RBM and LBM experiments, we manually chose learning rates and topologies.

Future work can incorporate our HPC findings here to find optimal hyperparameters for our deep learning networks before using a quantum annealer to further tune the networks. Our LBM experiment where we first trained an RBM on a classical machine before moving it to the annealer and adding intralyer connections seems particularly amenable to such a procedure.

5.2. Neuromorphic

The neuromorphic approach fits into the context of our overall project through its potential for low-power implementations of networks derived from the AQC and HPC portions of our work. AQC needs hardware to be cooled as much as possible and HPC needs thousands of CPUs/GPUs. The power consumption of either is far beyond what a neuromorphic solution requires to function.

For our neuromorphic comparison points we considered a two-phase experiment. The initial phase was to demonstrate the feasibility of a native spiking neuromorphic solution by implementing an SNN in a software-based simulation. The next phase was to collect energy estimates by simulating the characteristics of the corresponding SNN implemented on memristive neuromorphic hardware. In a previous work [28] for the MNIST task, we started by simulating a simple spiking neural network trained to classify MNIST images.

We used evolutionary optimization (EO) to generate an ensemble of networks that classified MNIST images with an accuracy of approximately 90%. The accuracy of the generated ensemble was comparable to some other non-convolutional spiking neural network approaches [27]. The network we considered for this experiment was one network in the ensemble. In particular, the network we chose is one that distinguishes between images of the digit 0 and images of other digit types. For the second phase of the experiment the energy consumption was also determined for a memristive implementation of this network. Here the synapses consisted of metal-oxide memristors and represented both a weight value and a delay value. Each synapse in the network had twin memristors to implement both positive and negative weights [31] and a synaptic buffer to control the delays and peripheral connections. The neurons used in the network are implemented using the mixed-signal integrate and fire approach.

The simulation of energy estimate leveraged the energy per spike values for each synapse and neuron phases gathered from low-level circuit simulation. The network was simulated with a clock speed of 16.67 MHz and the average power and energy calculated for the network was 304.3 mW and 18.26 nJ. We note that this estimate includes the digital programmable delays as well. However, if we consider the core analog neuromorphic logic, the energy per spike is 5.24 nJ and the average power was 87.43 mW, which is consistent with similar memristor-based neuromorphic systems [32]. In contrast, MNIST classification tasks on GPU, field-programmable gate arrays (FPGA), or even application-specific integrated circuit (ASIC) architectures were reported to be in the W range [33], far above neuromorphic implementations like the one we described or IBM's TrueNorth [34].

In previous work [35] we also applied this approach to estimating the energy usage of a memristive based implementation on the Fermi data. As opposed to the MNIST task in which we trained multiple SNNs to form an ensemble, we built a single SNN for the neutrino data with 50 input neurons and 11 output neurons where the 11 output neurons corresponded to the 11 class labels in the neutrino data. We used a single view of the data (the x-view) rather than all three views. Instead of interpreting the data as pixels in an image we utilized the time lattice of the data. In the time lattice each value in the x-view corresponds to the time at which the energy at that point exceeded a low threshold. We used these times to govern when spikes should appear as input in the SNN. This generated a natural encoding for SNN-style networks as opposed to the somewhat unnatural mapping of non-temporal data to an image format. We found a resulting network with 90 neurons and 86 synapses that reached approximately 80.63% accuracy on the testing set, comparable to the approximately 80.42% accuracy achieved by a CNN that was also restricted to the x-view [17]. We estimated the energy usage of a memristive based neuromorphic implementation of the network for the neutrino data to be approximately 1.66 µJ per classification. These results, more so than the MNIST results, demonstrate that leveraging the temporal nature of certain data may result in extremely efficient SNN solutions to certain tasks.

6. Discussion

We compared a standard benchmark problem, MNIST digit recognition, on three different platforms: quantum adiabatic optimization, HPC, and neuromorphic. Our results show each option offers a unique benefit. Quantum adiabatic computation opens up complex topologies for use in deep learning models that would normally prove intractable for classical machines. HPC allows us to optimize CNNs on a large scale to find an optimal topology with its associated parameters. Neuromorphic lets us implement low power neural network solutions derived from other platforms. Figure 12 provides a summary of these platforms and their associated qualities. However, it is also clear that the MNIST problem is not ideally suited to showcase the capabilities of either the quantum or neuromorphic systems because it has been essentially solved using CNNs.

For example, the greater representational power of the quantum LBM approach is likely better utilized on a more complex dataset. Similarly, spiking neuromorphic systems may be better suited for use on datasets that include temporal components. In Figure 13 we propose an architecture we believe provides the ability to leverage the strengths of each of these computing platforms for future, more complex data sets.

The goal of this study is to explore how to address some of the current limitations of deep learning, namely networks containing intralayer connections, automatically configuring the hyperparameters of a network, and natively implementing a deep learning model using energy efficient neuron and synapse hardware. We used quantum computing, high performance computing, and neuromorphic computing to address these issues using three different deep learning models (LBM, CNN, and SNN).

The quantum adiabatic computing approach allows deep learning network topologies to be much more complex than what is feasible with conventional von Neumann architecture computers. The results show training convergence with a high number of intralayer connections, thus opening the possibility of using much more complex topologies that can be trained on a quantum computer. There is no time-based performance penalty due to the addition of intralayer connections, though there may be a need to sample more often in order to reduce potential errors.

HPC allows us to automatically develop an optimal network topology and create a high performing network. Many popular topologies used today are developed through trial and error methods. This approach works well with standard research datasets because the research community can learn and publish the topologies that produce the highest accuracy networks for these data. However, when the dataset is relatively unknown or not well studied, the trial-and-error approach loses its effectiveness. The HPC approach provides a way to optimize the hyper-parameters of a CNN, saving significant amounts of time when working on new datasets, perhaps even bootstrapping under-studied datasets into the regular publish-and-review iterative process.

Memristor-based hardware provides an opportunity to natively implement a low-power SNN as part of a neuromorphic computing environment. Such a network has the potential to feature broader connectivity than a CNN and the ability to dynamically reconfigure itself over time. Neuromorphic computers' benefits, including robustness, low energy usage, and small device footprint, can prove useful in a real-world environment today if we develop a mechanism for finding good network solutions for deployment on memristor-based devices that do not rely on conversions from non-spiking neural network types.

Platform	Algorithm	Contribution	Significance
Quantum	Limited Boltzmann	Development of LBM for quantum computers	Increased representational ability for deep learning, potentially tractable for quantum computer
HPC	Convolutional Neural Network	Evolutionary optimization design CNN network	Rapid design of high performing network
Neuromorphic	Spiking Neural Network	Native implementation of SNN on neuromorphic hardware	Very low power implementation of SNN

Figure 12. A comparison of the platforms, deep learning approaches, contributions, and significance of the result from the MNIST experiment.

We can use the three different architectures together to create powerful deep learning systems to go beyond our current capabilities. For example, current quantum annealing hardware is limited in the size and scope of problems it can solve but does allow us to use more complex networks. We can turn this into an opportunity by using a complex network as a higher level layer in a CNN as seen in Figure 13. Higher layers typically combine rich features and can benefit from increase intralayer connectivity; they also have smaller-sized inputs than lower layers, easing the limited-scope issue of current quantum annealing hardware. Such an augmented CNN may improve overall accuracy.

The HPC approach of automatically finding optimal deep learning topologies is a fairly robust and scalable capability, though quite expensive in development and computer costs. The ability to use deep learning methods on new or under-studied datasets (such as the neutrino particle detection dataset) can provide huge time savings and analytical benefit to the scientific community.

The neuromorphic approach is limited by the lack of robust neuromorphic hardware and algorithms, but it holds the potential of analyzing complex data using temporal analysis using very low power hardware. One of the most compelling aspects of this approach is the combination of a SNN and neuromorphic hardware that can analyze the temporal aspects of data. The MNIST problem does not have a temporal component, but one can imagine a dataset that has both image and temporal aspects such as a video or our neutrino detection dataset. A CNN approach has been shown to perform well on the image side, so perhaps a SNN can provide increased accuracy by analyzing the temporal aspects as well. For example, a CNN could analyze an image to detect objects within the image and output the location and/or orientation of those objects. This output can be used as input for an SNN. As each video frame is processed independently by the CNN, the output can be fed into the SNN, which can aggregate information over time and make conclusions about what is occurring in the video or detect particular events that occur over time, all in an online fashion. In this example the CNN could be trained independently using the labeled frames of the video as input images while the SNN could be trained independently utilizing different objects with their locations and orientations as input.

These experiments provide valuable insights into deep learning by exploring the combination of three novel approaches to challenging deep learning problems. We believe that these three architectures can be combined to gain greater accuracy, flexibility, and insight into a deep learning approach. Figure 13 shows a possible configuration of the three approaches that addresses the three deep learning challenges we discussed above. The high performance computer is used to create a high performing CNN on image type data. The final layer or two is then processed by the quantum computer using an LBM network that contains greater complexity than a CNN. The temporal aspects of the data are modeled using an SNN, and the ensemble models are then merged and an output produced. Our belief is that this approach has the potential to yield greater accuracy than existing CNN models.

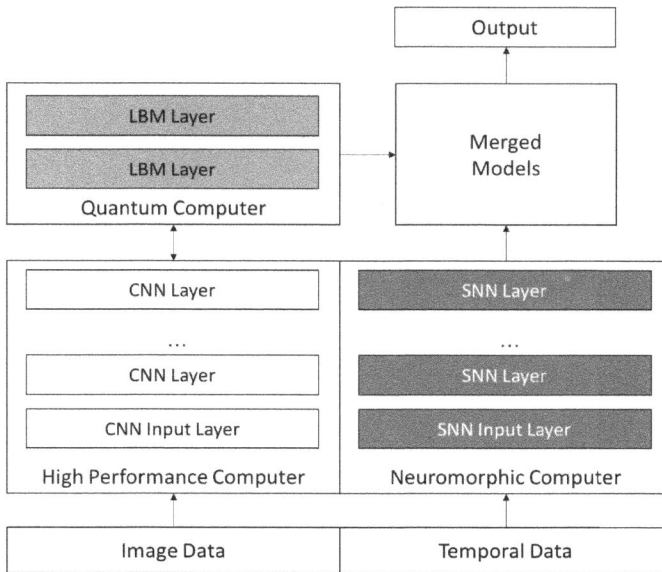

Figure 13. A proposed architecture that shows how the three approaches, quantum, HPC, and neuromorphic can be used to improve a deep learning approach. Image data can be analyzed using an HPC rapidly derived CNN with the top layers using an LBM on a quantum computer. The top layers have fewer inputs, and require greater representational capabilities which both play to the strength and limitations of a quantum approach. The temporal aspect of the data can be analyzed using an SNN. Finally, the image and temporal models will be merged to provide a richer and we believe a more accurate model, with an aim to be deployed in very low power neuromorphic hardware.

Future Work

We will test the proposed architecture to determine if it provides improved accuracy, flexibility, and insight into a dataset over methods derived from a traditional CNN approach. We will apply this to neutrino particle detection data and compare the proposed architecture against other contemporary methods.

We will also investigate how qubit mapping affects LBM results. Our experiment used a simple 1:1 mapping of hidden units to qubits by placing qubits in chimera cells in the order we defined them. However, this does not take advantage of locality within data; we will examine which methods of qubit mapping produce better results and see how they reveal patterns within our datasets.

7. Conclusions

Though inspired by biological neural models, deep learning networks make many simplifications to their connectivity topologies to enable efficient training algorithms and parallelization on GPUs. CNNs in particular have emerged as a standard high performance architecture on tasks such as object or facial recognition. While they are powerful tools, deep learning still has several limitations. First, we are restricted to relatively simple topologies; second, a significant portion of network tuning is done by hand; and third, we are still investigating how to implement low power, complex topologies in native hardware.

We chose three different computing environments to begin to address the issues respectively: quantum adiabatic computing, high performance computing clusters, and neuromorphic hardware. Because these environments are quite different, we chose to use different deep learning models for each. This includes Boltzmann machines in the quantum environment, CNNs in the HPC environment, and SNNs in the neuromorphic environment. We chose to use the well-understood MNIST hand-written digit dataset and a neutrino particle detection dataset.

Our results suggest these different architectures have the potential to address the identified deficiencies in complex deep learning networks that are inherent to the von Neumann CPU/memory architecture that is ubiquitous in computing.

The quantum annealing experiment showed that a complex neural network, namely one with intralayer connections, can be successfully trained on the MNIST digit recognition and neutrino particle detection tasks. The ability to train complex networks is a key advantage for a quantum annealing approach and opens the possibility of training networks with greater representational power than those currently used in deep learning trained on classical machines. High performance computing clusters can use such complex networks as building blocks to compare thousands of models to find the best performing networks for a given problem. Finally, the best performing neural network and its parameters can be implemented on a complex network of memristors to produce a low-power hardware device capable of solving difficult problems. This is a capability that is not feasible with a von Neumann architecture and holds the potential to solve much more complicated problems than can currently be solved with deep learning on classical machines.

We proposed a new deep learning architecture based on the unique capabilities of the quantum annealing, high performance computing, and neuromorphic approaches presented in this paper. This new architecture addresses three major limitations we see in current deep learning methods and holds the promise of higher classification accuracy, faster network creation times, and low power, native implementation in hardware.

Author Contributions: Robert M. Patton and Steven R. Young provided high performance computing results; Gangotree Chamka, Garrett S. Rose, and Catherine D. Schuman provided neuromorphic computing results; and Jeremy Liu, Federico M. Spedalieri, and Ke-Thia Yao provided quantum adiabatic computation results. Each group of authors was responsible for the methodology, software, resources, validation, investigation, and formal analysis associated with their research area. Conceptualization, funding acquisition, and supervision provided by Thomas E. Potok.Writing of original draft, review, and editing provided by Jeremy Liu.

Funding: This material is based upon work supported by the U.S. Department of Energy, Office of Science, Office of Advanced Scientific Computing Research, Robinson Pino, program manager, under contract number DE-AC05-00OR22725.

Acknowledgments: We would like to thank the MINERvA collaboration for the use of their simulated data and for many useful and stimulating conversations. MINERvA is supported by the Fermi National Accelerator Laboratory under US Department of Energy contract No. DE-AC02-07CH11359 which included the MINERvA construction project. MINERvA construction support was also granted by the United States National Science Foundation under Award PHY-0619727 and by the University of Rochester. Support for participating MINERvA physicists was provided by NSF and DOE (USA), by CAPES and CNPq (Brazil), by CoNaCyT (Mexico), by CONICYT (Chile), by CONCYTEC, DGI-PUCP and IDI/IGIUNI (Peru), and by Latin American Center for Physics (CLAF). This material is based upon work supported by the U.S. Department of Energy, Office of Science, Office of Advanced Scientific Computing Research, Robinson Pino, program manager, under contract number DE-AC05-00OR22725. This research used resources of the Oak Ridge Leadership Computing Facility, which is a DOE Office of Science User Facility supported under Contract DE-AC05-00OR22725. Notice: This manuscript has been authored by UT-Battelle, LLC under Contract No. DE-AC05-00OR22725 with the U.S. Department of Energy. The United States Government retains and the publisher, by accepting the article for publication, acknowledges that the United States Government retains a non-exclusive, paid-up, irrevocable, world-wide license to publish or reproduce the published form of this manuscript, or allow others to do so, for United States Government purposes. The Department of Energy will provide public access to these results of federally sponsored research in accordance with the DOE Public Access Plan (http://energy.gov/downloads/doe-public-access-plan).

Conflicts of Interest: The authors declare no conflict of interest. The founding sponsors had no role in the design of the study; in the collection, analyses, or interpretation of data; in the writing of the manuscript, and in the decision to publish the results.

Appendix A.

Appendix A.1. Related Works for High Performance Computing

Deep learning, being an early adopter of GPU technology, has benefited greatly from the speedup offered by these accelerated computing devices and has received great support from device manufacturers in the form of deep learning-specific GPU libraries. General purpose GPUs are the basic building blocks of today's HPC platforms and next generation machines will rely on them to an even greater degree. Thus, deep learning provides a great opportunity to fully utilize these machines, as they will have multiple GPUs per compute node. This leaves the question of how to best utilize thousands of GPUs for deep learning, as previous work has only utilized a maximum of 64 GPUs before encountering scaling problems when trying to exploit model parallelism to spread the weights of the network across multiple GPUs [36]. HPC provides the unique opportunity to address the problem of network specification. This refers to the problem of deciding upon the set of hyper-parameters needed to specify the network and training procedure in order to apply deep learning to a new dataset.

For convolutional neural networks, this could involve specifying parameters such as the number of layers, the number of hidden units, or the kernel size. For more general networks, such as RBMs, this could involve defining much more complicated connectivity between neurons.

Previously, it has been shown that HPC can be utilized to optimize the hyperparameters of a deep learning network [29]. This work utilized an evolutionary algorithm distributed across the nodes of Oak Ridge National Laboratory's (ORNL's) Titan supercomputer in order to optimize the performance of deep learning algorithms. We include the activation function used, the number of hidden units in a layer, the kernel size of a convolutional layer, and the learning rate of the solver as hyperparameters. As the size of the network grows, the hyper-parameter space grows increasingly larger. The size of deep learning networks used today have resulted in a hyper-parameter space that cannot be searched on a single machine or a small cluster. This is a result of the computational complexity of training and evaluating these networks. Without utilizing the computational capabilities provided by supercomputers, evaluating a sufficient number of hyper-parameter sets to search the enormous hyper-parameter space of these methods would be impossible.

Appendix A.2. Related Works for Neuromorphic Computing

There are two primary reasons that researchers have pursued the development of neuromorphic computing architectures: to develop custom hardware devices to accurately simulate biological neural systems with the goal of studying biological brains and to build computationally useful architectures that are inspired by the operation of biological brains and have some of their characteristics. In developing neuromorphic computing devices for computational purposes, there have been two main approaches: building devices based on spiking neural networks (SNNs), such as IBM's TrueNorth [37] or Darwin [38], and building devices based on traditional or convolutional neural networks, such as Google's Tensor Processing Unit [39] or Nervana's Nervana Engine [40], to serve as deep learning accelerators. The neuromorphic devices that have been built based on SNNs or built to simulate more biologically-accurate systems have vastly different characteristics than those that have been built based on deep learning networks, such as CNNs. The neurons in SNN-based systems are typically not organized in layers and have fewer restrictions on connectivity between neurons, allowing for more complex network topologies including recurrent networks. The neuron and synapse models also differ from those in convolutional neural networks and recurrent neural networks such as long short term memories (LSTMs) [41]. Specifically, in SNN-based neuromorphic systems, the neuron is typically some form of spiking neuron, such as a leaky-integrate-and-fire neuron, and the synapses usually have a delay value in addition to a weight value, thus introducing a temporal component to the processing of the network.

The primary computational issue associated with SNN-based systems is that few algorithms that train native networks for those systems have been developed. The key reason why algorithms have not been developed is the computational difficulty introduced by the broader connectivity in the network and the inclusion of the temporal component in both the neurons and synapses. One approach for training networks for neuromorphic computers has been to train a CNN offline and then create a mapping process from the CNN to the associated SNN-based neuromorphic hardware [42]. This mapping of an existing neural network trained with a well-studied algorithm (in this case, backpropagation) has been used for a variety of other neural network types beyond convolutional neural networks, such as spiking Hopfield networks and spiking restricted Boltzmann machines [43]. The algorithms that have been developed for spiking neuromorphic systems typically impose some sort of restriction for the network, or they have not yet been shown to be widely applicable. For example, a variation of back-propagation for spiking neural networks (SpikeProp) has been developed [44,45] but it is restricted to feed-forward networks and simply learns the weight values for the synapses. Learning rules based on spike-timing dependent plasticity or STDP have also been commonly used in spiking neural network architectures [46]. Though STDP has been shown to be useful on some tasks, including unsupervised tasks, the true impact of STDP on real applications has not yet been demonstrated. It is worth noting that STDP mechanisms have great potential to be used as unsupervised weight training method, but it may need to be used alongside a supervised algorithm that can help to determine network topology and parameters.

A key property of neuromorphic systems is their potential for more energy-efficient computation. To achieve energy-efficiency, we (and many others) have explored an implementation of a spiking neural network system utilizing memristors. Memristors are "memory resistors" in that their resistance can be altered depending on the magnitude of the voltage applied. When no voltage is applied across a memristor, the most recent resistance value is retained [47]. Memristors have similar behavior to biological synapses, and as such, have been frequently utilized to implement neuromorphic systems [48–50].

Appendix B.

Appendix B.1. Description of High Performance Computing

The high performance computer we are using is the ORNL's Titan computer with roughly 300,000 cores, and 18,000 GPUs. This is currently the fastest open science computer in the world.

Clearly a supercomputer is not needed to solve the MNIST problem; however, a supercomputer is extremely valuable in automatically finding an optimal deep learning topology for such a problem. Rather than using a trial-and-error method for finding a well performing network topology, we utilize an evolutionary optimization on Titan to evaluate tens of thousands of topologies [29]; therefore, systematically finding the best performing networks on this problem. If achievable, this would solve one of the major challenges in building deep learning networks.

For this project we used a CNN as our deep learning network since CNNs currently produce the top results. We approached the network topology problem of selecting optimal hyper-parameters as a massive search problem, where Titan can be used to quickly search the space.

We represented each individual within the population of the evolutionary algorithm (EA) as a single deep neural network or CNN. An individual consisted of a genome where the genes represented the various hyper-parameters that defined the network topology, i.e., the number of layers, type of layers (convolution, pooling, etc.), and order of the layers. We then applied parameters defined in the genes of the individual to construct and train a deep learning network on the MNIST dataset. The results of the network's performance in testing were then used as the "fitness" of the individual in the EA population, i.e., individual networks that had high accuracy were considered to be the most fit. Typically, generating the results for a single network on a small dataset like MNIST requires a modest amount of GPU/CPU time, and memory. However, creating, training, and evaluating tens of thousands networks requires a significant number of GPUs, like those in the Titan high performance computer.

After all the individuals in the population were evaluated, the top performing individuals were selected to generate a new population of individuals that represented the next generation of the EA. These new generations contained a mix of the well performing hyperparameters from the best performing networks in the population. Successive generations of individuals gradually led to an improved set of hyperparameters over time. This method is called Multi-node Evolutionary Neural Networks for Deep Learning (MENNDL) [29].

For this experiment, we were looking to automatically discover hyperparameters of a well performing deep learning network on the MNIST dataset. We used a simple EA that limited the search to the number of neurons per layer and the kernel size of convolutional layers.

The network architecture utilized was LeNet [4] and featured two convolutional layers, two pooling layers, and one hidden fully-connected layer. This is the network that is most often used with the MNIST dataset in the literature.

We showed that even with this widely studied MNIST dataset, better hyper-parameters could be found than those widely reported in the literature. An EA that can evolve the topology provides the opportunity for improved results and the ability to process more challenging datasets. Such an EA also provides the opportunity to meaningfully utilize the entirety of Titan's capacity. It provides challenging data management problems on a machine designed primarily for modeling and simulation, as opposed to these deep learning algorithms which require heavy amounts of data input in addition to heavy computation.

Appendix B.2. Description of Neuromorphic Computing

A spiking neuromorphic approach to the MNIST problem was not the ideal solution since there is not a temporal component in the task of recognizing a handwritten digit. In order to leverage the temporal processing capabilities of spiking neural networks, we added a temporal component to the task by using a streaming scan of the digits as input to the SNN such that columns in the input image were received over time rather than all at once. The SNN learned to recognize digits based on this scan pattern. For the results presented on the MNIST task, the goal was to understand the deployment benefits of using an SNN in memristive hardware as opposed to classification accuracy on this problem. For the neutrino data, where the data itself already had a temporal component, there was a more natural mapping to SNNs. Thus, classification accuracy may be a more accurate representation of potential performance of SNNs in general than for non-temporal data like MNIST.

As noted in Appendix A.2, there are not very many SNN training methods or neuromorphic training methods that can be applied to spiking neuromorphic networks and operate within the characteristics and constraints of a particular neuromorphic hardware implementation. To train both SNN models and neuromorphic networks we utilized an evolutionary optimization (EO) approach to determine the structure (e.g., number of neurons and synapses and how they are connected) and parameters (e.g., weight values of synapses and threshold values of neurons) [51].

The neuromorphic system we used to explore both the MNIST and neutrino detection problem was a memristive implementation of the neuroscience-inspired dynamic architectures (NIDA) system [52]. NIDA is a simple SNN model composed of integrate-and-fire neurons and synapses with delays and weights that are affected by processes similar to long-term potentiation and long-term depression in biological brains. The NIDA model allows us to study neuromorphic models in software and determine how restrictions different in hardware (such as weight resolution or connectivity) affect performance.

The EO approach for training networks in the MNIST problem was previously applied to the NIDA SNN [52]. An ensemble approach was utilized where each network in the ensemble was responsible for recognizing a particular digit type. For example, a network may be trained to recognize zeros, in which case the network will take the handwritten digit image as input and its output corresponds to either "yes, it is a zero" or "no, it is not a zero". Using this approach, ensembles that achieve around 90 percent accuracy were achieved.

The memristive device technology assumed for this simulation was characterized by a low resistance state (LRS) of 60 kΩ, about an order of magnitude larger than the resistance of a typical deep-submicron complementary metal–oxide–semiconductor (CMOS) transistor. This relatively high LRS for the memristor is desirable such that the CMOS channel resistance can effectively be neglected. The on-off ratio was assumed to be 10, providing a high resistance state (HRS) of 600 kΩ. Such characteristics for LRS, HRS and the associated on-off ratio have been observed for a range of memristive devices, including hafnium-oxide (HfO_2) [53], tantalum-oxide (TaO_2) [54], and titanium-oxide (TiO_2) [55]. All of these memristive material stacks consist of an oxide layer sandwiched between two metallic layers. Depending on the polarity and magnitude of an applied voltage bias, the oxide layer transitions between being less or more conductive, providing the switching characteristics desirable for representing synaptic weights.

Our memristive NIDA simulation setup also included analog integrate-and-fire neurons, implemented using a 65 nm CMOS process technology. Neuromorphic elements (neurons and synapses) were simulated using Cadence Spectre and system-level energy and power estimates were calculated using a high-level simulator written in C++. Specifically, we verified the high-level C++ model versus the circuit level implementation using small networks that were simulated using both Cadence Spectre and the high-level NIDA simulator. Larger networks, specifically MNIST, were simulated using the high-level NIDA simulator to determine neuron and synapse activity information.

The memristive NIDA simulation was based on two significant steps. Initially an evolutionary optimization training process was used to generate optimized networks for the low level simulation. At the same time, the transistor level simulation was done using Cadence Spectre simulator. Estimates were collected for the design components in different conditions (neuron accumulating but not firing, neuron firing, etc.). These "per component" energy estimates were used in conjunction with activity information from the high-level NIDA simulation to calculate the total energy consumed.

References

1. Ackley, D.H.; Hinton, G.E.; Sejnowski, T.J. A learning algorithm for Boltzmann machines. *Cogn. Sci.* **1985**, *9*, 147–169. [CrossRef]
2. Hinton, G.E.; Salakhutdinov, R.R. Reducing the Dimensionality of Data with Neural Networks. *Science* **2006**, *313*, 504–507. [CrossRef] [PubMed]
3. Hinton, G.E.; Osindero, S.; Teh, Y.W. A Fast Learning Algorithm for Deep Belief Nets. *Neural Comput.* **2006**, *18*, 1527–1554. [CrossRef] [PubMed]
4. LeCun, Y.; Bottou, L.; Bengio, Y.; Haffner, P. Gradient-based learning applied to document recognition. *Proc. IEEE* **1998**, *86*, 2278–2324. [CrossRef]
5. Scherer, D.; Müller, A.; Behnke, S. Evaluation of pooling operations in convolutional architectures for object recognition. In *International Conference on Artificial Neural Networks*; Springer: Berlin/Heidelberg, Germany, 2010; pp. 92–101.
6. Esser, S.K.; Merolla, P.A.; Arthur, J.V.; Cassidy, A.S.; Appuswamy, R.; Andreopoulos, A.; Berg, D.J.; McKinstry, J.L.; Melano, T.; Barch, D.R.; et al. Convolutional networks for fast, energy-efficient neuromorphic computing. *Proc. Natl. Acad. Sci. USA* **2016**, *113*, 11441–11446. [CrossRef] [PubMed]
7. Indiveri, G.; Corradi, F.; Qiao, N. Neuromorphic Architectures for Spiking Deep Neural Networks. In Proceedings of the IEEE International Electron Devices Meeting (IEDM), Washington, DC, USA, 7–9 December 2015.
8. Wiebe, N.; Kapoor, A.; Svore, K.M. Quantum deep learning. *Quantum Inf. Comput.* **2016**, *16*, 0541–0587.
9. LeCun, Y.; Cortes, C.; Burges, C.J. The MNIST Database of Handwritten Digits. 1998. Available online: http://yann.lecun.com/exdb/mnist/ (accessed on 17 May 2018).
10. Feynman, R.P. Simulating physics with computers. *Int. J. Theor. Phys.* **1982**, *21*, 467–488. [CrossRef]
11. Shor, P.W. Polynomial-Time Algorithms for Prime Factorization and Discrete Logarithms on a Quantum Computer. *SIAM J. Comput.* **1997**, *26*, 1484–1509. [CrossRef]
12. Raussendorf, R.; Briegel, H.J. A One-Way Quantum Computer. *Phys. Rev. Lett.* **2001**, *86*, 5188–5191. [CrossRef] [PubMed]
13. Farhi, E.; Goldstone, J.; Gutmann, S.; Sipser, M. Quantum Computation by Adiabatic Evolution. *arXiv* **2000**, arXiv:quant-ph/0001106.
14. Adachi, S.H.; Henderson, M.P. Application of Quantum Annealing to Training of Deep Neural Networks. *arXiv* **2015**, arXiv:1510.06356v1.
15. Benedetti, M.; Realpe-Gómez, J.; Biswas, R.; Perdomo-Ortiz, A. Estimation of effective temperatures in quantum annealers for sampling applications: A case study with possible applications in deep learning. *Phys. Rev. A* **2016**, *94*, 022308. [CrossRef]
16. Benedetti, M.; Realpe-Gómez, J.; Biswas, R.; Perdomo-Ortiz, A. Quantum-assisted learning of graphical models with arbitrary pairwise connectivity. *arXiv* **2016**, arXiv:1609.02542v2.
17. Terwilliger, A.M.; Perdue, G.N.; Isele, D.; Patton, R.M.; Young, S.R. Vertex Reconstruction of Neutrino Interactions Using Deep Learning. In Proceedings of the International Joint Conference on Neural Networks (IJCNN), Anchorage, AK, USA, 14–19 May 2017; pp. 2275–2281.
18. Barahona, F. On the computational complexity of Ising spin glass models. *J. Phys. A Math. Gen.* **1982**, *15*, 3241. [CrossRef]
19. Born, M.; Fock, V. Beweis des Adiabatensatzes. *Z. Phys.* **1928**, *51*, 165–180. [CrossRef]
20. Sarandy, M.S.; Wu, L.A.; Lidar, D.A. Consistency of the Adiabatic Theorem. *Quantum Inf. Process.* **2004**, *3*, 331–349. [CrossRef]
21. Boixo, S.; Somma, R.D. Necessary condition for the quantum adiabatic approximation. *Phys. Rev. A* **2010**, *81*, 032308. [CrossRef]
22. Somma, R.D.; Boixo, S. Spectral Gap Amplification. *arXiv* **2011**, arXiv:1110.2494.
23. Harris, R.; Johnson, M.W.; Lanting, T.; Berkley, A.J.; Johansson, J.; Bunyk, P.; Tolkacheva, E.; Ladizinsky, E.; Ladizinsky, N.; Oh, T.; et al. Experimental investigation of an eight-qubit unit cell in a superconducting optimization processor. *Phys. Rev. B* **2010**, *82*, 024511. [CrossRef]
24. Choi, V. Minor-embedding in adiabatic quantum computation: II. Minor-universal graph design. *arXiv* **2011**, arXiv:1001.3116.

25. Salakhutdinov, R.; Hinton, G. Deep boltzmann machines. In Proceedings of the Artificial Intelligence and Statistics, Beach, FL, USA, 16–18 April 2009; pp. 448–455.
26. Hinton, G. A practical guide to training restricted Boltzmann machines. *Momentum* **2010**, *9*, 926.
27. Diehl, P.U.; Neil, D.; Binas, J.; Cook, M.; Liu, S.C.; Pfeiffer, M. Fast-Classifying, High-Accuracy Spiking Deep Networks Through Weight and Threshold Balancing. In Proceedings of the International Joint Conference on Neural Networks (IJCNN), Killarney, Ireland, 12–16 July 2015; pp. 1–8.
28. Potok, T.E.; Schuman, C.; Young, S.; Patton, R.; Spedalieri, F.; Liu, J.; Yao, K.T.; Rose, G.; Chakma, G. A Study of Complex Deep Learning Networks on High Performance, Neuromorphic, and Quantum Computers. In Proceedings of the Machine Learning in HPC Environments, Salt Lake City, UT, USA, 30 January 2017; p. 47–55.
29. Young, S.R.; Rose, D.C.; Karnowski, T.P.; Lim, S.H.; Patton, R.M. Optimizing Deep Learning Hyper-Parameters Through an Evolutionary Algorithm. In Proceedings of the Workshop on Machine Learning in High-Performance Computing Environments, Austin, TX, USA, 15–20 November 2015; pp. 1–5.
30. Young, S.R.; Rose, D.C.; Johnston, T.; Heller, W.T.; Karnowski, T.P.; Potok, T.E.; Patton, R.M.; Perdue, G.; Miller, J. Evolving Deep Networks Using HPC. In Proceedings of the Machine Learning on HPC Environments, Denver, CO, USA, 12–17 November 2017; p. 7.
31. Sayyaparaju, S.; Chakma, G.; Amer, S.; Rose, G.S. Circuit Techniques for Online Learning of Memristive Synapses in CMOS-Memristor Neuromorphic Systems. In Proceedings of the Great Lakes Symposium on VLSI, Lake Louise, AL, Canada, 10–12 May 2017; pp. 479–482.
32. Liu, C.; Yang, Q.; Yan, B.; Yang, J.; Du, X.; Zhu, W.; Jiang, H.; Wu, Q.; Barnell, M.; Li, H. A Memristor Crossbar based Computing Engine Optimized for High Speed and Accuracy. In Proceedings of the IEEE Computer Society Annual Symposium on VLSI (ISVLSI), Pittsburgh, PA, USA, 11–13 July 2016; pp. 110–115.
33. Farabet, C.; Martini, B.; Akselrod, P.; Talay, S.; LeCun, Y.; Culurciello, E. Hardware accelerated convolutional neural networks for synthetic vision systems. In Proceedings of the 2010 IEEE International Symposium on Circuits and Systems, Vienna, Austria, 30 May–2 June 2010; pp. 257–260.
34. Yepes, A.; Tang, J.; Mashford, B.J. Improving classification accuracy of feedforward neural networks for spiking neuromorphic chips. *arXiv* **2017**, arXiv:1705.07755.
35. Schuman, C.D.; Potok, T.E.; Young, S.; Patton, R.; Perdue, G.; Chakma, G.; Wyer, A.; Rose, G.S. Neuromorphic Computing for Temporal Scientific Data Classification. In Proceedings of the Neuromorphic Computing Symposium, Knoxville, TN, USA, 17–19 July 2018.
36. Coates, A.; Huval, B.; Wang, T.; Wu, D.J.; Catanzaro, B.; Ng, A.Y. Deep Learning with COTS HPC Systems. In Proceedings of the 30th International Conference on Machine Learning (ICML), Atlanta, GA, USA, 16–21 June 2013; pp. 1337–1345.
37. Cassidy, A.S.; Merolla, P.; Arthur, J.V.; Esser, S.K.; Jackson, B.; Alvarez-Icaza, R.; Datta, P.; Sawada, J.; Wong, T.M.; Feldman, V.; et al. Cognitive Computing Building Block: A Versatile and Efficient Digital Neuron Model for Neurosynaptic Cores. In Proceedings of the International Joint Conference on Neural Networks (IJCNN), Killarney, Ireland, 12–17 July 2013; pp. 1–10.
38. Shen, J.; Ma, D.; Gu, Z.; Zhang, M.; Zhu, X.; Xu, X.; Xu, Q.; Shen, Y.; Pan, G. Darwin: A neuromorphic hardware co-processor based on Spiking Neural Networks. *Sci. China Inf. Sci.* **2016**, *59*, 1–5. [CrossRef]
39. Jouppi, N. Google Supercharges Machine Learning Tasks with TPU Custom Chip. 2016. Available online: https://cloudplatform.googleblog.com/2016/05/Google-supercharges-machine-learning-tasks-with-custom-chip.html (accessed on 17 May 2018).
40. Nervana. Nervana Engine. 2016. Available online: https://www.nervanasys.com/technology/engine/ (accessed on 30 January 2017).
41. Hochreiter, S.; Schmidhuber, J. Long short-term memory. *Neural Comput.* **1997**, *9*, 1735–1780. [CrossRef] [PubMed]
42. Esser, S.K.; Appuswamy, R.; Merolla, P.; Arthur, J.V.; Modha, D.S. Backpropagation for energy-efficient neuromorphic computing. In Proceedings of the Advances in Neural Information Processing Systems (NIPS), Montréal, QC, Canada, 7–12 December 2015; pp. 1117–1125.
43. Arthur, J.V.; Merolla, P.A.; Akopyan, F.; Alvarez, R.; Cassidy, A.; Chandra, S.; Esser, S.K.; Imam, N.; Risk, W.; Rubin, D.B.; et al. Building Block of a Programmable Neuromorphic Substrate: A Digital Neurosynaptic Core. In Proceedings of the International Joint Conference on Neural Networks (IJCNN), Brisbane, Australia, 10–15 June 2012; pp. 1–8.

44. Bohte, S.M.; Kok, J.N.; La Poutre, H. Error-backpropagation in temporally encoded networks of spiking neurons. *Neurocomputing* **2002**, *48*, 17–37. [CrossRef]
45. Schrauwen, B.; Van Campenhout, J. Extending Spikeprop. In Proceedings of the IEEE International Joint Conference on Neural Networks, Budapest, Hungary, 25–29 July 2004; Volume 1, pp. 471–475.
46. Song, S.; Miller, K.D.; Abbott, L.F. Competitive Hebbian learning through spike-timing-dependent synaptic plasticity. *Nat. Neurosci.* **2000**, *3*, 919–926. [CrossRef] [PubMed]
47. Williams, R.S. How We Found The Missing Memristor. *IEEE Spectr.* **2008**, *45*, 28–35. [CrossRef]
48. Jo, S.H.; Chang, T.; Ebong, I.; Bhadviya, B.B.; Mazumder, P.; Lu, W. Nanoscale memristor device as synapse in neuromorphic systems. *Nano Lett.* **2010**, *10*, 1297–1301. [CrossRef] [PubMed]
49. Kim, K.H.; Gaba, S.; Wheeler, D.; Cruz-Albrecht, J.M.; Hussain, T.; Srinivasa, N.; Lu, W. A functional hybrid memristor crossbar-array/CMOS system for data storage and neuromorphic applications. *Nano Lett.* **2011**, *12*, 389–395. [CrossRef] [PubMed]
50. Prezioso, M.; Merrikh-Bayat, F.; Hoskins, B.; Adam, G.; Likharev, K.K.; Strukov, D.B. Training and operation of an integrated neuromorphic network based on metal-oxide memristors. *Nature* **2015**, *521*, 61–64. [CrossRef] [PubMed]
51. Schuman, C.D.; Plank, J.S.; Disney, A.; Reynolds, J. An Evolutionary Optimization Framework for Neural Networks and Neuromorphic Architectures. In Proceedings of the International Joint Conference on Neural Networks (IJCNN), Vancouver, BC, USA, 24–29 July 2016; pp. 145–154.
52. Schuman, C.D.; Birdwell, J.D.; Dean, M.E. Spatiotemporal Classification Using Neuroscience-Inspired Dynamic Architectures. *Proc. Comput. Sci.* **2014**, *41*, 89–97. [CrossRef]
53. Cady, N.; Beckmann, K.; Manem, H.; Dean, M.; Rose, G.; Nostrand, J.V. Towards Memristive Dynamic Adaptive Neural Network Arrays. In Proceedings of the Government Microcircuit Applications and Critical Technology Conference (GOMACTech), Orlando, FL, USA, 14–17 March 2016.
54. Yang, J.J.; Zhang, M.; Strachan, J.P.; Miao, F.; Pickett, M.D.; Kelley, R.D.; Medeiros-Ribeiro, G.; Williams, R.S. High switching endurance in TaOx memristive devices. *Appl. Phys. Lett.* **2010**, *97*, 232102. [CrossRef]
55. Medeiros-Ribeiro, G.; Perner, F.; Carter, R.; Abdalla, H.; Pickett, M.D.; Williams, R.S. Lognormal switching times for titanium dioxide bipolar memristors: origin and resolution. *Nanotechnology* **2011**, *22*, 095702. [CrossRef] [PubMed]

MDPI

Article

Entropic Uncertainty Relations for Successive Measurements in the Presence of a Minimal Length

Alexey E. Rastegin

Department of Theoretical Physics, Irkutsk State University, Gagarin Bv. 20, 664003 Irkutsk, Russia; alexrastegin@mail.ru

Received: 30 March 2018; Accepted: 7 May 2018; Published: 9 May 2018

Abstract: We address the generalized uncertainty principle in scenarios of successive measurements. Uncertainties are characterized by means of generalized entropies of both the Rényi and Tsallis types. Here, specific features of measurements of observables with continuous spectra should be taken into account. First, we formulated uncertainty relations in terms of Shannon entropies. Since such relations involve a state-dependent correction term, they generally differ from preparation uncertainty relations. This difference is revealed when the position is measured by the first. In contrast, state-independent uncertainty relations in terms of Rényi and Tsallis entropies are obtained with the same lower bounds as in the preparation scenario. These bounds are explicitly dependent on the acceptance function of apparatuses in momentum measurements. Entropic uncertainty relations with binning are discussed as well.

Keywords: generalized uncertainty principle; successive measurements; minimal observable length; Rényi entropy; Tsallis entropy

---

1. Introduction

The Heisenberg uncertainty principle [1] is now avowed as a fundamental scientific concept. Heisenberg examined his thought experiment rather qualitatively. An explicit formal derivation appeared in [2]. This approach was later extended to arbitrary pairs of observables [3]. These traditional formulations are treated as preparation uncertainty relations [4], since repeated trials with the same quantum state are assumed here. This simple scenario differs from the situations typical in quantum information science. Since uncertainty relations are now examined not only conceptually, researchers often formulated them in information-theoretic terms. As was shown in [5], wave-particle duality can be interpreted on the basis of entropic uncertainty relations. Basic developments within the entropic approach to quantum uncertainty are reviewed in [6–8]. Interest in this approach has been stimulated by advances in using quantum systems as an informational resource [9–13]. Among more realistic cases, scenarios with successive measurements have been addressed in the literature [14–18]. Researchers are currently able to manipulate individual quantum systems [19,20]. In quantum information processing, our subsequent manipulations usually deal with an output of a latter stage. In effect, Heisenberg's thought experiment with microscope should rather be interpreted as related to uncertainties in successive measurements [21]. Uncertainty relations in the scenarios of successive measurements have received less attention than they deserve [15]. The authors of [15] also compared their findings with noise-disturbance relations given in [22]. Studies of scenarios with successive measurements allow us to understand whether preparation uncertainty relations are applicable to one or another question.

In principle, the Heisenberg uncertainty principle does not impose a restriction separately on spreads of position and momentum. It merely reveals that continuous trajectories are unspeakable in standard quantum mechanics, although such principles remain valid within Bohmian mechanics [23]. The generalized uncertainty principle is aimed to involve the existence of a minimal observable length. The latter is naturally connected with efforts to describe quantum gravity [24]. Some advances in merging quantum mechanics and general relativity are summarized in [25]. It is believed that quantum gravitational effects begin to be apparent at the scale corresponding to the Planck length $\ell_P = \sqrt{G\hbar/c^3} \approx 1.616 \times 10^{-35}$ m. Below this scale, the very structure of space-time is an open problem [26]. In addition, Heisenberg's principle is assumed to be converted into the generalized uncertainty principle (GUP) [27–29]. There exist proposals to test observable effects of the minimal length, including astronomical observations [30,31] and experimental schemes feasible within current technology [32,33]. The GUP case connects to many aspects that are currently the subject of active research [34–38]. The generalized uncertainty principle declares a non-zero lower bound on position spread. To reach such a model, the canonical commutation relation should be modified. Deformed forms of the commutation relation were recently studied from several viewpoints. On the other hand, the connections of the GUP with the real world represent an open question. In the context of non-relativistic quantum mechanics, the corresponding formalism was proposed in [39]. Another approach to representation of the used observables was suggested in [40]. This way is very convenient in extending entropic uncertainty relations to the GUP case [41].

In this paper, we aim to consider entropic uncertainty relations for successive measurements in the presence of a minimal observable length. Of course, our presentation is essentially based on mathematical relations given by Beckner [42] and by Białynicki-Birula and Mycielski [43]. This direction was initially inspired by Hirschman [44]. For observables with finite spectra, basic developments appeared due to [45–47]. We will largely use the results reported in [48,49]. The work in [48] is devoted to formulating entropic uncertainty relations for successive measurements of canonically conjugate observables. The case of position and momentum was addressed therein as a particular example of the scheme developed in [50,51]. Entropic uncertainty relations in the presence of a minimal length were examined in [49], and mainly focused on those points that were not considered in this context previously. Combining these two aspects finally led to the generalized uncertainty principle in scenarios of successive measurements. This paper is organized as follows. In Section 2, we review preliminary material, including properties of used information-theoretic measures. In Section 3, we briefly discuss successive quantum measurements in general. The main results of this paper are presented in Section 4. Both of the typical scenarios of successive measurements will be examined. In particular, we will see how formulating lower entropic bounds depends on the actual order in which measurements of position and momentum have been performed. In Section 5, we conclude the paper with a summary of the obtained results.

2. Preliminaries

In this section, we review the required material and fix the notations. To characterize measurement uncertainties, we use entropies of the Rényi and Tsallis types. Let us begin with the case of probability distributions with a discrete label. For the given probability distribution $\mathbf{p} = \{p_i\}$, its Rényi entropy of order α is defined as [52]

$$R_\alpha(\mathbf{p}) := \frac{1}{1-\alpha} \ln\left(\sum_i p_i^\alpha\right), \tag{1}$$

where $0 < \alpha \neq 1$. For $0 < \alpha < 1$, the Rényi α-entropy is a concave function of the probability distribution. For $\alpha > 1$, it is neither purely convex nor purely concave [53]. In the limit $\alpha \to 1$, the formula (1) gives the standard Shannon entropy

$$H_1(\mathbf{p}) = -\sum_i p_i \ln p_i. \tag{2}$$

For the given probability distribution $\mathbf{p} = \{p_i\}$ and $0 < \alpha \neq 1$, the Tsallis α-entropy is defined as [54]

$$H_\alpha(\mathbf{p}) := \frac{1}{1-\alpha}\left(\sum_i p_i^\alpha - 1\right) = -\sum_i p_i^\alpha \ln_\alpha(p_i). \tag{3}$$

Here, we use the α-logarithm expressed as $\ln_\alpha(y) := (y^{1-\alpha} - 1)/(1 - \alpha)$ for positive variable y and $0 < \alpha \neq 1$. When $\alpha \to 1$, the α-logarithm reduces to the usual one. Then, the α-entropy (3) also leads to the Shannon entropy (2). An axiomatic approach to generalized information-theoretic quantities is reviewed in [55]. In more detail, properties and applications of generalized entropies in physics are discussed in [56]. In the present paper, we will deal only with entropies of probability distributions. Quantum entropies of very general family were thoroughly examined in [57,58]. Quantum Rényi and Tsallis entropies are both particular representatives of this family.

Let $w(x)$ be a probability density function defined for all real x. Then, the differential Shannon entropy is introduced as

$$H_1(w) := - \int_{-\infty}^{+\infty} w(x) \ln w(x) \, dx. \tag{4}$$

Similarly, we determine entropies for other continuous variables of interest. For $0 < \alpha \neq 1$, the differential Rényi α-entropy is defined as

$$R_\alpha(w) := \frac{1}{1 - \alpha} \ln \left(\int_{-\infty}^{+\infty} w(x)^\alpha \, dx \right). \tag{5}$$

In contrast to entropies of a discrete probability distribution, differential entropies are not positive definite in general. To quantify an amount of uncertainty, we often tend to deal with positive entropic functions. One possible approach is such that the continuous axis of interest is divided into a set of non-intersecting bins. Preparation uncertainty relations with binning were derived in terms of the Shannon [59] and Rényi entropies [60]. To reach a good exposition, the size of these bins should be sufficiently small in comparison with a scale of considerable changes of $w(x)$. Keeping an obtained discrete distribution, we further calculate entropies of the forms (1) and (3).

The generalized uncertainty principle declares the deformed commutation relation for the position and momentum operators [39]. For convenience, we will use the wavenumber operator $\hat{k}$ instead of the momentum operator $\hbar\hat{k}$. It is helpful to rewrite this relation as

$$[\hat{x}, \hat{k}] = \mathrm{i}\left(1 + \beta\hat{k}^2\right). \tag{6}$$

Here, the positive parameter β is assumed to be rescaled by factor $\hbar^2$ from its usual sense. With the limit $\beta \to 0$, the formula (6) gives the standard commutation relation of ordinary quantum mechanics. Due to the Robertson formulation [3], the standard deviations in the pre-measurement state $\hat{\rho}$ satisfy

$$\Delta\hat{A}\,\Delta\hat{B} \geq \left| \frac{1}{2} \langle [\hat{A}, \hat{B}] \rangle_{\hat{\rho}} \right|. \tag{7}$$

By $\langle\hat{A}\rangle_{\hat{\rho}} = \mathrm{Tr}(\hat{A}\,\hat{\rho})$, we mean the quantum-mechanical expectation value. Combining (6) with (7) then gives

$$\Delta\hat{x}\,\Delta\hat{k} \geq \frac{1}{2}\left(1 + \beta\langle\hat{k}^2\rangle_{\hat{\rho}}\right) \geq \frac{1}{2}\left(1 + \beta(\Delta\hat{k})^2\right). \tag{8}$$

The principal parameter β is positive and independent of $\Delta\hat{x}$ and $\Delta\hat{k}$ [39]. It directly follows from (8) that $\Delta\hat{x}$ is not less than the square root of β. As was shown in [40], the auxiliary wavenumber operator $\hat{q}$ allows us to mediate between (6) and the standard commutation relation. Let $\hat{x}$ and $\hat{q}$ be self-adjoint operators satisfying $[\hat{x}, \hat{q}] = \mathrm{i}$. In the q-space, the action of $\hat{q}$ results in multiplying a wave function $\varphi(q)$ by q, whereas $\hat{x}\varphi(q) = \mathrm{i}\,d\varphi/dq$. Then, the wavenumber $\hat{k}$ can be represented as [40]

$$\hat{k} = \frac{1}{\sqrt{\beta}} \tan\left(\sqrt{\beta}\hat{q}\right). \tag{9}$$

The auxiliary wavenumber obeys the standard commutation relation but ranges between $\pm q_0(\beta) = \pm \pi/(2\sqrt{\beta})$. The function $q \mapsto k = \tan(\sqrt{\beta}q)/\sqrt{\beta}$ gives a one-to-one correspondence between $q \in (-q_0; +q_0)$ and $k \in (-\infty; +\infty)$. Hence, the eigenvalues of $\hat{k}$ fully cover the real axis. Further details of the above representation are examined in [40].

For any pure state, we will deal with three wave functions $\phi(k)$, $\varphi(q)$, and $\psi(x)$. The formalism of [40] is convenient in the sense that it explicitly describes the space of acceptable wave packets. In the q-space, these states should have wave functions that vanish for $|q| > q_0(\beta)$. Here, the auxiliary wave function $\varphi(q)$ is a useful tool related to $\psi(x)$ via the Fourier transform. In the q-space, the eigenfunctions of $\hat{x}$ appear as $\exp(-iqx)/\sqrt{2\pi}$. Thus, any wave function in the coordinate space is expressed as

$$\psi(x) = \frac{1}{\sqrt{2\pi}} \int_{-q_0}^{+q_0} \exp(+iqx)\,\varphi(q)\,dq\,. \tag{10}$$

Wave functions in the q- and x-spaces are connected by the Fourier transform [40],

$$\varphi(q) = \frac{1}{\sqrt{2\pi}} \int_{-\infty}^{+\infty} \exp(-iqx)\,\psi(x)\,dx\,. \tag{11}$$

The distinction from ordinary quantum mechanics is that wave functions in the q-space should be formally treated as 0 for all $|q| > q_0(\beta)$.

Using the above connection, the author of [41] affirmed the following. The uncertainty relation given in [42,43] is still valid in the GUP case. However, wave functions in the q-space are actually auxiliary. In the GUP case, the physically legitimate wavenumber and momentum involved in the relation (6) are described by wavefunctions in the k-space. A real distribution of physical wavenumber values is determined with respect to $\phi(k)$ instead of $\varphi(q)$. Let us examine the probability that momentum lies between two prescribed values. In view of the bijection between the intervals $(k_1; k_2)$ and $(q_1; q_2)$, this probability is expressed as

$$\int_{k_1}^{k_2} |\phi(k)|^2\,dk = \int_{q_1}^{q_2} |\varphi(q)|^2\,dq\,, \tag{12}$$

so that $|\phi(k)|^2\,dk = |\varphi(q)|^2\,dq$. Hence, two probability density functions $u(k)$ and $v(q)$ are connected as $u(k)\,dk = v(q)\,dq$, in another form

$$u(k) = \frac{v(q)}{1 + \beta k^2}\,. \tag{13}$$

For pure states, when $u(k) = |\phi(k)|^2$ and $v(q) = |\varphi(q)|^2$, the formula (13) is obvious. It can be extended to mixed states due to the spectral decomposition. However, one is actually unable to obtain the probability density functions $u(k)$ and $w(x)$ immediately.

In reality, any measurement apparatus is inevitably of a finite size. Devices with a finite extension need a finite amount of energy. Hence, one cannot ask for a state in which the measurement of an observable gives exactly one particular value of position. In more detail, measurements of coordinates of a microparticle are considered by Blokhintsev ([61], Chapter II). The generalized uncertainty principle imposes another limitation for position measurements. Although eigenstates of position and momentum are often considered explicitly, they are rather convenient tools of mathematical technique. The corresponding kets are not elements of the Hilbert space, but can be treated in the context of rigged Hilbert spaces [62]. Instead, we aim to use narrow distributions of a finite but small width. Measuring or preparing some state with the particular value ζ of position, one has to be affected by a neighborhood of ζ. Therefore, we treat each concrete result only as an estimation compatible with the GUP.

Thus, we cannot directly obtain probability density functions of the form $u(k)$ and $w(x)$. Here, a finiteness of detector resolution should be addressed [15,48]. Measuring or preparing a state with the particular value ζ of position, one is affected by some vicinity of ζ. In this way, we refer to generalized quantum measurements. Let the eigenkets $|x\rangle$ be normalized through Dirac's delta function. As was already mentioned, such kets cannot be treated as physical states even within ordinary quantum mechanics. In a finite-resolution measurement of position, the set $\mathcal{X} = \{|x\rangle\langle x|\}$ is replaced with some set $\mathcal{N}$ of operators of the form

$$\hat{N}(\xi) := \int_{-\infty}^{+\infty} dx\, g(\xi - x)\, |x\rangle\langle x|\,. \tag{14}$$

An acceptance function $\xi \mapsto g(\xi)$ satisfies the condition $\int_{-\infty}^{+\infty} |g(\xi)|^2\, d\xi = 1$. Then operators of the form (14) lead to a generalized resolution of the identity,

$$\int_{-\infty}^{+\infty} d\xi\, \hat{N}(\xi)^{\dagger}\hat{N}(\xi) = 1\,, \tag{15}$$

where the right-hand side is treated as the identity operator. For the pre-measurement state $\hat{\rho}$, the measurement leads to the probability density function

$$W_{\hat{\rho}}(\xi) = \operatorname{Tr}\!\big(\hat{N}(\xi)^{\dagger}\hat{N}(\xi)\hat{\rho}\big) = \int_{-\infty}^{+\infty} |g(\xi - x)|^2\, w_{\hat{\rho}}(x)\, dx\,. \tag{16}$$

This should be used instead of $w_{\hat{\rho}}(x) = \langle x|\hat{\rho}|x\rangle$. When the acceptance function is sufficiently narrow, we will obtain a good "footprint" of $w_{\hat{\rho}}(x)$. Let $\zeta \mapsto f(\zeta)$ be another acceptance function that also obeys the normalization condition. A finite-resolution measurement of the legitimate wavenumber is described by some set $\mathcal{N}$ of operators

$$\hat{M}(\zeta) := \int_{-\infty}^{+\infty} dk\, f(\zeta - k)\, |k\rangle\langle k|\,. \tag{17}$$

Here, the initial resolution $\mathcal{K} = \{|k\rangle\langle k|\}$ is replaced with $\mathcal{M} = \{\hat{M}(\zeta)\}$. Instead of $u_{\hat{\rho}}(k) = \langle k|\hat{\rho}|k\rangle$, we actually deal with the probability density function

$$U_{\hat{\rho}}(\zeta) = \operatorname{Tr}\!\big(\hat{M}(\zeta)^{\dagger}\hat{M}(\zeta)\hat{\rho}\big) = \int_{-\infty}^{+\infty} |f(\zeta - k)|^2\, u_{\hat{\rho}}(k)\, dk\,, \tag{18}$$

For good acceptance functions, a distortion of statistics will be small. The Gaussian distribution is a typical form of such functions [15]. We will assume that a behavior of acceptance functions is qualitatively similar.

3. On Successive Measurements of Observables in General

In this section, we generally formulate the question with respect to two successive measurements of observables with continuous spectra. It is more sophisticated than an intuitive obvious treatment of successive measurements on a finite-dimensional system. The latter allows us to deal with projective measurements, since all observables have a discrete spectrum. Such an approach is not meaningful for the case of position and momentum. On the other hand, the finite-dimensional case is important for understanding basic formulations related to continuous observables. To motivate our approach, we briefly review entropic uncertainty relations for successive projective measurements. Further, we will present a suitable reformulation for the case of position and momentum. Together with the entropic formulation, other approaches to express uncertainties in quantum measurements are of interest. In particular, modern investigations are based on the sum of variances [63,64], majorization relations [65–69], and the method of effective anticommutators [70]. The authors of [71] discussed some surprising results that may occur in application of entropic measures to quantify uncertainties in quantum measurements. These questions are beyond the scope of our consideration.

Scenarios with successive measurements are of interest for several reasons. The concept of wave function reduction assumes that we perform at least two successive measurements on a system (see for example Section 5.5 of [72]). By $\hat{\Lambda}_a \in \mathcal{A}$, we denote a projector onto the a-th eigenspace of finite-dimensional observable $\hat{A}$. For the pre-measurement state $\hat{\rho}$, the probability of outcome a is written as $\operatorname{Tr}(\hat{\Lambda}_a\hat{\rho})$. Such probabilities form a discrete distribution, from which we calculate quantities of interest. By $R_\alpha(\mathcal{A}; \hat{\rho})$ and $H_\alpha(\mathcal{A}; \hat{\rho})$, we further mean the entropies (1) and (3) calculated with the probabilities $\mathrm{Tr}(\hat{\Lambda}_a \hat{\rho})$. After the measurement of $\hat{A}$, we measure another observable $\hat{B}$. It is actually described by the set $\mathcal{B} = \{\hat{\Pi}_b\}$. Note that subsequent measurements are assumed to be performed with a new ensemble of states. The latter differs from traditional uncertainty relations in the preparation scenario. Scenarios with successive measurement are fixed by the used form of post-first-measurement states [16].

In the first scenario, the second measurement is performed on the state immediately following the first measurement with completely erased information. Here, the pre-measurement state of the second measurement is expressed as [14]

$$\mathsf{Y}_{\mathcal{A}}(\hat{\rho}) = \sum_a \hat{\Lambda}_a \hat{\rho} \hat{\Lambda}_a \,. \tag{19}$$

To characterize the amount of uncertainty in two successive measurements, we will use quantities of the form

$$R_\alpha(\mathcal{A}; \hat{\rho}) + R_\gamma(\mathcal{B}; \mathsf{Y}_{\mathcal{A}}(\hat{\rho})) \,, \tag{20}$$

and similarly with the corresponding Tsallis entropies. In the second scenario of successive measurements, we assume that the result of the first measurement is maintained. A focus on actual measurement outcomes is typical for the so-called selective measurements. For example, incoherent selective measurements are used in the formulation of monotonicity of coherence measures [73]. Coherence quantifiers can be defined with entropic functions of the Tsallis [74] and Rényi types [75]. In effect, the second measurement will be performed on the post-first-measurement state selected with respect to the actual outcome [16,17]. Due to the Lüders reduction rule [76], this state is written as

$$\hat{\tau}_a = \left(\mathrm{Tr}(\hat{\Lambda}_a \hat{\rho})\right)^{-1} \hat{\Lambda}_a \hat{\rho} \hat{\Lambda}_a \,, \tag{21}$$

whenever $\mathrm{Tr}(\hat{\Lambda}_a \hat{\rho}) \neq 0$. Measuring the observable $\hat{B}$ in each $\hat{\tau}_a$, we obtain the corresponding entropy $R_\gamma(\mathcal{B}; \hat{\tau}_a)$. Averaging over all a, we introduce the quantity

$$\sum_a \mathrm{Tr}(\hat{\Lambda}_a \hat{\rho}) \, R_\gamma(\mathcal{B}; \hat{\tau}_a) = \sum_a \mathrm{Tr}(\hat{\Lambda}_a \hat{\rho}) \, R_\alpha(\mathcal{A}; \hat{\tau}_a) + \sum_a \mathrm{Tr}(\hat{\Lambda}_a \hat{\rho}) \, R_\gamma(\mathcal{B}; \hat{\tau}_a) \,. \tag{22}$$

Of course, the first sum in the right-hand side of (22) vanishes. Measuring $\hat{A}$ in its eigenstate leads to a deterministic probability distribution, whence $R_\alpha(\mathcal{A}; \hat{\tau}_a) = 0$ for all a. It is for this reason that only the left-hand side of (22) is used in studies of uncertainties in successive measurements of finite-dimensional observables. In a similar manner, we can rewrite (20) and (22) with the use of Tsallis' entropies. For $\alpha = \gamma = 1$, the quantity (22) becomes the Shannon entropy averaged over all a. The authors of [16] utilized the latter as a measure of uncertainties in successive measurements. Uncertainty relations for successive projective measurements in terms of Rényi's entropies were analyzed in [17]. Formally, the sums involved in (22) are similar to one of several existing definitions of conditional Rényi's entropy. In more detail, these definitions are discussed [77]. The simplest of them just leads to expressions of the form (22). Moreover, the two kinds of conditional Tsallis entropy are known in the literature [78,79]. More properties of generalized conditional entropies are discussed in [80].

Let us proceed to exact formulations for successive measurements of position and momentum. One cannot provide states in which the measurement of position or momentum gives exactly one particular value. Instead, we deal with well localized states of finite or even small scales. Following [48], the right-hand side of (22) will be used in extending the second scenario to the position-momentum case in the presence of a minimal length. Suppose that the first applied measurement aims to measure momentum. The authors of [15] mentioned how the post-first-measurement state should be posed. In our notation, we write

$$\Phi_{\mathcal{M}}(\hat{\rho}) = \int_{-\infty}^{+\infty} \mathrm{d}\zeta \, \hat{M}(\zeta)\hat{\rho}\hat{M}(\zeta)^{\dagger}.$$ (23)

This expression replaces the formula (19) suitable for observables with a purely discrete spectrum. The following important fact should be pointed out. If we again measure momentum, but now with the state (23), then it will result in the same probability distribution function. It can be derived from (17) that

$$\langle k|\hat{\rho}|k\rangle = \langle k|\Phi_{\mathcal{M}}(\hat{\rho})|k\rangle, \qquad U_{\hat{\rho}}(\zeta) = U_{\Phi_{\mathcal{M}}(\hat{\rho})}(\zeta).$$ (24)

Such relations may be interpreted as a mild version of the repeatability concept. For strictly positive $\alpha \neq 1$, the Rényi α-entropy $R_{\alpha}(\mathcal{M};\hat{\rho})$ is given by substituting $U_{\hat{\rho}}(\zeta)$ into (5). The standard differential entropy $H_1(\mathcal{M};\hat{\rho})$ can be obtained within the limit $\alpha \to 1$. Also, the Rényi α-entropy $R_{\alpha}(\mathrm{p}_{\mathcal{M}}^{(\delta)};\hat{\rho})$ is defined by (1) by substituting probabilities defined through a discretization of the ζ-axis. When the first measurement is described by the set $\mathcal{N}$, the post-first-measurement state is specified as

$$\Phi_{\mathcal{N}}(\hat{\rho}) = \int_{-\infty}^{+\infty} \mathrm{d}\xi \, \hat{N}(\xi)\hat{\rho}\hat{N}(\xi)^{\dagger}.$$ (25)

Let $\hat{\rho}$ denote the state right before the sequence of successive measurements. In the first scenario of successive measurements, we will characterize uncertainties by entropic quantities of the form

$$R_{\alpha}(\mathcal{M};\hat{\rho}) + R_{\gamma}(\mathcal{N};\Phi_{\mathcal{M}}(\hat{\rho})), \qquad R_{\alpha}(\mathcal{M};\Phi_{\mathcal{N}}(\hat{\rho})) + R_{\gamma}(\mathcal{N};\hat{\rho}).$$ (26)

The former of the two sums concerns the case in which momentum is measured. Another useful approach is to calculate entropies with binning. For instance, sampling of the function (18) into bins between marks ζ_j gives a discrete probability distribution $\mathrm{p}_{\mathcal{M}}^{(\delta)}$. In the second measurement, entropies can be taken with binning between some marks ξ_k. By $\mathrm{p}_{\mathcal{N}}^{(\delta)}$, we mean the corresponding probability distribution. This approach leads to the characteristic quantities

$$R_{\alpha}(\mathrm{p}_{\mathcal{M}}^{(\delta)};\hat{\rho}) + R_{\gamma}(\mathrm{p}_{\mathcal{N}}^{(\delta)};\Phi_{\mathcal{M}}(\hat{\rho})), \qquad R_{\alpha}(\mathrm{p}_{\mathcal{M}}^{(\delta)};\Phi_{\mathcal{N}}(\hat{\rho})) + R_{\gamma}(\mathrm{p}_{\mathcal{N}}^{(\delta)};\hat{\rho}).$$ (27)

In a similar manner, we formulate entropic measures of the Tsallis type. As was already mentioned, such entropies will be taken only with binning.

The second scenario of successive measurements prescribes that each actual result of the first measurement should be retained. Assuming $U_{\hat{\rho}}(\zeta) \neq 0$ in the corresponding domain, we now consider the normalized output state

$$\hat{\varrho}(\zeta) = U_{\hat{\rho}}(\zeta)^{-1}\hat{M}(\zeta)\hat{\rho}\hat{M}(\zeta)^{\dagger}.$$ (28)

Each $\hat{\varrho}(\zeta)$ is used as one of possible pre-measurement states in the second measurement. Similarly to (22), we then consider the quantity

$$\int_{-\infty}^{+\infty} R_{\alpha}(\mathcal{M};\hat{\varrho}(\zeta))U_{\hat{\rho}}(\zeta)\,\mathrm{d}\zeta + \int_{-\infty}^{+\infty} R_{\gamma}(\mathcal{N};\hat{\varrho}(\zeta))U_{\hat{\rho}}(\zeta)\,\mathrm{d}\zeta.$$ (29)

When position is measured first, particular outputs are of the form

$$\hat{\sigma}(\xi) = W_{\hat{\rho}}(\xi)^{-1}\hat{N}(\xi)\hat{\rho}\hat{N}(\xi)^{\dagger}.$$ (30)

To describe the amount of uncertainty here, we rewrite (29) with $\hat{\sigma}(\xi)$ instead of $\hat{\varrho}(\zeta)$ and $W_{\hat{\rho}}(\xi)$ instead of $U_{\hat{\rho}}(\zeta)$. We will also utilize entropic uncertainty relations with binning. Here, one replaces (29) with

$$\int_{-\infty}^{+\infty} R_\alpha\big(\mathrm{p}_{\mathcal{M}}^{(\delta)};\hat{\varrho}(\zeta)\big) U_\beta(\zeta)\,\mathrm{d}\zeta + \int_{-\infty}^{+\infty} R_\gamma\big(\mathrm{p}_{\mathcal{N}}^{(\delta)};\hat{\varrho}(\zeta)\big) U_\beta(\zeta)\,\mathrm{d}\zeta\,, \tag{31}$$

and similarly with the Tsallis entropies. Quantities of the form (31) concern successive measurements, in which position is measured after momentum. When position is measured by the first, we rewrite such expressions with $\hat{\sigma}(\xi)$ and $W_\beta(\xi)$. In the paper [48], the above treatment of successive measurements was considered for general canonically conjugate operators. This approach to the concept of canonical conjugacy is based on the Pegg–Barnett formalism [50]. The Pegg–Barnett formalism was originally proposed to explain a Hermitian phase operator [81,82]. Entropic uncertainty relations on the base of this formalism were examined in [51,83,84].

4. Main Results

In this section, we shall formulate entropic uncertainty relations for successive measurements within the GUP case. For this case, preparation uncertainty relations with a correction term were derived in [49]. For the convenience of further calculations, the prepared pre-measurement state will be denoted by $\hat{\omega}$. Due to [49], we have

$$H_1(\mathcal{M};\hat{\omega}) + H_1(\mathcal{N};\hat{\omega}) \geq H_1(\mathcal{K};\hat{\omega}) + H_1(\mathcal{X};\hat{\omega}) \geq \ln(e\pi) + \big\langle \ln(1+\beta\hat{k}^2)\big\rangle_{\hat{\omega}}. \tag{32}$$

The well-known bound $\ln(e\pi)$ corresponds to the entropic uncertainty relation of Beckner [42] and Białynicki-Birula and Mycielski [43]. The second term in the right-hand side of (32) reflects the fact that the legitimate momentum of the commutation relation (6) is given by $\hbar\hat{k}$. Here, the wavenumber operator $\hat{q}$ plays an auxiliary role. Note that this correction term depends on the pre-measurement state. As some numerical results in [85] later showed, the presented correction is sufficiently tight. It is similar to the correction term obtained in the Robertson formulation (8). However, the inequality (32) is a preparation uncertainty relation.

Suppose now that we measure momentum by the first and position by the second. In the first scenario, the pre-measurement state $\hat{\rho}$ leads to the post-first-measurement state $\Phi_{\mathcal{M}}(\hat{\rho})$. Due to (24), we immediately write

$$H_1\big(\mathcal{M};\hat{\rho}\big) = H_1\big(\mathcal{M};\Phi_{\mathcal{M}}(\hat{\rho})\big)\,, \qquad \big\langle \ln(1+\beta\hat{k}^2)\big\rangle_{\hat{\rho}} = \big\langle \ln(1+\beta\hat{k}^2)\big\rangle_{\Phi_{\mathcal{M}}(\hat{\rho})}. \tag{33}$$

Substituting $\hat{\omega} = \Phi_{\mathcal{M}}(\hat{\rho})$ into (32) and using (33), we easily get

$$H_1(\mathcal{M};\hat{\rho}) + H_1\big(\mathcal{N};\Phi_{\mathcal{M}}(\hat{\rho})\big) \geq \ln(e\pi) + \big\langle \ln(1+\beta\hat{k}^2)\big\rangle_{\hat{\rho}}. \tag{34}$$

This is an entropic uncertainty relation in the first scenario of successive measurements such that momentum is measured by the first. The corresponding lower bound is the same as in the preparation scenario. It is not the case, when we measure position by the first and momentum by the second. Putting $\hat{\omega} = \Phi_{\mathcal{N}}(\hat{\rho})$ into (32) finally gives

$$H_1(\mathcal{N};\hat{\rho}) + H_1\big(\mathcal{M};\Phi_{\mathcal{N}}(\hat{\rho})\big) \geq \ln(e\pi) + \big\langle \ln(1+\beta\hat{k}^2)\big\rangle_{\Phi_{\mathcal{N}}(\hat{\rho})}. \tag{35}$$

The correction term in the right-hand side of (35) is similar in form but should be calculated with the post-first-measurement state $\Phi_{\mathcal{N}}(\hat{\rho})$. Taking $\beta = 0$, the above entropic bounds for successive measurements do not differ from the bound in the preparation scenario. Here, we see a manifestation of the deformed commutation relation (6). The latter disturbs a certain symmetry between position and momentum.

Let us proceed to the second scenario of successive measurements. Suppose again that momentum is measured by the first. Substituting $\hat{\omega} = \hat{\varrho}(\zeta)$ into (32), we multiply it by $U_\beta(\zeta)$ and then integrate with respect to ζ. This results in the inequality

$$\int_{-\infty}^{+\infty} H_1\big(\mathcal{M}; \hat{\varrho}(\zeta)\big) U_\rho(\zeta)\, d\zeta + \int_{-\infty}^{+\infty} H_1\big(\mathcal{N}; \hat{\varrho}(\zeta)\big) U_\rho(\zeta)\, d\zeta$$

$$\geq \ln(e\pi) + \int_{-\infty}^{+\infty} \big\langle \ln(1 + \beta\hat{k}^2) \big\rangle_{\hat{\varrho}(\zeta)} U_\rho(\zeta)\, d\zeta . \tag{36}$$

Using (28), the second term in the right-hand side of (36) can be simplified, viz.,

$$\int_{-\infty}^{+\infty} d\zeta\, U_\rho(\zeta) \int_{-\infty}^{+\infty} dk\, \ln(1 + \beta k^2)\, \langle k|\hat{\varrho}(\zeta)|k\rangle$$

$$= \int_{-\infty}^{+\infty} d\zeta \int_{-\infty}^{+\infty} dk\, \ln(1 + \beta k^2)\, \langle k|\hat{M}(\zeta)\hat{\rho}\hat{M}(\zeta)^\dagger|k\rangle \tag{37}$$

$$= \int_{-\infty}^{+\infty} dk\, \ln(1 + \beta k^2)\, \langle k|\hat{\rho}|k\rangle \int_{-\infty}^{+\infty} d\zeta\, |f(\zeta - k)|^2 .$$

In the right-hand side of (37), the last integral with respect to ζ is equal to 1. For the second scenario of successive measurements, we obtain

$$\int_{-\infty}^{+\infty} H_1\big(\mathcal{M}; \hat{\varrho}(\zeta)\big) U_\rho(\zeta)\, d\zeta + \int_{-\infty}^{+\infty} H_1\big(\mathcal{N}; \hat{\varrho}(\zeta)\big) U_\rho(\zeta)\, d\zeta \geq \ln(e\pi) + \big\langle \ln(1 + \beta\hat{k}^2) \big\rangle_\rho . \tag{38}$$

Hence, entropic uncertainty relations (34) and (38) are obtained with the same lower bound calculated with the pre-measurement state. Let us consider the case when position is measured by the first. Substituting $\hat{\omega} = \hat{\sigma}(\zeta)$ into (32), we multiply it by $W_\rho(\xi)$ and integrate with respect to ξ, whence

$$\int_{-\infty}^{+\infty} H_1\big(\mathcal{M}; \hat{\sigma}(\xi)\big) W_\rho(\xi)\, d\xi + \int_{-\infty}^{+\infty} H_1\big(\mathcal{N}; \hat{\sigma}(\xi)\big) W_\rho(\xi)\, d\xi$$

$$\geq \ln(e\pi) + \int_{-\infty}^{+\infty} \big\langle \ln(1 + \beta\hat{k}^2) \big\rangle_{\hat{\sigma}(\xi)} W_\rho(\xi)\, d\xi . \tag{39}$$

In the right-hand side of (39), the second integral is a correction term averaged over particular outputs $\hat{\sigma}(\xi)$. In general, an expression for this term cannot be simplified without additional assumptions. We have already seen how the relation (35) differs from (34). The formula (39) differs from (38) in a similar vein. In the presence of a minimal length, the preparation uncertainty relation (32) remains valid for successive measurements, when momentum is measured by the first. Otherwise, it should be reformulated.

Entropic uncertainty relations with binning can be treated in a similar manner. Using some discretization of axes, we take into account sufficiently typical setup. This approach also leads to entropic functions with only positive values. In contrast, differential entropies can generally have arbitrary signs. In the case of momentum measurements, values ζ_i denote the ends of intervals $\delta\zeta_i = \zeta_{i+1} - \zeta_i$. For the prepared state $\hat{\omega}$, we deal with probabilities

$$p_i^{(\delta)} := \int_{\zeta_i}^{\zeta_{i+1}} U_{\hat{\omega}}(\zeta)\, d\zeta , \tag{40}$$

which form the discrete distribution $\mathrm{p}_\mathcal{M}^{(\delta)}$. Using (40), one calculates the Shannon entropy $H_1\big(\mathrm{p}_\mathcal{M}^{(\delta)}; \hat{\omega}\big)$. In a similar way, we discretize the ξ-axes into bins $\delta\xi_j = \xi_{j+1} - \xi_j$ with the resulting distribution $\mathrm{p}_\mathcal{N}^{(\delta)}$. It can be shown that

$$H_1\big(\mathrm{p}_\mathcal{M}^{(\delta)}; \hat{\omega}\big) + H_1\big(\mathrm{p}_\mathcal{N}^{(\delta)}; \hat{\omega}\big) \geq \ln\left(\frac{e\pi}{\delta\zeta\, \delta\xi}\right) + \big\langle \ln(1 + \beta\hat{k}^2) \big\rangle_{\hat{\omega}} , \tag{41}$$

where $\delta\zeta = \max \delta\zeta_i$ and $\delta\xi = \max \delta\xi_j$. The formula (41) gives a preparation uncertainty relation with binning. It involves the same correction term due to the existence of a minimal length. To pose entropic uncertainty relations in the first scenario of successive measurements, we again use reasons that have lead to (34) and (35). Finally, one gets

$$H_1\left(\mathsf{p}_{\mathcal{M}}^{(\delta)}; \hat{\rho}\right) + H_1\left(\mathsf{p}_{\mathcal{N}}^{(\delta)}; \Phi_{\mathcal{M}}(\hat{\rho})\right) \geq \ln\left(\frac{e\pi}{\delta\zeta\,\delta\xi}\right) + \left\langle \ln(1 + \beta\hat{k}^2) \right\rangle_{\hat{\rho}},\tag{42}$$

$$H_1\left(\mathsf{p}_{\mathcal{N}}^{(\delta)}; \hat{\rho}\right) + H_1\left(\mathsf{p}_{\mathcal{M}}^{(\delta)}; \Phi_{\mathcal{N}}(\hat{\rho})\right) \geq \ln\left(\frac{e\pi}{\delta\xi\,\delta\zeta}\right) + \left\langle \ln(1 + \beta\hat{k}^2) \right\rangle_{\Phi_{\mathcal{N}}(\hat{\rho})}.\tag{43}$$

In the second scenario of successive measurements, entropic uncertainty relations with binning are obtained in the form

$$\int_{-\infty}^{+\infty} H_1\left(\mathsf{p}_{\mathcal{M}}^{(\delta)}; \hat{\varrho}(\zeta)\right) U_{\hat{\rho}}(\zeta)\,d\zeta + \int_{-\infty}^{+\infty} H_1\left(\mathsf{p}_{\mathcal{N}}^{(\delta)}; \hat{\varrho}(\zeta)\right) U_{\hat{\rho}}(\zeta)\,d\zeta \geq \ln\left(\frac{e\pi}{\delta\zeta\,\delta\xi}\right) + \left\langle \ln(1 + \beta\hat{k}^2) \right\rangle_{\hat{\rho}}\tag{44}$$

$$\int_{-\infty}^{+\infty} H_1\left(\mathsf{p}_{\mathcal{N}}^{(\delta)}; \hat{\sigma}(\xi)\right) W_{\hat{\rho}}(\xi)\,d\xi + \int_{-\infty}^{+\infty} H_1\left(\mathsf{p}_{\mathcal{M}}^{(\delta)}; \hat{\sigma}(\xi)\right) W_{\hat{\rho}}(\xi)\,d\xi \geq \ln\left(\frac{e\pi}{\delta\xi\,\delta\zeta}\right) \\ + \int_{-\infty}^{+\infty} \left\langle \ln(1 + \beta\hat{k}^2) \right\rangle_{\hat{\sigma}(\xi)} W_{\hat{\rho}}(\xi)\,d\xi.\tag{45}$$

In the presence of a minimal length, distinctions of (43) and (45) from the corresponding preparation relations are concentrated in correction terms. In effect, these terms are not state-independent. On the other hand, entropic bounds of preparation uncertainty relations remain valid when momentum is measured by the first. The author of [49] also reported on state-independent entropic uncertainty relations in the presence of a minimal length. Such relations were posed in terms of the Rényi and Tsallis entropies with binning. An alteration of statistics due to a finite resolution of the measurements is also taken into account. When acceptance functions of measurement apparatuses are sufficiently spread, they lead to an increase of entropic lower bounds. To pose uncertainty relations formally, we introduce the following quantity [49],

$$S_f := \sup_\zeta \int_{-\infty}^{+\infty} \frac{|f(\zeta - k)|^2}{1 + \beta k^2}\,dk,\tag{46}$$

where the acceptance function $\zeta \mapsto f(\zeta)$ corresponds to momentum measurements. Let $\hat{\omega}$ represent the prepared state. As was shown in [49], the existence of a minimal length leads to preparation uncertainty relations of the form

$$R_\alpha(\mathcal{M}; \hat{\omega}) + R_\gamma(\mathcal{N}; \hat{\omega}) \geq \ln\left(\frac{\varkappa\pi}{S_f}\right).\tag{47}$$

Here, positive entropic parameters obey $1/\alpha + 1/\gamma = 2$ and

$$\varkappa^2 = \alpha^{1/(\alpha-1)}\gamma^{1/(\gamma-1)}.\tag{48}$$

In the limit $\alpha \to 1$, the parameter $\varkappa$ becomes equal to e. When $\beta = 0$, we clearly have $S_f = 1$, so that the right-hand side of (47) reduces to $\ln(\varkappa\pi)$. The latter is a known entropic bound for the case of usual position and momentum. For $\beta > 0$ and physically reasonable acceptance functions, we obtain an improved lower due to $S_f < 1$. It is important that the quantity (46) depends only on β and the actual acceptance function in momentum measurements. Preparation entropic uncertainty relations with binning are posed as follows [49]. Let probability density functions $U_{\hat{\omega}}(\zeta)$ and $W_{\hat{\omega}}(\xi)$ be sampled into discrete probability distributions. Then the corresponding Rényi and Tsallis entropies satisfy

$$R_\alpha\left(\mathrm{p}_{\mathcal{M}}^{(\delta)};\hat{\omega}\right) + R_\gamma\left(\mathrm{p}_{\mathcal{N}}^{(\delta)};\hat{\omega}\right) \geq \ln\left(\frac{\varkappa\pi}{S_f\,\delta\zeta\,\delta\xi}\right), \tag{49}$$

$$H_\alpha\left(\mathrm{p}_{\mathcal{M}}^{(\delta)};\hat{\omega}\right) + H_\gamma\left(\mathrm{p}_{\mathcal{N}}^{(\delta)};\hat{\omega}\right) \geq \ln_v\left(\frac{\varkappa\pi}{S_f\,\delta\zeta\,\delta\xi}\right), \tag{50}$$

where $1/\alpha + 1/\gamma = 2$ and $v = \max\{\alpha,\gamma\}$.

Due to equalities of the form (24), the preparation uncertainty relations (47), (49), and (50) are immediately converted into relations for successive measurements. In the first scenario, we obtain

$$R_\alpha(\mathcal{M};\hat{\rho}) + R_\gamma(\mathcal{N};\Phi_{\mathcal{M}}(\hat{\rho})) \geq \ln\left(\frac{\varkappa\pi}{S_f}\right), \tag{51}$$

where $1/\alpha + 1/\gamma = 2$ and the momentum measurement is assumed to be made by the first. When position is measured by the first, we replace $\hat{\rho}$ with $\Phi_{\mathcal{N}}(\hat{\rho})$ and $\Phi_{\mathcal{M}}(\hat{\rho})$ with $\hat{\rho}$ in the left-hand side of (51). For $1/\alpha + 1/\gamma = 2$ and $v = \max\{\alpha,\gamma\}$, entropic uncertainty relations with binning are written as

$$R_\alpha\left(\mathrm{p}_{\mathcal{M}}^{(\delta)};\hat{\rho}\right) + R_\gamma\left(\mathrm{p}_{\mathcal{N}}^{(\delta)};\Phi_{\mathcal{M}}(\hat{\rho})\right) \geq \ln\left(\frac{\varkappa\pi}{S_f\,\delta\zeta\,\delta\xi}\right), \tag{52}$$

$$H_\alpha\left(\mathrm{p}_{\mathcal{M}}^{(\delta)};\hat{\rho}\right) + H_\gamma\left(\mathrm{p}_{\mathcal{N}}^{(\delta)};\Phi_{\mathcal{M}}(\hat{\rho})\right) \geq \ln_v\left(\frac{\varkappa\pi}{S_f\,\delta\zeta\,\delta\xi}\right). \tag{53}$$

The same entropic lower bounds hold, when position is measured by the first. We refrain from presenting the details here. In the second scenario of successive measurements, one immediately gets

$$\int_{-\infty}^{+\infty} R_\alpha\left(\mathcal{M};\hat{\varrho}(\zeta)\right)U_{\hat{\rho}}(\zeta)\,\mathrm{d}\zeta + \int_{-\infty}^{+\infty} R_\gamma\left(\mathcal{N};\hat{\varrho}(\zeta)\right)U_{\hat{\rho}}(\zeta)\,\mathrm{d}\zeta \geq \ln\left(\frac{\varkappa\pi}{S_f}\right), \tag{54}$$

where $1/\alpha + 1/\gamma = 2$ and the momentum measurement is assumed to be made by the first. Replacing $\hat{\varrho}(\zeta)$ with $\hat{\sigma}(\xi)$ and $U_{\hat{\rho}}(\zeta)$ with $W_{\hat{\rho}}(\xi)$, we resolve the case when position is measured by the first. For $1/\alpha + 1/\gamma = 2$ and $v = \max\{\alpha,\gamma\}$, entropic uncertainty relations with binning are expressed as

$$\int_{-\infty}^{+\infty} R_\alpha\left(\mathrm{p}_{\mathcal{M}}^{(\delta)};\hat{\varrho}(\zeta)\right)U_{\hat{\rho}}(\zeta)\,\mathrm{d}\zeta + \int_{-\infty}^{+\infty} R_\gamma\left(\mathrm{p}_{\mathcal{N}}^{(\delta)};\hat{\varrho}(\zeta)\right)U_{\hat{\rho}}(\zeta)\,\mathrm{d}\zeta \geq \ln\left(\frac{\varkappa\pi}{S_f\,\delta\zeta\,\delta\xi}\right), \tag{55}$$

$$\int_{-\infty}^{+\infty} H_\alpha\left(\mathrm{p}_{\mathcal{M}}^{(\delta)};\hat{\varrho}(\zeta)\right)U_{\hat{\rho}}(\zeta)\,\mathrm{d}\zeta + \int_{-\infty}^{+\infty} H_\gamma\left(\mathrm{p}_{\mathcal{N}}^{(\delta)};\hat{\varrho}(\zeta)\right)U_{\hat{\rho}}(\zeta)\,\mathrm{d}\zeta \geq \ln_v\left(\frac{\varkappa\pi}{S_f\,\delta\zeta\,\delta\xi}\right). \tag{56}$$

When position is measured first, we merely replace here $\hat{\varrho}(\zeta)$ with $\hat{\sigma}(\xi)$ and $U_{\hat{\rho}}(\zeta)$ with $W_{\hat{\rho}}(\xi)$. Thus, state-independent entropic lower bounds of preparation uncertainty relation remain valid for scenarios with successive measurements. The existence of a minimal length is taken into account due to the quantity (46). In the case $\alpha = \gamma = 1$, the above relations are expressed via the Shannon entropies. We have also obtained state-dependent entropic uncertainty relations such as (35), (39), (43), and (45). Their formulations differ from preparation uncertainty relations since they depend on the quantum state immediately following the first measurement.

5. Conclusions

We have formulated entropic uncertainty relations for successive measurements in the presence of a minimal length. The presented formulation is explicitly dependent on the order of the measurements, though the bounds themselves may not be optimal. The problem of a minimal observable length is related to efforts to describe gravitation at the quantum level. In effect, the generalized uncertainty principle restricts the space of acceptable wave packets. Scenarios with successive measurements are interesting for several reasons. The traditional scenario of preparation uncertainty relations is insufficient even from the viewpoint of Heisenberg's thought experiment [1]. Successive measurements of position and momentum cannot be treated as projective even within ordinary quantum mechanics. The GUP case implies additional limitation for a spatial width of the acceptance function in position measurements. Thus, entropic measures of uncertainty should be formulated differently from the finite-dimensional case. One of distinctions concerns a proper form of the state immediately following the first measurement. The post-first-measurement state was chosen according to the two possible scenarios. Uncertainty relations in terms of Shannon entropies contain a state-dependent correction term. Hence, entropic lower bounds for successive measurements generally differ from lower bounds involved into preparation uncertainty relations. We also formulated state-independent uncertainty bounds in terms of Rényi entropies and, with binning, in terms of Tsallis entropies. In the presence of a minimal length, state-independent entropic lower bounds of preparation uncertainty relations remain valid for scenarios with successive measurements. When acceptance functions of measurement apparatuses are sufficiently well spread, the existing entropic lower bounds are improved.

Conflicts of Interest: The author declares no conflict of interest.

References

1. Heisenberg, W. Über den anschaulichen inhalt der quanten theoretischen kinematik und mechanik. *Z. Phys.* **1927**, *43*, 172–198. (In German) [CrossRef]
2. Kennard, E.H. Zur quantenmechanik einfacher bewegungstypen. *Z. Phys.* **1927**, *44*, 326–352. (In German) [CrossRef]
3. Robertson, H.P. The uncertainty principle. *Phys. Rev.* **1929**, *34*, 163–164. [CrossRef]
4. Rozpędek, F.; Kaniewski, J.; Coles, P.J.; Wehner, S. Quantum preparation uncertainty and lack of information. *New J. Phys.* **2017**, *19*, 023038. [CrossRef]
5. Coles, P.J.; Kaniewski, J.; Wehner, S. Equivalence of wave-particle duality to entropic uncertainty. *Nat. Commun.* **2014**, *5*, 5814. [CrossRef] [PubMed]
6. Wehner, S.; Winter, A. Entropic uncertainty relations—A survey. *New J. Phys.* **2010**, *12*, 025009. [CrossRef]
7. Białynicki-Birula, I.; Rudnicki, Ł. Entropic Uncertainty Relations in Quantum Physics. In *Statistical Complexity*; Springer: Berlin, Germany, 2011; pp. 1–34.
8. Coles, P.J.; Berta, M.; Tomamichel, M.; Wehner, S. Entropic uncertainty relations and their applications. *Rev. Mod. Phys.* **2017**, *89*, 015002. [CrossRef]
9. Berta, M.; Christandl, M.; Colbeck, R.; Renes, J.M.; Renner, R. The uncertainty principle in the presence of quantum memory. *Nat. Phys.* **2010**, *6*, 659–662. [CrossRef]
10. Tomamichel, M.; Renner, R. Uncertainty relation for smooth entropies. *Phys. Rev. Lett.* **2011**, *106*, 110506. [CrossRef] [PubMed]
11. Ng, N.H.Y.; Berta, M.; Wehner, S. Min-entropy uncertainty relation for finite-size cryptography. *Phys. Rev. A* **2012**, *86*, 042315. [CrossRef]
12. Furrer, F. Reverse-reconciliation continuous-variable quantum key distribution based on the uncertainty principle. *Phys. Rev. A* **2014**, *90*, 042325. [CrossRef]
13. Li, J.; Fei, S.-M. Uncertainty relation based on Wigner–Yanase–Dyson skew information with quantum memory. *Entropy* **2018**, *20*, 132. [CrossRef]
14. Srinivas, M.D. Optimal entropic uncertainty relation for successive measurements in quantum information theory. *Pramana J. Phys.* **2003**, *60*, 1137–1152. [CrossRef]
15. Distler, J.; Paban, S. Uncertainties in successive measurements. *Phys. Rev. A* **2013**, *87*, 062112. [CrossRef]

16. Baek, K.; Farrow, T.; Son, W. Optimized entropic uncertainty for successive projective measurements. *Phys. Rev. A* **2014**, *89*, 032108. [CrossRef]
17. Zhang, J.; Zhang, Y.; Yu, C.-S. Rényi entropy uncertainty relation for successive projective measurements. *Quantum Inf. Process.* **2015**, *14*, 2239–2253. [CrossRef]
18. Baek, K.; Son, W. Entropic uncertainty relations for successive generalized measurements. *Mathematics* **2016**, *4*, 41. [CrossRef]
19. Wineland, D. Superposition, entanglement, and raising Schrödinger's cat. *Ann. Phys.* **2013**, *525*, 739–752. [CrossRef]
20. Haroche, S. Controlling photons in a box and exploring the quantum to classical boundary. *Ann. Phys.* **2013**, *525*, 753–776. [CrossRef]
21. Busch, P.; Lahti, P.; Werner, R.F. Proof of Heisenberg's error-disturbance relation. *Phys. Rev. Lett.* **2013**, *111*, 160405. [CrossRef] [PubMed]
22. Ozawa, M. Uncertainty relations for noise and disturbance in generalized quantum measurements. *Ann. Phys.* **2004**, *311*, 350–416. [CrossRef]
23. Gisin, N. Why Bohmian Mechanics? One- and two-time position measurements, Bell inequalities, philosophy, and physics. *Entropy* **2018**, *20*, 105. [CrossRef]
24. Hossenfelder, S. Minimal length scale scenarios for quantum gravity. *Living Rev. Relativ.* **2013**, *16*, 2. [CrossRef] [PubMed]
25. Rovelli, C. *Quantum Gravity*; Cambridge University Press: Cambridge, UK, 2004.
26. Amati, D.; Ciafaloni, M.; Veneziano, G. Can spacetime be probed below the string size? *Phys. Lett. B* **1989**, *216*, 41–47. [CrossRef]
27. Scardigli, F. Generalized uncertainty principle in quantum gravity from micro-black hole gedanken experiment. *Phys. Lett. B* **1999**, *452*, 39–44. [CrossRef]
28. Bambi, C. A revision of the generalized uncertainty principle. *Class. Quantum Grav.* **2008**, *25*, 105003. [CrossRef]
29. Tawfik, A.N.; Diab, A.M. A review of the generalized uncertainty principle. *Rep. Prog. Phys.* **2015**, *78*, 126001. [CrossRef] [PubMed]
30. Amelino-Camelia, G.; Ellis, J.; Mavromatos, N.E.; Nanopoulos, D.V.; Sarkar, S. Tests of quantum gravity from observations of γ-ray bursts. *Nature* **1998**, *393*, 763–765. [CrossRef]
31. Jacob, U.; Piran, T. Neutrinos from gamma-ray bursts as a tool to explore quantum-gravity-induced Lorentz violation. *Nat. Phys.* **2007**, *3*, 87–90. [CrossRef]
32. Pikovski, I.; Vanner, M.R.; Aspelmeyer, M.; Kim, M.; Brukner, Č. Probing Planck-scale physics with quantum optics. *Nat. Phys.* **2012**, *8*, 393–397. [CrossRef]
33. Marin, F.; Marino, F.; Bonaldi, M.; Cerdonio, M.; Conti, L.; Falferi, P.; Mezzena, R.; Ortolan, A.; Prodi, G.A.; Taffarello, L.; et al. Gravitational bar detectors set limits to Planck-scale physics on macroscopic variables. *Nat. Phys.* **2013**, *9*, 71–73. [CrossRef]
34. Tawfik, A. Impacts of generalized uncertainty principle on black hole thermodynamics and Salecker–Wigner inequalities. *JCAP* **2013**, *7*, 040. [CrossRef]
35. Dey, S.; Fring, A.; Khantoul, B. Hermitian versus non-Hermitian representations for minimal length uncertainty relations. *J. Phys. A Math. Theor.* **2013**, *46*, 335304. [CrossRef]
36. Tawfik, A.; Diab, A. Generalized uncertainty principle: Approaches and applications. *Int. J. Mod. Phys. A* **2014**, *23*, 1430025. [CrossRef]
37. Faizal, M.; Majumder, B. Incorporation of generalized uncertainty principle into Lifshitz field theories. *Ann. Phys.* **2015**, *357*, 49–58. [CrossRef]
38. Masood, S.; Faizal, M.; Zaz, Z.; Ali, A.F.; Raza, J.; Shah, M.B. The most general form of deformation of the Heisenberg algebra from the generalized uncertainty principle. *Phys. Lett. B* **2016**, *763*, 218–227. [CrossRef]
39. Kempf, A.; Mangano, G.; Mann, R.B. Hilbert space representation of the minitial length uncertainty relation. *Phys. Rev. D* **1995**, *52*, 1108–1118. [CrossRef]
40. Pedram, P. New approach to nonperturbative quantum mechanics with minimal length uncertainty. *Phys. Rev. D* **2012**, *85*, 024016. [CrossRef]
41. Pedram, P. The minimal length and the Shannon entropic uncertainty relation. *Adv. High Energy Phys.* **2016**, *2016*, 5101389. [CrossRef]
42. Beckner, W. Inequalities in Fourier analysis. *Ann. Math.* **1975**, *102*, 159–182. [CrossRef]

43. Białynicki-Birula, I.; Mycielski, J. Uncertainty relations for information entropy in wave mechanics. *Commun. Math. Phys.* **1975**, *44*, 129–132. [CrossRef]

44. Hirschman, I.I. A note on entropy. *Am. J. Math.* **1957**, *79*, 152–156. [CrossRef]

45. Deutsch, D. Uncertainty in quantum measurements. *Phys. Rev. Lett.* **1983**, *50*, 631–633. [CrossRef]

46. Kraus, K. Complementary observables and uncertainty relations. *Phys. Rev. D* **1987**, *35*, 3070–3075. [CrossRef]

47. Maassen, H.; Uffink, J.B.M. Generalized entropic uncertainty relations. *Phys. Rev. Lett.* **1988**, *60*, 1103–1106. [CrossRef] [PubMed]

48. Rastegin, A.E. Entropic uncertainty relations for successive measurements of canonically conjugate observables. *Ann. Phys.* **2016**, *528*, 835–844. [CrossRef]

49. Rastegin, A.E. On entropic uncertainty relations in the presence of a minimal length. *Ann. Phys.* **2017**, *382*, 170–180. [CrossRef]

50. Pegg, D.T.; Vaccaro, J.A.; Barnett, S.M. Quantum-optical phase and canonical conjugation. *J. Mod. Opt.* **1990**, *37*, 1703–1710. [CrossRef]

51. Gonzalez, A.R.; Vaccaro, J.A.; Barnett, S.M. Entropic uncertainty relations for canonically conjugate operators. *Phys. Lett. A* **1995**, *205*, 247–254. [CrossRef]

52. Rényi, A. On measures of entropy and information. In *Proceedings of the 4th Berkeley Symposium on Mathematical Statistics and Probability*; University of California Press: Berkeley, CA, USA, 1961; pp. 547–561.

53. Jizba, P.; Arimitsu, T. The world according to Rényi: Thermodynamics of multifractal systems. *Ann. Phys.* **2004**, *312*, 17–59. [CrossRef]

54. Tsallis, C. Possible generalization of Boltzmann–Gibbs statistics. *J. Stat. Phys.* **1988**, *52*, 479–487. [CrossRef]

55. Csiszár, I. Axiomatic characterizations of information measures. *Entropy* **2008**, *10*, 261–273. [CrossRef]

56. Bengtsson, I.; Życzkowski, K. *Geometry of Quantum States: An Introduction to Quantum Entanglement*; Cambridge University Press: Cambridge, UK, 2006.

57. Holik, F.; Bosyk, G.M.; Bellomo, G. Quantum information as a non-Kolmogorovian generalization of Shannon's theory. *Entropy* **2015**, *17*, 7349–7373. [CrossRef]

58. Bosyk, G.M.; Zozor, S.; Holik, F.; Portesi, M.; Lamberti, P.W. A family of generalized quantum entropies: Definition and properties. *Quantum Inf. Process.* **2016**, *15*, 3393–3420. [CrossRef]

59. Białynicki-Birula, I. Entropic uncertainty relations. *Phys. Lett. A* **1984**, *103*, 253–254. [CrossRef]

60. Białynicki-Birula, I. Formulation of the uncertainty relations in terms of the Rényi entropies. *Phys. Rev. A* **2006**, *74*, 052101. [CrossRef]

61. Blokhintsev, D.I. *Space and Time in the Microworld*; D. Reidel Publishing Company: Dordrecht, The Netherlands, 1973.

62. De la Madrid, R.; Bohm, A.; Gadella, M. Rigged Hilbert space treatment of continuous spectrum. *Fortschr. Phys.* **2002**, *50*, 185–216. [CrossRef]

63. Huang, Y. Variance-based uncertainty relations. *Phys. Rev. A* **2012**, *86*, 024101. [CrossRef]

64. Maccone, L.; Pati, A.K. Stronger uncertainty relations for all incompatible observables. *Phys. Rev. Lett.* **2014**, *113*, 260401. [CrossRef] [PubMed]

65. Puchała, Z.; Rudnicki, Ł.; Życzkowski, K. Majorization entropic uncertainty relations. *J. Phys. A Math. Theor.* **2013**, *46*, 272002. [CrossRef]

66. Friedland, S.; Gheorghiu, V.; Gour, G. Universal uncertainty relations. *Phys. Rev. Lett.* **2013**, *111*, 230401. [CrossRef] [PubMed]

67. Rudnicki, Ł.; Puchała, Z.; Życzkowski, K. Strong majorization entropic uncertainty relations. *Phys. Rev. A* **2014**, *89*, 052115. [CrossRef]

68. Rudnicki, Ł. Majorization approach to entropic uncertainty relations for coarse-grained observables. *Phys. Rev. A* **2015**, *91*, 032123. [CrossRef]

69. Rastegin, A.E.; Życzkowski, K. Majorization entropic uncertainty relations for quantum operations. *J. Phys. A Math. Theor.* **2016**, *49*, 355301. [CrossRef]

70. Kaniewski, J.; Tomamichel, M.; Wehner, S. Entropic uncertainty from effective anticommutators. *Phys. Rev. A* **2014**, *90*, 012332. [CrossRef]

71. Luis, A.; Bosyk, G.M.; Portesi, M. Entropic measures of joint uncertainty: Effects of lack of majorization. *Physica A* **2016**, *444*, 905–913. [CrossRef]

72. Basdevant, J.-L.; Dalibard, J. *Quantum Mechanics*; Springer: Berlin, Germany, 2002.

73. Baumgratz, T.; Cramer, M.; Plenio, M.B. Quantifying coherence. *Phys. Rev. Lett.* **2014**, *113*, 140401. [CrossRef] [PubMed]

74. Rastegin, A.E. Quantum-coherence quantifiers based on the Tsallis relative α entropies. *Phys. Rev. A* **2016**, *93*, 032136. [CrossRef]

75. Zhu, H.; Hayashi, M; Chen, L. Coherence and entanglement measures based on Rényi relative entropies. *J. Phys. A Math. Theor.* **2017**, *50*, 475303. [CrossRef]

76. Lüders, G. Über die zustandsänderung durch den meßprozeß. *Ann. Phys.* **1950**, *443*, 322–328. (In German) [CrossRef]

77. Teixeira, A.; Matos, A.; Antunes, L. Conditional Rényi entropies. *IEEE Trans. Inf. Theory* **2012**, *58*, 4273–4277. [CrossRef]

78. Furuichi, S. Information-theoretical properties of Tsallis entropies. *J. Math. Phys.* **2006**, *47*, 023302. [CrossRef]

79. Rastegin, A.E. Convexity inequalities for estimating generalized conditional entropies from below. *Kybernetika* **2012**, *48*, 242–253.

80. Rastegin, A.E. Further results on generalized conditional entropies. *RAIRO Theor. Inf. Appl.* **2015**, *49*, 67–92. [CrossRef]

81. Barnett, S.M.; Pegg, D.T. On the Hermitian optical phase operator. *J. Mod. Opt.* **1989**, *36*, 7–19. [CrossRef]

82. Pegg, D.T.; Barnett, S.M. Phase properties of the quantized single-mode electromagnetic field. *Phys. Rev. A* **1989**, *39*, 1665–1675. [CrossRef]

83. Abe, S. Information-entropic uncertainty in the measurements of photon number and phase in optical states. *Phys. Lett. A* **1992**, *166*, 163–167. [CrossRef]

84. Rastegin, A.E. Number-phase uncertainty relations in terms of generalized entropies. *Quantum Inf. Comput.* **2012**, *12*, 0743–0762.

85. Hsu, L.-Y.; Kawamoto, S.; Wen, W.-Y. Entropic uncertainty relation based on generalized uncertainty principle. *Mod. Phys. Lett. A* **2017**, *32*, 1750145. [CrossRef]

entropy

MDPI

Article

Quantization and Bifurcation beyond Square-Integrable Wavefunctions

Ciann-Dong Yang * and Chung-Hsuan Kuo

Department of Aeronautics and Astronautics, National Cheng Kung University, No. 1, University Road, Tainan 701, Taiwan; genieskuo@gmail.com
* Correspondence: cdyang@mail.ncku.edu.tw

Received: 27 March 2018; Accepted: 26 April 2018; Published: 29 April 2018

Abstract: Probability interpretation is the cornerstone of standard quantum mechanics. To ensure the validity of the probability interpretation, wavefunctions have to satisfy the square-integrable (SI) condition, which gives rise to the well-known phenomenon of energy quantization in confined quantum systems. On the other hand, nonsquare-integrable (NSI) solutions to the Schrödinger equation are usually ruled out and have long been believed to be irrelevant to energy quantization. This paper proposes a quantum-trajectory approach to energy quantization by releasing the SI condition and considering both SI and NSI solutions to the Schrödinger equation. Contrary to our common belief, we find that both SI and NSI wavefunctions contribute to energy quantization. SI wavefunctions help to locate the bifurcation points at which energy has a step jump, while NSI wavefunctions form the flat parts of the stair-like distribution of the quantized energies. The consideration of NSI wavefunctions furthermore reveals a new quantum phenomenon regarding the synchronicity between the energy quantization process and the center-saddle bifurcation process.

Keywords: square integrable; energy quantization; Quantum Hamilton-Jacobi Formalism; quantum trajectory

---

1. Introduction

In the statistical formulation of quantum mechanics, a wavefunction ψ has to be square integrable (SI) to ensure the qualification of $\psi^*\psi$ as a probability density function. SI solutions to the Schrödinger equation can be used to determine the energy levels in a confined system. Nonsquare-integrable (NSI) solutions to the Schrödinger equation otherwise are ruled out and their role has been unknown till now. To investigate the role of NSI wavefunctions, we need a formulation of quantum mechanics, which does not require the SI condition. Among the nine different formulations of quantum mechanics [1], there is a formulation known as the quantum Hamilton-Jacobi (H-J) formalism [2,3], which meets our purpose. The quantum H-J formalism has been developed since the inception of quantum mechanics along the line of Jordan [4], Dirac [5] and Schwinger [6]. The main advantage of the classical H-J formalism is to give the frequencies of a periodic motion directly without solving the equations of motion. Analogous to its classical counterpart, the advantage of quantum H-J formalism is recognized as a method of finding energy eigenvalues directly without solving the related Schrödinger equation.

Based on the quantum H-J equation, Leacock and Padgett [2,3] proposed an ingenious method to evaluate energy eigenvalues by contour integral. This approach to energy eigenvalues E_n is entirely independent of whether the related wavefunction is SI or not, and allows us to examine the participation of the NSI wavefunctions in the process of energy quantization. Apart from providing energy eigenvalues, quantum H-J formalism like its classical counterpart produces quantum Hamilton dynamics [7], from which complex quantum trajectories can be solved to describe the quantum motion associated with a given wavefunction. Probability interpretation isolates SI wavefunctions from NSI wavefunctions; on the contrary, under the quantum H-J formalism SI and NSI wavefunctions are indivisible with continuously connected quantum trajectories. Because NSI wavefunctions ψ fail to serve as probability density functions, we need an alternative operation to replace the expectation (assemble average) $\langle \psi | \hat{\Omega} | \psi \rangle$ of a quantum observable Ω. The complex quantum trajectory method developed from the quantum H-J formalism can provide the time average $\langle \Omega \rangle_T$ to substitute for the usual assemble average $\langle \psi | \hat{\Omega} | \psi \rangle$.

Based on the time-average operation $\langle \Omega \rangle_T$, which applies to both SI and NSI wavefunctions, we can derive quantization laws more general than those based on the assemble average $\langle \psi | \hat{\Omega} | \psi \rangle$, which applies only to SI wavefunctions. One of the general results shows that as the total energy E of a confined system increases monotonically, the time-average kinetic energy $\langle E_k \rangle_T$ of a confined particle exhibits a stair-like distribution in such a way that the step jumps occur as E equal to one of the energy eigenvalue E_n and the flat part of the distribution is formed over the interval $E_n \leq E < E_{n+1}$. During the process as E increases continuously from E_n to the next energy eigenvalue E_{n+1}, we note that all the corresponding wavefunctions are NSI, but they all yield the same value of $\langle E_k \rangle_T$ and form the flat part of the stair-like energy distribution. In other words, the transition from the eigenstate ψ_n to the next eigenstate ψ_{n+1} can be connected smoothly by the NSI wavefunctions ψ_E with $E_n < E < E_{n+1}$, which otherwise have been ruled out in standard quantum mechanics.

Compared to the complex quantum trajectory derived from the quantum H-J formalism, de Broglie-Bohm (dBB) quantum trajectory [8–10] is real-valued. The equivalence between dBB trajectory interpretation and probability interpretation of quantum mechanics has been well developed over the last several decades. Although under the dBB formulation, particles follow continuous trajectories with well-defined two-time position correlations, a recent paper by Gisin [11] pointed out that Bohmian mechanics makes the same predictions as standard quantum mechanics: the violation of Bell inequalities. The studies of dBB formulation of quantum mechanics [12–14] revealed that like the way that thermal probabilities arise in ordinary statistical mechanics, the quantum probabilities $|\psi(x,t)|^2$ arise dynamically in a similar way that a simple initial ensemble with a non-equilibrium distribution $P(x,0) \neq |\psi(x,0)|^2$ of particle positions evolves towards the equilibrium distribution via the relaxation process $P(x,t) \to |\psi(x,t)|^2$. Meanwhile, the speed of the convergence of $P(x,t)$ to $|\psi(x,t)|^2$ was found to correlate with the degree of chaos of the involved Bohmian trajectories [15–17], which in turn was shown to be related to the vortex dynamics generated by nodal points in the wavefunction $\psi(x,t)$ [18,19]. The degree of chaos produced by many interacting vortices ultimately depends on the number and spatial distribution of the nodal points in the configuration space [20,21].

Parallel to the development of real-valued dBB trajectories, the study of complex-valued quantum trajectories has evolved into a quantum trajectory method, which integrates the hydrodynamic equations *on the fly* to synthesize the probability density by evolving ensembles of complex quantum trajectories [22–24]. Due to the additional degree of freedom given by the imaginary part of a complex trajectory, it is possible to synthesize the quantum probability $|\psi(x,t)|^2$ by a single complex-valued trajectory, instead of an ensemble of real-valued or complex-valued trajectories [25].

To date, the trajectory approaches to quantum mechanics, either using real-valued trajectories based on dBB formulation or using complex-valued trajectories based on H-J formulation, mainly deal with SI wavefunctions in order to show their consistency with the probability interpretation. Here we will go beyond SI wavefunctions to find out what will happen, when statistical interpretation is not applicable. On one hand we will use complex trajectories to demonstrate the energy quantization process after releasing the SI condition, and on the other we will apply the quantum Hamilton dynamics to demonstrate the sequential center-saddle bifurcations as the energy quantization proceeds. The combined result manifests a new quantum phenomenon regarding the synchronicity between the energy quantization process and the center-saddle bifurcation process.

NSI wavefunctions not only participate in the quantization and bifurcation process, but also in the formation of spin degree of freedom. In spite of their distinct statistical properties, SI and NSI wavefunctions have similar velocity fields with the only difference in their directions of rotation on the complex plane. As the third goal of the paper, we will contrast quantum trajectories of SI wavefunctions with those of NSI wavefunctions to manifest the invisible spin degree of freedom as a rotational motion on the complex plane.

The remainder of this paper is organized as follows: Section 2 presents quantum H-J formalism and the related method of determining energy eigenvalues. In Section 3, time average operation for NSI wavefunctions along a complex quantum trajectory is developed from the quantum H-J formalism to replace the ensemble average. The proposed time average operation is then used in Section 4 to derive the universal quantization laws regarding the kinetic energy and the quantum potential. Section 5 demonstrates the participation of NSI wavefunctions in the energy quantization process for a harmonic oscillator. Section 6 proposes a quantum dynamic description of energy quantization, in terms of which a new phenomenon regarding the synchronicity between quantization and bifurcation is revealed. Finally, both SI and NSI solutions to the Schrödinger equation are considered in Section 7 and their relations to spin degree of freedom are explained.

2. Quantum Hamilton-Jacobi Formalism

While the quantum H-J theory is general, here we consider its application to bound states, which have quantized energy levels and closed quantum trajectories. The quantum H-J approach to determining energy eigenvalues can be conceived of as an extension of the Wilson-Sommerfeld quantization rule [26]. In this approach, the quantum energy levels are given exactly by setting the quantum action variable equal to an integer multiple of Planck constant:

$$J(E) = \frac{1}{2\pi} \oint_c p(x)dx = n\hbar, \; n = 0, \, 1, \, 2, \, \cdots , \tag{1}$$

where $p(x)$ is called quantum momentum function (QMF) and the contour c is defined on the complex plane with the integer n being the number of poles of $p(x)$ enclosed by c. The QMF $p(x)$ is related to the quantum action function S and the wavefunction ψ as:

$$p(x) = \frac{\partial S}{\partial x} = -i\hbar \frac{\partial \ln \psi}{\partial x}, \tag{2}$$

with S satisfying the quantum H-J equation:

$$\frac{\partial S}{\partial t} + H(t, x, p)|_{p=\partial S/\partial x} = \frac{\partial S}{\partial t} + \left[\frac{p^2}{2m} + V - \frac{i\hbar}{2m} \frac{\partial p}{\partial x} \right]_{p=\partial S/\partial x} = 0, \tag{3}$$

and with ψ satisfying the Schrödinger equation:

$$i\hbar \frac{\partial \psi}{\partial t} = -\frac{\hbar^2}{2m} \frac{\partial^2 \psi}{\partial x^2} + V\psi. \tag{4}$$

It appears that the quantum H-J Equation (3) and the Schrödinger Equation (4) are equivalent expressions via the relation $S = -i\hbar \ln \psi$.

Leacock and Padgett [2,3] proposed an ingenious method to evaluate $J(E)$ without actually solving $p(x)$ from the quantum H-J Equation (3). They showed that for a given potential $V(x)$, $J(E)$ can be computed simply by a suitable deformation of the complex contour c and the change of variables in Equation (1). Once $J(E)$ is found, the energy eigenvalues E_n can be determined by solving E in terms of the integer n via the relation $J(E) = n\hbar$.

The quantum H-J approach to determining energy eigenvalues has two significant implications. Firstly, this approach suggests that the energy eigenvalue E_n stems from the quantization of the action variable J, rather than from the quantization of the total energy E itself. Precisely speaking, the energy eigenvalue E_n is the specific energy E at which the action variable $J(E)$ happens to be an integer multiple of $\hbar$, i.e., $J(E_n) = n\hbar$. Inspired by this implication, the first goal of this paper is to reveal the internal mechanism causing the quantization of the action variable J and find out its relation to the energy quantization.

Secondly, the quantum H-J approach implies that the SI condition is not required throughout the process of determining energy eigenvalues, which means that whether wavefunctions are SI or not is unconcerned upon evaluating eigen energies. Based on this observation, our second goal here is to expose how SI and NSI wavefunctions cooperate to form the observed energy levels within a confining potential. For a given wavefunction $\psi(t,x)$ either SI or NSI, the associated quantum dynamics can be described by the quantum Hamilton equations with the quantum Hamiltonian H given by Equation (3):

$$\frac{dx}{dt} = \frac{\partial H}{\partial p} = \frac{p}{m}, \ x \in \mathbb{C}, \tag{5a}$$

$$\frac{dp}{dt} = -\frac{\partial H}{\partial x} = -\frac{\partial}{\partial x}(V(x) + Q(t,x)), \ p \in \mathbb{C}, \tag{5b}$$

where $Q(x)$ is the complex quantum potential defined by:

$$Q(t,x) = -\frac{i\hbar}{2m}\frac{\partial p}{\partial x}\Big|_{p=\partial S/\partial x} = -\frac{i\hbar}{2m}\frac{\partial^2 S}{\partial x^2} = -\frac{\hbar^2}{2m}\frac{\partial^2 \ln \psi(t,x)}{\partial x^2}. \tag{6}$$

Quantum potential $Q(t,x)$ is intrinsic to the quantum state $\psi(t,x)$ and is independent of the externally applied potential $V(x)$. The quantum Hamilton Equations (5) are distinct from the classical ones in two aspects: the complex nature and the state-dependent nature. The complex nature is a consequence of the fact that the canonical variables (x,p) solved from Equation (5) are, in general, complex variables. The state-dependent nature means that Equation (5) governs the quantum motion specifically in the quantum state described by ψ. The Hamilton Equations (5), which is usually regarded as the complex-extension of Bohmian mechanics, can be derived independently by the optimal stochastic control theory [27].

For a given wavefunction $\psi(t,x)$, the complex contour c traced by $x(t)$ can be solved from Equation (5a), along which the contour integral in Equation (1) then can be evaluated. The second Hamilton Equation (5b) is an alternative expression of the Schrödinger Equation (4) as can be shown by substituting $p(t,x)$ from Equation (2) and $Q(t,x)$ from Equation (6).

3. Time Average along a Complex Quantum Trajectory

The necessity of considering time average along a complex quantum trajectory comes from the fact that the action variable J introduced in Equation (1) is equal to the time-average kinetic energy, as will be shown below. For a particle confined by a time-independent potential $V(x)$, we have wavefunction $\psi(t,x) = e^{-iEt/\hbar}\psi_E(x)$ and quantum action function $S(t,x) = -Et - i\hbar \ln \psi_E(x)$, with which the quantum H-J Equation (3) can be recast into the following form:

$$H(x,p) = \frac{p^2}{2m} + V(x) + Q(x) = -\frac{\partial S}{\partial t} = E. \tag{7}$$

This is the energy conservation law in the quantum H-J formalism, indicating that the conserved total energy E comprises three terms: the kinetic energy $E_k = p^2/2m$, the applied potential $V(x)$, and the quantum potential $Q(x)$. When expressed in terms of the wavefunction $\psi_E(x)$, Equation (7) and Equation (4) become the time-independent Schrödinger equation:

$$\frac{\hbar^2}{2m}\frac{d^2\psi_E}{dx^2} + (E - V(x))\psi_E = 0. \tag{8}$$

The energy conservation law (7) is valid for any solution $\psi_E(x)$ to the Schrödinger Equation (8), either SI or NSI. The Schrodinger Equation (8) has a continuum of solutions, unless it is supplemented with appropriate boundary conditions. Without loss of generality, we consider $V(x)$ in the form of a potential well with the property $V(x) \to \infty$, as $x \to \pm\infty$. Due to the presence of the infinite potential, the probability of finding the particle at infinity is zero, i.e.,

$$\psi_E(x) \to 0, \text{ as } x \to \pm\infty. \tag{9}$$

This boundary condition gets rid of most of the solutions to Equation (8) and selects out only a discrete set of ψ_E and E. Consequently, it is the boundary conditions in standard quantum mechanics that actually enforce the quantization. The boundary condition (9) originates from the fundamental requirement that wavefunctions must be SI, i.e.,

$$\int_{-\infty}^{+\infty} \psi_E^*(x)\psi_E(x)dx < \infty, \tag{10}$$

which allows the normalization of the total probability to unity. If the SI condition (10) or the boundary condition (9) is released, the total energy E will be still conserved, but no longer quantized, because the participation of NSI wavefunctions $\psi_E(x)$ in Equation (7) will result in an arbitrary total energy E other than E_n. However, even if the total energy E is allowed to be varied continuously, there exist intrinsic quantization laws from which the energy eigenvalue E_n can be recovered. In other words, probability interpretation with the accompanying SI condition is not the only way to arrive at the quantization. This issue was first addressed by Leacock and Padgett [2,3] and demonstrated in detail by Bhalla [28,29].

Although NSI wavefunctions ψ fail to serve as probability density functions in the assemble average $\langle \Omega(x,p) \rangle_\psi = \langle \psi | \Omega(\hat{x}, \hat{p}) | \psi \rangle$, complex quantum trajectories for NSI wavefunction still exist, along which time average of $\Omega(x,p)$ can be defined to substitute for the assemble average $\langle \Omega(x,p) \rangle_\psi$. The complex quantum trajectory describing the particle's motion in a confined system can be solved from Equations (5a) and (2), which together with $\psi(t,x) = e^{-iEt/\hbar}\psi_E(x)$ gives the governing equation as:

$$\frac{dx}{dt} = -\frac{i\hbar}{m}\frac{\psi_E'(x)}{\psi_E(x)}, \tag{11}$$

where $\psi_E(x)$ is a general solution to Equation (8) with given energy E. The resulting complex trajectory $x(t)$ serves as a physical realization of the complex contour c appearing in Equation (1), and allows the contour integral to be evaluated along the particle's path of motion.

By treating the quantum Hamiltonian $H(x,p)$ defined in Equation (7) as a Lyapunov function, the energy conservation law $dH/dt = 0$ implies that the autonomous nonlinear system (11) is Lyapunov stable (neutrally stable) with equilibrium points in the form of centers, irrespective of whether ψ_E is SI or not. The trajectory solved from Equation (11) coincides with the Lyapunov contour lines defined by $H(x,p) = E = $ constant, which are concentric curves surrounding equilibrium points.

The time average of $\Omega(x,p)$ along the particle's trajectory $x(t)$ is defined as:

$$\langle \Omega(x,p) \rangle_T = \frac{1}{T}\int_0^T \Omega(x(t), p(t))dt, \tag{12}$$

where T is the period of oscillation of $x(t)$. The quantum action variable J defined in Equation (1) is a ready example of taking time average along a complex contour. Letting c be a closed trajectory solved from Equation (11), we can rewrite the contour integral (1) in terms of the time-average kinetic energy as:

$$J = \frac{1}{2\pi}\oint_c p(x)dx = \frac{1}{2\pi m}\int_0^T p^2 dt = \frac{2}{\omega}\langle E_k \rangle_T, \tag{13}$$

where $\omega = 2\pi/T$ is the angular frequency of the periodic motion. Therefore, the Wilson-Sommerfeld quantization law $J = n\hbar$ is simply an alternative expression of the energy quantization law $\langle E_k \rangle_T = n(\hbar\omega/2)$.

In general, the time average of an arbitrary function $\Omega(x, p)$ can be expressed in terms of a contour integral by using Equations (11) and (12):

$$\langle \Omega(x, p) \rangle_T = i \frac{m\omega}{2\pi\hbar} \oint_c \Omega(x) \frac{\psi_E(x)}{\psi_E'(x)} dx, \tag{14}$$

where c is the closed contour traced by $x(t)$ on the complex plane, and the symbol "prime" denotes the differentiation with respect to x. Since QMF $p(x)$ can be expressed as a function of x, we simply write $\Omega(x, p)$ as $\Omega(x)$ in the integrand. According to the residue theorem, the value of $\langle \Omega(x, p) \rangle_T$ is determined only by the poles of the integrand enclosed by the contour c and is independent of the actual form of c. We will see below that the discrete change of the number of poles in the integrant leads to the quantization of $\langle \Omega(x, p) \rangle_T$.

4. General Quantization Laws without SI Condition

Let $\psi_E(x)$ be a general solution to the Schrödinger Equation (8) with a given energy E. We can treat the time average $\langle \Omega(x, p) \rangle_T$ as a function of the total energy E by noting that $\langle \Omega(x, p) \rangle_T$ is computed by Equation (14) with wavefunction $\psi_E(x)$, which in turn depends on the energy E. The time average $\langle \Omega(x, p) \rangle_T$ is said to be quantized, if its value manifests a stair-like distribution as the total energy E increases monotonically. We will derive several energy quantization laws originating from such a stair-like behavior of $\langle \Omega(x, p) \rangle_T$, which are universal for all confined quantum systems. The energy quantization defined here denotes the discrete change of the considered energy, which is different from the definition in standard quantum mechanics, where energy quantization denotes the discrete energies satisfying the SI condition (10).

Firstly, we consider the quantization of the time-average kinetic energy. By substituting $\Omega(x, p) = p^2/2m$ into Equation (14), we obtain:

$$\langle E_k \rangle_T = \frac{1}{T} \int_0^T \frac{1}{2m} p^2 dt = \frac{\hbar\omega}{4\pi i} \oint_c \frac{\psi_E'(x)}{\psi_E(x)} dx. \tag{15}$$

To evaluate the above contour integral, we recall a formula from the residue theorem:

$$\oint_c \frac{\Omega'(x)}{\Omega(x)} dx = 2\pi i \left(Z_f - P_f \right), \tag{16}$$

where Z_f and P_f are, respectively, the numbers of zero and pole of $\Omega(x)$ enclosed by the contour c. Using this formula in Equation (15) yields:

$$\langle E_k \rangle_T = \frac{\hbar\omega}{2} \left(Z_\psi - P_\psi \right) = \frac{\hbar\omega}{2} n_\psi, \tag{17}$$

where the integer $n_\psi = Z_\psi - P_\psi$ is the difference between the numbers of zero and pole of $\psi_E(x)$. It appears that the time average of the particle's kinetic energy in a confined potential is an integer multiple of $\hbar\omega/2$. This is an universal quantization law independent of the confining potential $V(x)$. Using Equation (17) in Equation (13), we recover the Wilson-Sommerfeld quantization law $J = n_\psi \hbar$.

The other quantized energy is the quantum potential Q. The evaluation of Equation (14) with $\Omega(x, p) = Q(x)$ gives:

$$\langle Q \rangle_T = \frac{1}{T} \int_0^T Q dt = \frac{\hbar}{2miT} \int_0^T \frac{dp}{dx} dt = \frac{\hbar\omega}{4\pi i} \oint_c \frac{p'(x)}{p(x)} dx. \tag{18}$$

Applying Formula (16) once again, we arrive at the second energy quantization law:

$$\langle Q \rangle_T = \frac{\hbar\omega}{2} \left(Z_p - P_p \right) = \frac{\hbar\omega}{2} n_p, \tag{19}$$

where integer $n_p = Z_p - P_p$ is the difference between the numbers of zero and pole of $p(x)$. Like the quantization of $\langle E_k \rangle_T$, Equation (19) reveals that the value of $\langle Q \rangle_T$ is an integer multiple of $\hbar\omega/2$, irrespective of the confining potential $V(x)$.

The Kinetic energy E_k and the quantum potential energy Q, individually, are quantized quantities, and their combination leads to another quantization law. This can be verified from the combination of Equations (15) and (18):

$$\langle E_k + Q \rangle_T = \frac{\hbar\omega}{4\pi i} \oint_c \left[\frac{\psi'_E(x)}{\psi_E(x)} + \frac{p'(x)}{p(x)} \right] dx, \tag{20}$$

where in the integrand can be simplified further as:

$$\frac{\psi'_E(x)}{\psi_E(x)} + \frac{p'(x)}{p(x)} = \frac{d}{dx} \ln[p(x)\psi_E(x)] = \frac{d}{dx} \ln \psi'_E(x).$$

With the above simplification and the Formula (16), Equation (20) yields a new quantization law:

$$\langle E_k + Q \rangle_T = \frac{\hbar\omega}{4\pi i} \oint_c \frac{\psi''_E(x)}{\psi'_E(x)} dx = \frac{\hbar\omega}{2} n_{\psi'}, \tag{21}$$

where integer $n_{\psi'} = Z_{\psi'} - P_{\psi'}$ is the difference between the numbers of zero and pole of $\psi'_E(x)$.

The three integers, n_{ψ}, n_p and $n_{\psi'}$, are solely determined by the wavefunction ψ_E, which in turn is solved from Equation (8) with a prescribed energy E. As E increases, the three integers can only change discretely in response to the continuous change of E. Let $E_0 < \cdots < E_{n-1} < E_n < \cdots$ be the sequence of specific energies at which the integer $n_{\psi'}$ experiences a step jump, $n - 1 \to n$. With increasing E, the value of $\langle E_k + Q \rangle_T$ then assumes a stair-like distribution described by:

$$\langle E_k + Q \rangle_T = \frac{\hbar\omega}{2} n, \quad E_{n-1} < E \leq E_n, \quad n = 1, 2, \cdots. \tag{22}$$

The values of $\langle E_k \rangle_T$ and $\langle Q \rangle_T$ have a similar distribution. It is noted that the wavefunction $\psi_E(x)$ solved from Equation (8) with an energy E in the interval $E_{n-1} < E \leq E_n$ is generally NSI. Our next task is to clarify the roles of these NSI wavefunctions in the quantization process of $\langle E_k \rangle_T$ and $\langle Q \rangle_T$.

5. Energy Quantization beyond SI Wavefunctions

To elucidate how SI and NSI wavefunctions cooperate to form the observed quantization levels, we consider the typical quantum motion under a quadratic confining potential $V(x) = x^2/2$. The related Schrödinger equation in dimensionless form is:

$$\frac{d^2\psi_E}{dx^2} + \left(2E - x^2 \right) \psi_E = 0, \tag{23}$$

where the total energy E is allowed to be any positive real number. A general solution to the Schrödinger Equation (23), which takes into account NSI wavefunctions, can be expressed in terms of the Whittaker function $W(k, m, z)$ as:

$$\psi_E(x) = \frac{C_1}{\sqrt{x}} W\left(\frac{E}{2}, \frac{1}{4}, x^2 \right) \tag{24a}$$

$$= C_1 e^{-x^2/2} \left[F\left(\frac{1}{4} - \frac{E}{2}, \frac{1}{2}, x^2 \right) / \Gamma\left(\frac{3}{4} - \frac{E}{2} \right) - 2xF\left(\frac{3}{4} - \frac{E}{2}, \frac{3}{2}, x^2 \right) / \Gamma\left(\frac{1}{4} - \frac{E}{2} \right) \right], \tag{24b}$$

where $F(\alpha, \beta, z)$ is the hypergeometric function, and $\Gamma(\alpha)$ is the Gamma function. Detailed discussions on the above-mentioned special functions can be found in standard textbooks of physical mathematics [30]. For a given energy E, the obtained solution $\psi_E(x)$ is generally NSI, except for the energy eigenvalues $E_n = n + 1/2$, $n = 0, 1, 2, \cdots$, at which Equation (24b) becomes:

$$\psi_n(x) = C_1 e^{-x^2/2} \left[F\left(-\frac{n}{2}, \frac{1}{2}, x^2\right) / \Gamma\left(\frac{1}{2} - \frac{n}{2}\right) - 2x F\left(\frac{1}{2} - \frac{n}{2}, \frac{3}{2}, x^2\right) / \Gamma\left(-\frac{n}{2}\right) \right]. \tag{25}$$

Depending on whether n is even or odd, simplification of $\psi_n(x)$ is given respectively by:

- $n = 2m$:

$$\psi_m(x) = C_1 e^{-x^2/2} F\left(-m, 1/2, x^2\right) / \Gamma(1/2 - m) = C_1 e^{-x^2/2} H_{2m}(x). \tag{26a}$$

- $n = 2m + 1$:

$$\psi_m(x) = C_1 e^{-x^2/2} F\left(-m, 3/2, x^2\right) / \Gamma(-1/2 - m) = C_1 e^{-x^2/2} H_{2m+1}(x). \tag{26b}$$

where we note $1/\Gamma(-m) = 0$ in Equation (25) for negative integer $-m$. Combining the above two equations yields the eigenfunctions $\psi_n(x) = C_1 e^{-x^2/2} H_n(x)$ for the quantum harmonic oscillator. The eigenfunctions $\psi_n(x)$ are the only solutions to the Schrödinger Equation (23), satisfying the boundary condition (9) and the SI condition (10).

All the existing discussions on energy quantization in the harmonic oscillator focus on the SI eigenfunctions and their linear combinations. Here we are interested in the energy quantization related to the NSI wavefunctions described by Equation (24) with $E \neq n + 1/2$. According to Equations (17) and (21), the quantization of $\langle E_k \rangle_T$ and $\langle E_k + Q \rangle_T$ is determined by the numbers of zero and pole of $\psi_E(x)$ and $\psi_E'(x)$. Examining the expression for $\psi_E(x)$ given by Equation (24b), we find that $\psi_E(x)$ and $\psi_E'(x)$ do not have any pole over the entire complex plane, because the hypergeometric function $F(\alpha, \beta, z)$ and its derivative are analytic functions for any $z \in \mathbb{C}$. Accordingly, we have $P_\psi = P_{\psi'} = 0$, and:

$$n_\psi = Z_\psi - P_\psi = Z_\psi, \ n_{\psi'} = Z_{\psi'} - P_{\psi'} = Z_{\psi'}, \ n_p = Z_p - P_p = Z_{\psi'} - Z_\psi. \tag{27}$$

Hence the three quantum numbers, n_ψ, n_p and $n_{\psi'}$, can be determined by the two independent integers: Z_ψ and $Z_{\psi'}$, the numbers of zero of $\psi_E(x)$ and $\psi_E'(x)$, respectively. Regarding the computation of Z_ψ, we can find the zero of ψ_E by solving the roots of the Whittaker function according to Equation (24a):

$$W\left(\frac{E}{2}, \frac{1}{4}, x^2\right) = 0, \ x \in \mathbb{C}, \ E \in \mathbb{R}^+. \tag{28}$$

For a given energy E, the resulting root is denoted by $x_s(E)$, and Z_ψ is the number of $x_s(E)$ satisfying Equation (28). The blue line in Figure 1 illustrates the variation of Z_ψ with respect to the energy E. Similarly, $Z_{\psi'}$ can be found by solving the roots of $\psi_E'(x) = 0$:

$$\left(2E + 1 - 2x^2\right) \cdot W\left(\frac{E}{2}, \frac{1}{4}, x^2\right) + 4 \cdot W\left(\frac{E}{2} + 1, \frac{1}{4}, x^2\right) = 0, \ x \in \mathbb{C}, \ E \in \mathbb{R}^+. \tag{29}$$

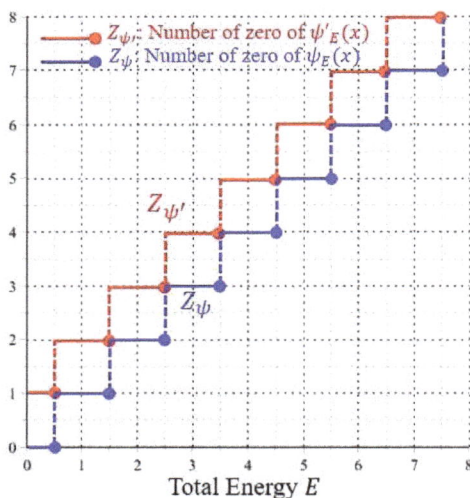

Figure 1. The stair-like distributions of the numbers of zero of $\psi_E(x)$ and $\psi'_E(x)$, as the total energy E changes continuously

The resulting root is denoted by $x_{eq}(E)$ and the number of $x_{eq}(E)$ satisfying Equation (29) for a given energy E gives the value of $Z_{\psi'}$. The red line in Figure 1 illustrates the variation of $Z_{\psi'}$ with respect to the energy E.

As can be seen from Figure 1, when the total energy E increases monotonically, $Z_{\psi'}$ and Z_{ψ} exhibit a stair-like distribution in the form of:

$$Z_{\psi'} = n_{\psi'} = n+1, \; Z_{\psi} = n_{\psi} = n, \; n - \frac{1}{2} < E \leq n + \frac{1}{2}, \; n = 1, \, 2, \, \cdots. \tag{30}$$

and $Z_{\psi'} = 1$, $Z_{\psi} = 0$, as $0 < E \leq 1/2$. Based on the above distributions of $Z_{\psi'}$ and Z_{ψ}, the quantization laws derived in Equations (17), (19) and (21) now become:

$$\langle E_k \rangle_T = \frac{n_{\psi}}{2} = \frac{n}{2}, \; \langle E_k + Q \rangle_T = \frac{n_{\psi'}}{2} = \frac{n+1}{2}, \; \langle Q \rangle_T = \frac{1}{2} \left(n_{\psi'} - n_{\psi} \right) = \frac{1}{2}, \tag{31}$$

when the total energy E falls in the interval $n - 1/2 < E \leq n + 1/2$. All the energies in Equation (31) have been expressed in terms of the multiples of $\hbar\omega$. Consequently, as we increase the total energy E monotonically, $\langle E_k \rangle_T$ and $\langle E_k + Q \rangle_T$ increase in a stair-like manner with the step levels given by Equation (31), as shown in Figure 2. Up to this stage, the two components E_k and Q in the energy conservation law (7) have been found to be quantized, while the third component, i.e., the externally applied potential $V(x)$, is not a quantized quantity, which otherwise changes continuously with E via the relation:

$$\langle V(x) \rangle_T = E - E_k + Q_T = E - \frac{n+1}{2}. \tag{32}$$

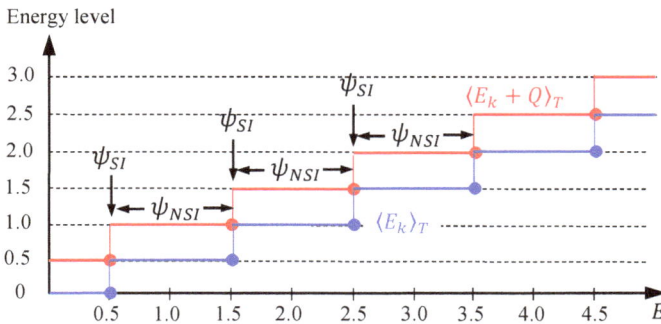

Figure 2. The step changes of $\langle E_k \rangle_T$ and $\langle E_k + Q \rangle_T$ occur at the SI wavefunctions ψ_E with $E = n + 1/2$, as the total energy E changes continuously in a harmonic oscillator. The flat parts of $\langle E_k \rangle_T$ and $\langle E_k + Q \rangle_T$ are constituted by the NSI ψ_E with $E \neq n + 1/2$.

The most noticeable point is that the step change of $\langle E_k \rangle_T$ and $\langle E_k + Q \rangle_T$ occurs at the specific energies $E_n = n + 1/2$, which coincide with the energy eigenvalues of the harmonic oscillator. In other words, the role of the SI condition amounts to determining the discrete energy E_n at which the numbers of zero of $\psi_E(x)$ and $\psi'_E(x)$ exhibit a step jump, while the role of the NSI wavefunctions $\psi_E(x)$ with $E \neq E_n$ is to form the flat parts of the stair-like distribution as shown in Figures 1 and 2, where the numbers of zero of $\psi_E(x)$ and $\psi'_E(x)$, or equivalently the time-average energies $\langle E_k \rangle_T$ and $\langle E_k + Q \rangle_T$, keep unchanged.

6. Quantum Bifurcation beyond SI Wavefunctions

As the total energy E increases, the wavefunction $\psi_E(x)$ transits repeatedly from NSI states to a SI state, once E coincides with an energy eigenvalue E_n. In this section, we will show that the encounter with an energy eigenvalue not only causes a step jump of $\langle E_k \rangle_T$ and $\langle E_k + Q \rangle_T$, but also causes a nonlinear phenomenon - quantum bifurcation, where the number of equilibrium points of the quantum dynamics experiences an instantaneous change.

With $\psi_E(x)$ given by Equation (24a), the quantum dynamics (11) assumes the following dimensionless form:

$$\frac{dx}{dt} = -i \frac{\psi'_E(x)}{\psi_E(x)} = \frac{i}{2x} \left(2E + 1 - 2x^2 \right) + \frac{2i}{x} \frac{W(E/2 + 1, 1/4, x^2)}{W(E/2, 1/4, x^2)}, \tag{33}$$

where the total energy E is treated as a free parameter, whose critical values for the occurrence of bifurcation are to be identified. The quantum trajectories $x(t)$ solved from Equation (33) provide us with a quantitative comparison between SI and NSI wavefunctions, which otherwise cannot be compared under the probability interpretation of $\psi_E(x)$.

As can be seen from Equation (33), the equilibrium point x_{eq} of the quantum dynamics is equal to the zero of $\psi'_E(x)$, while the singular point x_s is just the zero of $\psi_E(x)$. Hence the step changes of Z_ψ and $Z_{\psi'}$ shown in Figure 1 also imply the step changes of the numbers of the equilibrium points x_{eq} and the singular points x_s, respectively. In other words, we can say that the following two processes occur synchronously as E increases monotonically: one process is the quantization of $\langle E_k \rangle_T$ and $\langle E_k + Q \rangle_T$ regarding the step changes of Z_ψ and $Z_{\psi'}$ as discussed previously, and the other is the bifurcation of the quantum dynamics (33) regarding the step changes of the equilibrium points and singular points, as to be discussed below.

(1) SI wavefunctions:

Firstly, we consider the special cases that the total energy E happens to be one of the eigen energies: $E_0 = 1/2$, $E_1 = 3/2$, and $E_2 = 5/2$. The related eigenfunctions and the eigen-dynamics derived from Equation (33) are given by:

$$\psi_0(x) = e^{-x^2/2} : \frac{dx}{dt} = ix, \tag{34a}$$

$$\psi_1(x) = 2xe^{-x^2/2} : \frac{dx}{dt} = i\frac{x^2 - 1}{x}, \tag{34b}$$

$$\psi_2(x) = 2\left(2x^2 - 1\right)e^{-x^2/2} : \frac{dx}{dt} = i\frac{x\left(x^2 - 5/2\right)}{x^2 - 1/2}. \tag{34c}$$

These three equations describe the velocity fields and their solutions give the eigen-trajectories for the first three SI states of the harmonic oscillator. It can be shown that the equilibrium points of Equation (34) are centers, while their singular points are saddles. For instance, $\dot{x} = ix$ in Equation (34a) has an equilibrium point at the origin with solution given by $x(t) = ce^{it}$, whose trajectories on the complex plane are concentric circles around the equilibrium point, showing that $x = 0$ is a center. To show singular points in Equation (34) are saddles, we consider the following complex-valued system with a singular point at the origin:

$$\dot{x} = f(x) = \frac{g(x)}{x}, \; x \in \mathbb{C}, \tag{35}$$

where $g(x)$ is analytic at $x = 0$. In a neighborhood of the origin, Equation (35) can be approximated by $\dot{x} = \lambda/x$, where $\lambda = g(0)$ is the residue of $f(x)$ evaluated at $x = 0$. The substitution of $x = x_R + ix_I$ and $\lambda = \alpha + i\beta$ into $\dot{x} = \lambda/x$ leads to the equivalent real-valued nonlinear system:

$$x_R \dot{x}_R - x_I \dot{x}_I = \alpha, \; x_R \dot{x}_I + x_I \dot{x}_R = \beta. \tag{36}$$

Its solution is a set of hyperbolas expressed by $x_R^2 - 2(\alpha/\beta)x_R x_I - x_I^2 = C$, showing that the singular point $x = 0$ in Equation (35) are saddles. The centers and saddles of Equation (34) generated by the SI wavefunctions $\psi_E(x)$ with $E_0 = 1/2$, $E_1 = 3/2$, and $E_2 = 5/2$ are illustrated in Figure 3, which displays the distribution and movement of the centers and saddles of the quantum dynamics (33) on the horizontal x axis, as the total energy E changes continuously along the vertical axis. Detailed discussions on the quantum trajectories of the SI wavefunctions $\psi_n(x)$ for a harmonic oscillator were reported in the literature [31,32]. Here our concern is the quantum trajectories of the NSI wavefunctions $\psi_E(x)$ with $E \neq n + 1/2$. Quantum trajectories in the first three quantization intervals of E will be examined below, from which a global picture of center-saddle bifurcation can be drawn.

(2) NSI wavefunctions $\psi_E(x)$ with $0 < E \leq 1/2$:

The wavefunction $\psi_E(x)$ in this range of energy is NSI, except for $E = 1/2$. It seems to be a reasonable conjecture that NSI wavefunctions naturally give rise to unbound quantum trajectories; however, this is not the case. Figure 4 illustrates the quantum trajectories solved from Equation (33) for the NSI wavefunctions $\psi_E(x)$ with $E = 0.1$, $E = 0.49$, and $E = 0.51$. In spite of being generated by NSI wavefunctions, the resulting quantum trajectories are bound with slight deviations from the eigen-trajectories of $E = 1/2$, which are concentric circles around the equilibrium point at the origin, as described by Equation (34a). It appears that SI eigenfunctions are not isolated from the neighboring NSI wavefunctions, because their quantum trajectories can be deformed continuously into each other.

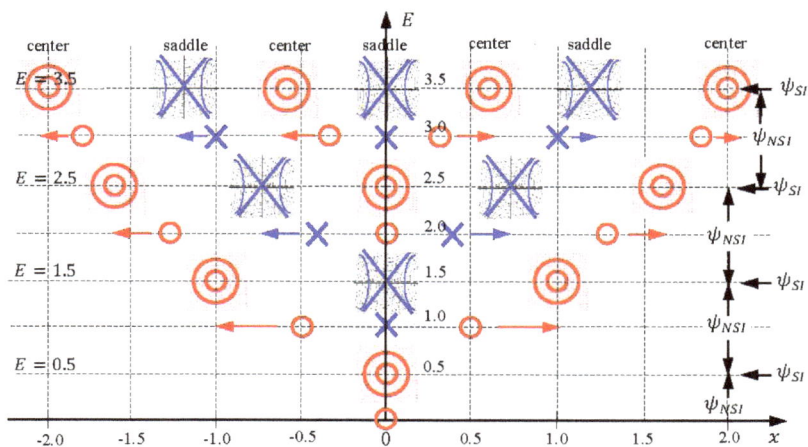

Figure 3. The distribution and movement of the centers and saddles over the horizontal x axis, as the total energy E changes continuously along the vertical axis.

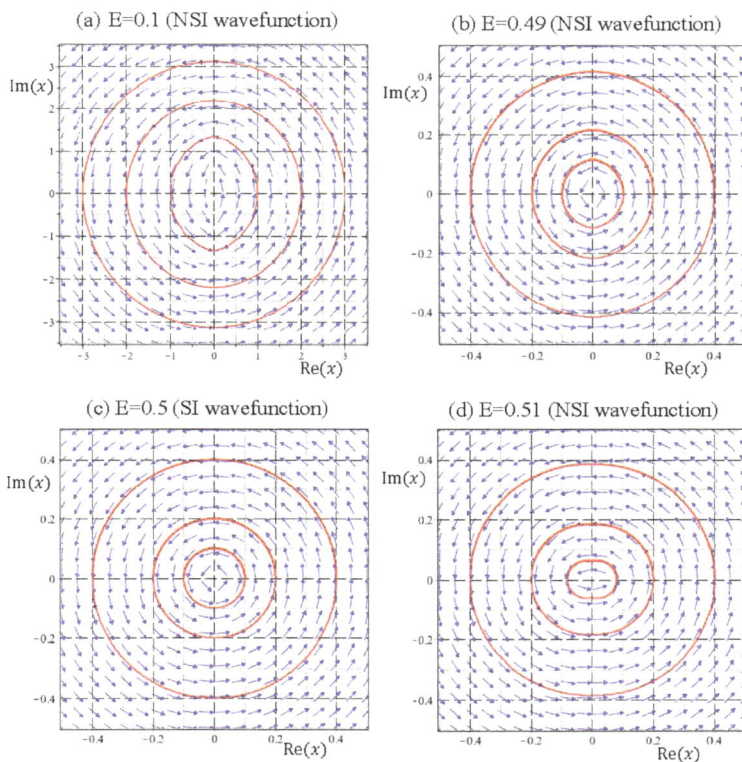

Figure 4. The wavefunctions corresponding to $E = 0.1$, $E = 0.49$, and $E = 0.51$ are NSI, but their quantum trajectories are bound and closely connected to the eigen trajectories of $E = 0.5$, which are concentric circles around the equilibrium point at the origin.

113

(3) NSI wavefunctions $\psi_E(x)$ with $1/2 < E \leq 3/2$:

According to Figure 1, the number of equilibrium points $Z_{\psi'}$ of the quantum dynamics (33), jumps from one to two as E across the energy eigenvalue $E = 1/2$. This bifurcation phenomenon is illustrated in Figure 5. There is only one equilibrium point at the origin in the energy interval $0 < E \leq 0.5$, while beyond the bifurcation point $E = 1/2$, two equilibrium points come out from the origin. Particular attention is paid to the quantum trajectories of $E = 0.51$ depicted in Figure 4d. At first glance, it looks like that the quantum trajectories of $E = 0.51$ have a single equilibrium point at the origin. However, the enlargement of Figure 4d near the origin as illustrated in Figure 5a indicates that the single equilibrium point at the origin for $E = 1/2$ splits into two equilibrium points as E increases to 0.51. When E increases to $3/2$, the two equilibrium points (two centers) move further to $x_{eq} = \pm 1$, as described by the quantum dynamics (34b) and illustrated in Figure 5d.

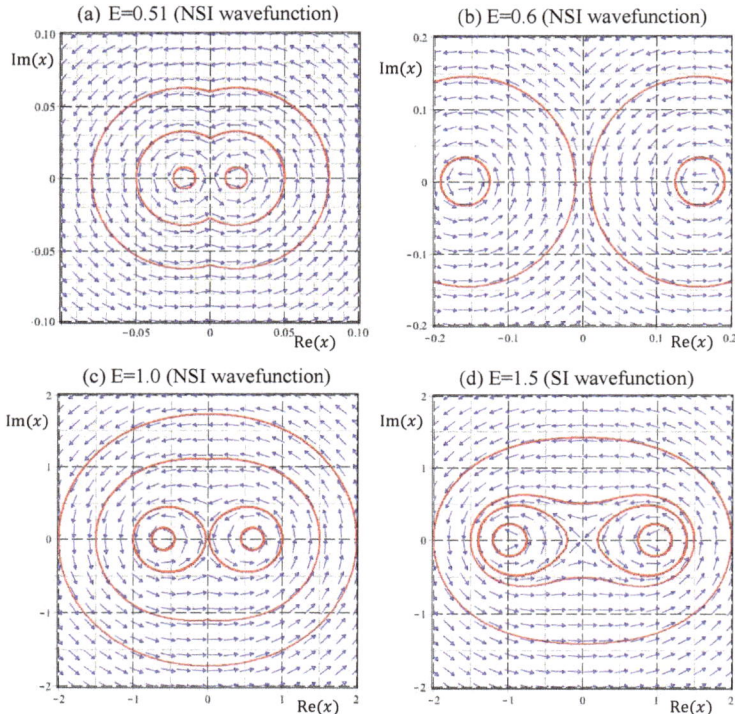

Figure 5. Velocity fields and quantum trajectories in the energy interval $0.5 < E \leq 1.5$ show the quantum bifurcation that the number of equilibrium points jumps from one to two as energy across $E = 0.5$. Part (**a**) is the enlargement of Figure 4d near the origin to illustrate the split of the single equilibrium point at the origin into a pair of equilibrium points as E increases from 0.5 to 0.51. In this energy interval, there are two equilibrium points and one singular point at the origin, corresponding to the energy levels $Z_{\psi'} = 2$ and $Z_{\psi} = 1$ as shown in Figure 1.

Coincident with the splitting of the equilibrium point at the bifurcation point $E = 1/2$, a singular point emerges from the origin in the form of a saddle point. The resulting saddle point pattern in the vicinity of the origin is clearly manifested in Figure 5. It turns out that at the bifurcation energy $E = 1/2$, two kinds of bifurcation occur simultaneously: one bifurcation regards the change of the number of equilibrium points from a single center at the origin into a pair of centers moving apart along the positive and negative real axis as E increases from $1/2$ to $3/2$, and the other bifurcation regards the change from a center into a saddle at the origin.

(4) NSI wavefunctions $\psi_E(x)$ with $3/2 < E \leq 5/2$:

At the energy eigenvalue $E = 3/2$, the value of $Z_{\psi'}$ experiences the second step jump, and a new bifurcation is expected to form here. This prediction is confirmed in Figure 6a, where the enlargement of the velocity field near the origin shows that the saddle-point singularity at the origin for $E = 3/2$ now transforms into a center for $E = 1.51$. As E increases further to $E = 1.6$ and $E = 2$, flow circulation around the origin as a center becomes more apparent. Counting the new equilibrium point emerging from the origin and the already existing pair of centers, the number of equilibrium points increases from two to three as E across $E = 3/2$, and remains three in the interval $3/2 < E \leq 5/2$. Coincident with the emergence of a new center from the origin at $E = 3/2$, the singular saddle point previously residing at the origin now splits into a pair of saddles with their separation increasing with E. The two saddles move to $x_s = \pm\sqrt{1/2}$ when E increases to $5/2$, as described by Equation (34c) and illustrated in Figure 6d.

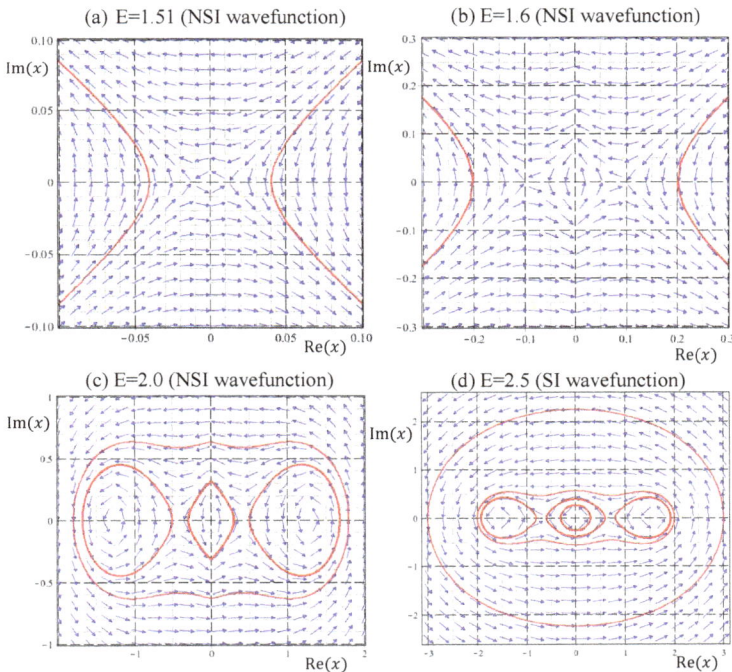

Figure 6. Velocity fields and quantum trajectories in the energy interval $1.5 < E \leq 2.5$ show the quantum bifurcation that the number of equilibrium points jumps from two to three as energy across $E = 1.5$. Part (**a**,**b**) are the enlargements of the velocity field near the origin to illustrate the emergence of a new equilibrium point (a center). In this energy interval, there are three equilibrium points and two singular points, corresponding to the energy levels $Z_{\psi'} = 3$ and $Z_\psi = 2$ as shown in Figure 1. Part (**d**) plots the eigen trajectories for $E = 2.5$ to show the three equilibrium points at $x_{eq} = 0$, $\pm\sqrt{5/2}$ and two singular points at $x_s = \pm\sqrt{1/2}$.

(5) Center-Saddle Bifurcation

When we proceed further, the bifurcations of the equilibrium centers x_{eq} and the singular saddles x_s of the quantum dynamics (33) occur alternatively as E increases. To gain a global picture of the bifurcation pattern, we solve the equilibrium points $x_{eq}(E)$ and the singular points $x_s(E)$ from Equations (29) and (28), respectively, and then plot them as functions of E. The resulting plots generate two sequences of pitchfork bifurcation diagram as shown in Figures 7 and 8 for $x_{eq}(E)$ and $x_s(E)$, respectively. It can be seen that the bifurcations of $x_{eq}(E)$ and $x_s(E)$ occur alternatively at the critical energies $E_n = n + 1/2$ in such a way that the branches of $x_{eq}(E)$ bifurcate sequentially at $E = 0 + (1/2), 2 + (1/2), 4 + (1/2), \cdots$, while the branches of $x_s(E)$ bifurcate sequentially at $E = 1 + (1/2), 3 + (1/2), 5 + (1/2), \cdots$.

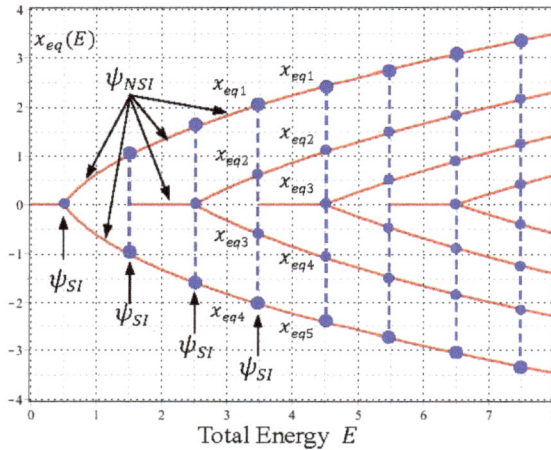

Figure 7. A sequence of pitchfork bifurcation curves shows the variation of equilibrium points $x_{eq}(E)$ with respect to the total energy E. The number of $x_{eq}(E)$ at each E, denoted by the blue dots, is equal to the energy level Z_ψ as plotted in Figure 1. The branches of the $x_{eq}(E)$ curves start sequentially at $E = 0, 3/2, 7/2, \cdots$, and bifurcate sequentially at $E = 1/2, 5/2, 9/2, \cdots$. Except for the bifurcation points (the blue dots), the entire sequential bifurcation diagram is formed by the NSI wavefunctions $\psi_E(x)$ with $E \neq n + 1/2$.

Furthermore, it is worth noting that except for the bifurcation points (the blue dots in Figure 7 and the red dots in Figure 8), the sequential bifurcation diagram is constructed entirely by the NSI wavefunctions $\psi_E(x)$ with $E \neq n + 1/2$. Without the participation of the NSI wavefunctions, adjacent eigenfunctions lose their interconnection and a continuous description of the bifurcation sequence becomes impossible. The other perspective of center-saddle bifurcation can be gained from Figure 3, where we can see that centers and saddles appear alternatively at the origin as the total energy E increases monotonically along the vertical axis.

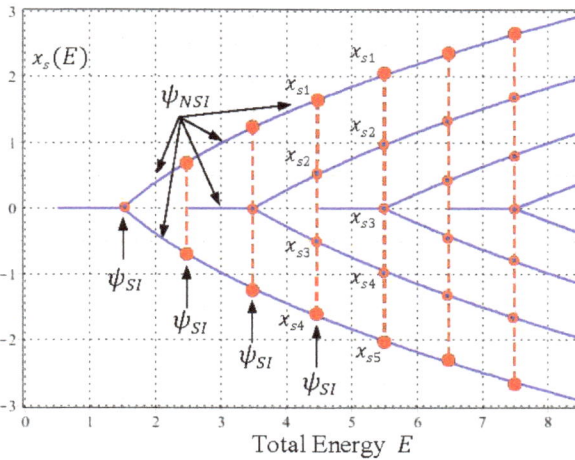

Figure 8. A sequence of pitchfork bifurcation curves shows the variation of singular points $x_s(E)$ with respect to the total energy E. The number of $x_s(E)$ at each E, denoted by the red dots, is equal to the energy level Z_ψ as plotted in Figure 1. The branches of the $x_s(E)$ curves start sequentially at $E = 1/2,\ 5/2,\ 9/2, \cdots$, and bifurcate sequentially at $E = 3/2,\ 7/2,\ 11/2, \cdots$. Except for the bifurcation points (the red dots), the entire sequential bifurcation diagram is formed by the NSI wavefunctions $\psi_E(x)$ with $E \neq n + 1/2$.

(6) Synchronicity between quantization and bifurcation

We recall that the number of $x_{eq}(E)$ at each E is just the number of zero of $\psi'_E(x)$, which gives the quantization level of $\langle E_k + Q \rangle_T$. This relation indicates that the bifurcation of the equilibrium center point $x_{eq}(E)$ and the quantization of $\langle E_k + Q \rangle_T$ occur synchronously. Similarly, because the number of $x_s(E)$ at each E is the number of zero of $\psi_E(x)$, which gives the quantization level of $\langle E_k \rangle_T$, the bifurcation of the singular saddle point $x_s(E)$ is thus synchronous with the quantization of $\langle E_k \rangle_T$. Table 1 lists the numbers of saddles and centers, and the energy levels of $\langle E_k + Q \rangle_T$ and $\langle E_k \rangle_T$ for the first several energy intervals. As can be seen, the numbers of saddles and centers change synchronously with the change of $\langle E_k + Q \rangle_T$ and $\langle E_k \rangle_T$. It is noted that as energy E varies continuously during the quantization and bifurcation processes, the instantaneous changes of energy levels and equilibrium points are triggered by the SI condition $E = n + 1/2$.

Table 1. The step changes of $\langle E_k \rangle_T$ and $\langle E_k + Q \rangle_T$ are synchronous with the changes of saddles and centers.

Quantized Items	$0 < E < \frac{1}{2}$	$E = \frac{1}{2}$	$\frac{1}{2} < E < \frac{3}{2}$	$E = \frac{3}{2}$	$\frac{3}{2} < E < \frac{5}{2}$	$E = \frac{5}{2}$	$\frac{5}{2} < E < \frac{7}{2}$
Wavefunctions	NSI	SI	NSI	SI	NSI	SI	NSI
Zeros of $\psi_E(x)$	0	0	1	1	2	2	3
Number of saddles	0	0	1	1	2	2	3
Levels of $\langle E_k \rangle_T$	0	0	1/2	1/2	1	1	3/2
Zeros of $\psi'_E(x)$	1	1	2	2	3	3	4
Number of centers	1	1	2	2	3	3	4
Levels of $\langle E_k + Q \rangle_T$	1/2	1/2	1	1	3/2	3/2	2

7. Spin Degree of Freedom beyond SI Wavefunctions

The role of the Schrödinger equation has long been considered as describing spinless particles only, because the Schrödinger charge current for the s-states of a hydrogen-like atom vanishes and produces no intrinsic angular momentum. However, based on the observation that in the absence of a magnetic field, the Pauli equation reduces to the Schrödinger equation, it was pointed out [33] that the Schrödinger equation must be regarded as describing an electron in an eigenstate of spin and not, as universally supposed, an electron without spin. According to the dBB trajectory approach [34,35], spin is interpreted as a dynamical property of electron motion and is attributed to a circulating movement of a point, i.e., to a pure orbital motion, but not to an extended spinning object. To be consistent with the Dirac theory and with the condition of Lorentz invariance, Holland [34] proposed that the Schrödinger charge current must be supplemented by a spin magnetization current, which is generated by a circulating flow of energy in the wave field of the electron [36,37].

The NSI solutions to the Schrödinger equation considered in the present paper might give an alternative explanation for the origin of particle's spin motion. The Schrödinger Equation (8) with given energy E actually has two independent solutions. It is surprising to find that the quantum trajectories generated by the two independent solutions are indistinguishable, except for their directions of rotation. Inspecting the quantum trajectories shown in Figures 4–6, it appears that all the trajectories, either generated by SI or NSI wavefunctions, rotate counterclockwise (CCW). In fact, all the trajectories produced by the general wavefunctions given by Equation (24) rotate in the same direction, because Equation (24) only gives one of the independent solutions. A complete general solution to the Schrödinger Equation (23) comprises two independent parts:

$$\psi_E(x) = \psi_{CCW}(x) + \psi_{CW}(x) = \frac{C_1}{\sqrt{x}} W\left(\frac{E}{2}, \frac{1}{4}, x^2\right) + \frac{C_2}{\sqrt{x}} W\left(-\frac{E}{2}, \frac{1}{4}, -x^2\right), \tag{37}$$

where $\psi_{CCW}(x)$ is the solution considered previously and $\psi_{CW}(x)$ is the other independent solution, whose quantum trajectories rotate clockwise (CW). The wavefunction $\psi_{CW}(x)$ represents the second half of solutions to the Schrödinger Equation (23), which is NSI for any energy E and we usually take the neglect of it as granted.

The consideration of the wavefunctions $\psi_{CW}(x)$ helps to identify the additional degree of freedom independent of the particle's orbital motion. To highlight the difference between $\psi_{CCW}(x)$ and $\psi_{CW}(x)$, their velocity fields computed by Equation (11) with $E = 1/2$ are illustrated in Figures 9a,b, respectively. The velocity field of $\psi_{CCW}(x)$ is identical to Figure 4c, which shows circular flows surrounding the origin counterclockwise. By contrast, the velocity field of $\psi_{CW}(x)$ depicted in Figure 9b appears to be clockwise circulation around the origin. The quantum trajectories generated by $\psi_{CW}(x)$ are almost indistinguishable from those generated by $\psi_{CCW}(x)$, and the only difference between them is the directions of rotation. In addition to the orbital motion, the new degree of freedom manifested in the combination of $\psi_{CCW}(x)$ and $\psi_{CW}(x)$ is the dual directions of rotation, which is otherwise invisible along a single trajectory generated by either $\psi_{CCW}(x)$ or $\psi_{CW}(x)$. Due to their same spatial motion with dual directions, $\psi_{CCW}(x)$ and $\psi_{CW}(x)$ can be reasonably recognized as the same spatial solution to the Schrödinger equation but with opposing directions of spin.

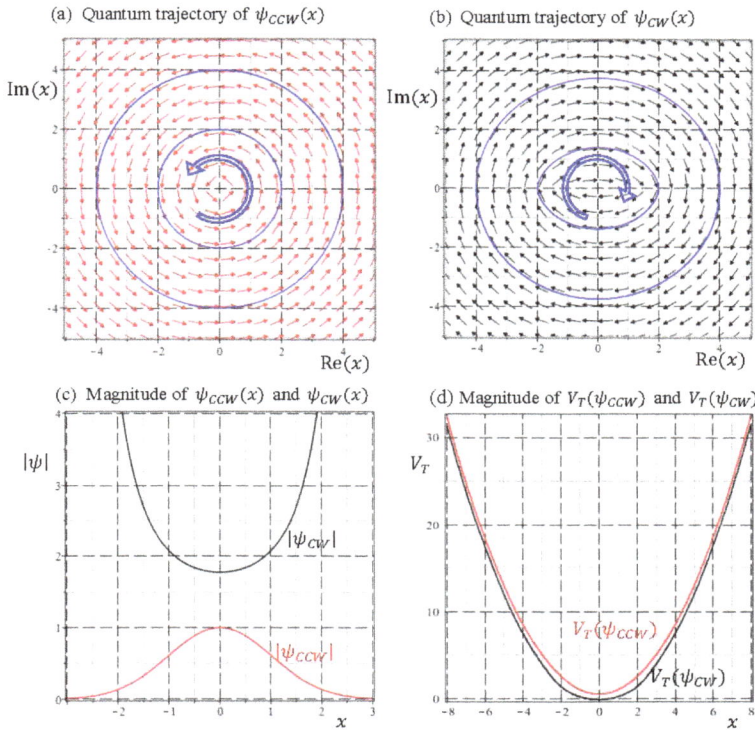

Figure 9. The dual-rotation solutions to the Schrödinger equation with $E = 1/2$: (**a**) quantum trajectory of the wavefunction $\psi_{CCW}(x)$; (**b**) quantum trajectory of the wavefunction $\psi_{CW}(x)$; (**c**) comparison between $|\psi_{CCW}|$ and $|\psi_{CW}|$, and (**d**) comparison between total potential $V_T(\psi_{CCW})$ and $V_T(\psi_{CW})$.

The reason underlying the opposite rotation of $\psi_{CCW}(x)$ and $\psi_{CW}(x)$ can be explained by using the asymptotic expansion property for the Whittaker function:

$$W(\pm k, m, \pm z) = e^{\mp z/2}(\pm z)^{\pm k}\left[1 + \mathcal{O}\left(z^{-1}\right)\right]. \tag{38}$$

With this property, the asymptotic expansions of $\psi_{CCW}(x)$ and $\psi_{CW}(x)$ take the following forms:

$$\psi_{CCW}(x) = \frac{C_1}{\sqrt{x}}W\left(\frac{E}{2}, \frac{1}{4}, x^2\right) \approx C_1 e^{-x^2/2} x^{E-1/2}, \tag{39a}$$

$$\psi_{CW}(x) = \frac{C_2}{\sqrt{x}}W\left(-\frac{E}{2}, \frac{1}{4}, -x^2\right) \approx C_2 e^{x^2/2} x^{-E-1/2}. \tag{39b}$$

The asymptotic quantum dynamics of $\psi_{CCW}(x)$ and $\psi_{CW}(x)$ then can be derived as:

$$\frac{dx}{dt} = -i\frac{d}{dx}\ln\psi_{CCW}(x) = ix - i\frac{E-1/2}{x} \approx ix, \; |x| \gg 0, \tag{40a}$$

$$\frac{dx}{dt} = -i\frac{d}{dx}\ln\psi_{CW}(x) = -ix + i\frac{E+1/2}{x} \approx -ix, \; |x| \gg 0. \tag{40b}$$

Irrespective to the energy E, both of the asymptotic quantum dynamics converge to the ground-state quantum dynamics (34a) with the only difference in their rotation direction. This similarity in the velocity field of $\psi_{CCW}(x)$ and $\psi_{CW}(x)$ has been ignored in the literature. Probability interpretation is only concerned with the square-integrable condition of $\psi_{CCW}(x)$ and $\psi_{CW}(x)$, as shown in Figure 9c, which obviously fails to explain the similarity in the quantum dynamics of $\psi_{CCW}(x)$ and $\psi_{CW}(x)$.

The factor dominating the similarity in the quantum dynamics of $\psi_{CCW}(x)$ and $\psi_{CW}(x)$ can be explained by Equation (5b):

$$\frac{dp}{dt} = -\frac{\partial}{\partial x} V_T. \tag{41}$$

The total potential $V_T = V + Q$, comprising the applied potential V and the quantum potential Q, dominates the time evolution of the QMF p. For the case of a harmonic oscillator, V_T turns out to be (in dimensionless form):

$$V_T(\psi) = \frac{1}{2}x^2 - \frac{1}{2}\frac{d^2}{dx^2}\ln\psi(x). \tag{42}$$

The evaluations of $V_T(\psi)$ at $\psi = \psi_{CCW}$ and $\psi = \psi_{CW}$ for $E = 1/2$ are plotted in Figure 9d. Despite of the adverse nature between SI wavefunction $\psi_{CCW}(x)$ and NSI wavefunction $\psi_{CW}(x)$, their total potentials exhibit a high degree of resemblance, which explains the observed similarity in the velocity fields of Figure 9a,b. The comparisons regarding the magnitudes of ψ and the total potential $V_T(\psi)$ for $E = 3/2$ and $E = 5/2$ are shown in Figure 10, where we observe that except for the region neighboring the origin, $V_T(\psi_{CCW})$ is close to $V_T(\psi_{CW})$ and they become identical as $|x| \to \infty$. This asymptotic identity leads to the dual velocity fields derived in Equation (40).

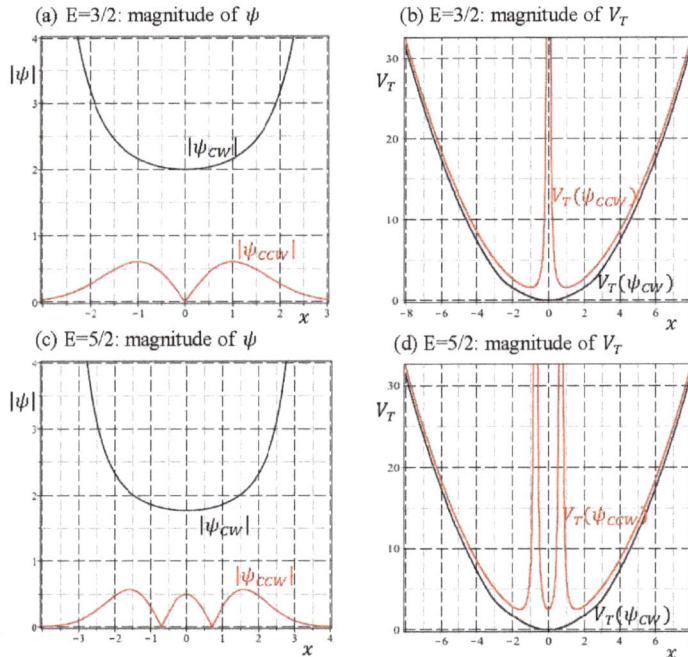

Figure 10. The comparisons between CCW solution $\psi_{CCW}(x)$ and CW solution $\psi_{CW}(x)$ regarding the magnitudes of ψ and the total potential $V_T(\psi)$ for $E = 3/2$ and $E = 5/2$, respectively.

Schrödinger equation is a second-order differential equation with respect to its spatial coordinates, and its complete solution should be composed of two independent solutions. The common belief that

Schrödinger equation is unable to describe spin motion seems to stem from our disregard of one of the independent solution. While probability interpretation of wavefunctions excludes the NSI solution $\psi_{CW}(x)$ from the general solution $\psi_E(x)$, the spin degree of freedom is removed at the same time. Under the framework of quantum H-J formalism, we have seen that by incorporating $\psi_{CW}(x)$ with $\psi_{CCW}(x)$ to form a general wavefunction as expressed by Equation (37), both spatial and spin motion can be described by the Schrödinger equation, and even for one-dimensional quantum motion, the spin degree of freedom can be manifested as the dual rotations on the complex x plane.

8. Conclusions

Standard approach to energy quantization in confined quantum systems is to seek for the allowable energies E_n for which the time-independent Schrödinger equation has square-integrable solutions. The obtained energy eigenvalue E_n is recognized as the quantization level of the system's total energy. In this paper we have given a renewed interpretation for E_n and considered NSI solutions $\psi_E(x)$ to the Schrödinger equation by releasing the SI requirement. The release of this requirement leads to several new findings as summarized in the following points:

- Universal quantization laws: The total energy $E = E_k + Q(x) + V(x)$ derived from the time-independent Schrödinger equation is shown to be conserved, but not quantized. Regardless of the confining potential $V(x)$, quantization always occurs in the kinetic energy $\langle E_k \rangle_T$ and the quantum potential $\langle Q \rangle_T$, whose values can only change by an integer multiple of $\hbar\omega/2$.

- Renewed meaning of the energy eigenvalues: The energy eigenvalues E_n derived conventionally from the SI condition are shown to be the special energies at which the quantization levels of $\langle E_k \rangle_T$ and $\langle E_k + Q \rangle_T$ experience a step jump.

- The origin of energy quantization: Energy quantization in a confined system originates from the discrete change of the numbers of zero of $\psi_E(x)$ and $\psi_E'(x)$, whose values determine the quantization levels of $\langle E_k \rangle_T$ and $\langle E_k + Q \rangle_T$.

- Concurrence of Quantization and bifurcation: Bifurcations of equilibrium center points and singular saddle points of the quantum dynamics are shown to be synchronous, respectively, with the quantization process of $\langle E_k \rangle_T$ and $\langle E_k + Q \rangle_T$, as the total energy E increases monotonically.

- Undivided SI and NSI wavefunctions: probability interpretation isolates SI wavefunctions from NSI wavefunctions; however, under the quantum H-J formalism, SI and NSI wavefunctions are indivisible with continuously connected velocity field and quantum trajectories.

- The role of NSI wavefunctions in energy quantization: Both SI and NSI wavefunctions contribute to the energy quantization. SI wavefunctions help to locate the bifurcation points at which $\langle E_k \rangle_T$ and $\langle E_k + Q \rangle_T$ have a step jump, while NSI wavefunctions form the flat parts of the stair-like distribution of the quantized energies.

- The role of NSI wavefunctions in spin: The second-order Schrödinger equation generally contains two independent solutions with opposite rotation on the complex plane. The inclusion of both solutions allows the Schrödinger equation to describe the spatial motion as well as the spin motion.

At the present stage of this research, we can only say that if the SI requirement is released tentatively, above information can be gained from NSI wavefunctions. We need further experimental results to support the existence of NSI wavefunctions in confined quantum systems. The result of this paper gives a clue to the design of such an experiment. Based on the SI condition, what we consider to be quantized is the particle's total energy. On the other hand, if the SI condition is released, what to be quantized is the particle's time-average kinetic energy. Experiments on particle's motion in confining potentials can be performed to verify which prediction is correct.

Author Contributions: C.D.Y. developed the main theory of nonsquare-integrable wavefunctions. C.H.K. performed the numerical computations.

Acknowledgments: This wok was financially supported by Ministry of Science and Technology, Taiwan, with contract number: MOST 104-2221-E-006-143-MY3.

Conflicts of Interest: The authors declare no conflict of interest.

References

1. Styer, D.F.; Balkin, M.S.; Becker, K.M.; Burns, M.R.; Dudley, C.E.; Forth, S.T.; Gaumer, J.S.; Kramer, M.A.; Oertel, D.C.; Park, L.H.; et al. Nine formulations of quantum mechanics. *Am. J. Phys.* **2002**, *70*, 288–297. [CrossRef]
2. Leacock, R.A.; Padgett, M.J. Hamilton-Jacobi Theory and the Quantum Action Variable. *Phys. Rev. Lett.* **1983**, *50*, 3–6. [CrossRef]
3. Leacock, R.A.; Padgett, M.J. Hamilton-Jacobi/Action-Angle Quantum Mechanics. *Phys. Rev. D* **1983**, *28*, 2491–2502. [CrossRef]
4. Jordan, P. Über kanonische Transformationen in der Quantenmechanik. *Z. Phys.* **1926**, *37*, 383–386. (In German) [CrossRef]
5. Dirac, P.A.M. *The Principles of Quantum Mechanics*; Oxford University Press: London, UK, 1958.
6. Schwinger, J. *Quantum Kinematics and Dynamics*; Benjamin-Cummings: Menlo Park, CA, USA, 1970.
7. Yang, C.D. Quantum Hamilton Mechanics: Hamilton Equations of Quantum Motion, Origin of Quantum Operators, and Proof of Quantization Axiom. *Ann. Phys.* **2006**, *321*, 2876–2926. [CrossRef]
8. Bohm, D. A Suggested Interpretation of the Quantum Theory in Terms of Hidden Variables. *Phys. Rev.* **1952**, *85*, 166–193. [CrossRef]
9. Bohm, D.; Hiley, B.J. *The Undivided Universe*; Routledge: London, UK, 1993.
10. Holland, P.R. *The Quantum Theory of Motion*; Cambridge University Press: Cambridge, UK, 1993.
11. Gisin, N. Why Bohmian Mechanics? One- and Two-Time Position Measurements, Bell Inequalities, Philosophy, and Physics. *Entropy* **2018**, *20*, 105. [CrossRef]
12. Valentini, A. Signal-locality, Uncertainty, and the Sub-quantum H-theorem, Part I. *Phys. Lett. A* **1991**, *156*, 5–11. [CrossRef]
13. Valentini, A.; Westman, H. Dynamical Origin of Quantum Probabilities. *Proc. R. Soc. A* **2005**, *461*, 253–272. [CrossRef]
14. Colin, S.; Valentini, A. Instability of Quantum Equilibrium in Bohm's Dynamics. *Proc. R. Soc. A* **2014**.
15. Bohm, D.; Vigier, J.P. Model of the Causal Interpretation of Quantum Theory in Terms of a Fluid with Irregular Fluctuations. *Phys. Rev.* **1954**, *96*, 208–216. [CrossRef]
16. Efthymiopoulos, C.; Contopoulos, G. Chaos in Bohmian Quantum Mechanics. *J. Phys. A* **2006**, *39*, 1819–1852. [CrossRef]
17. Contopoulos, G.; Delis, N.; Efthymiopoulos, C. Order in de Broglie-Bohm Quantum Mechanics. *J. Phys. A* **2012**, *45*, 165301. [CrossRef]
18. Tzemos, C.; Efthymiopoulos, C.; Contopoulosz, G. Origin of chaos near three-dimensional quantum vortices: A general Bohmian theory. *Phys. Rev. E* **2018**, *97*, 042201. [CrossRef]
19. Borondo, F.; Luque, A.; Villanueva, J.; Wisniacki, D.A. A dynamical systems approach to Bohmian trajectories in a 2D harmonic oscillator. *J. Phys. A* **2009**, *42*, 495103. [CrossRef]
20. Wisniacki, D.A.; Pujals, E.R.; Borondo, F. Vortex interaction, chaos and quantum probabilities. *Europhys. Lett.* **2006**, *73*, 671. [CrossRef]
21. Wisniacki, D.A.; Pujals, E.R.; Borondo, F. Vortex dynamics and their interactions in quantum trajectories. *J. Phys. A* **2007**, *40*, 14353. [CrossRef]
22. Wyatt, R.E.; Rowland, B.A. Quantum Trajectories in Complex Phase Space: Multidimensional Barrier Transmission. *J. Chem. Phys.* **2007**, *127*, 044103. [CrossRef] [PubMed]
23. Chou, C.C.; Sanz, A.S.; Miret-Artes, S.; Wyatt, R.E. Hydrodynamic View of Wave Packet Interference: Quantum Caves. *Phys. Rev. Lett.* **2009**, *102*, 250401. [CrossRef] [PubMed]
24. Goldfarb, Y.; Degani, I.; Tannor, D.J. Bohmian Mechanics with Complex Action: A New Trajectory-based Formulation of Quantum Mechanics. *J. Chem. Phys.* **2006**, *125*, 231103. [CrossRef] [PubMed]
25. Yang, C.D.; Wei, C.H. Synthesizing Quantum Probability by a Single Chaotic Complex-Valued Trajectory. *Int. J. Quantum Chem.* **2016**, *116*, 428–437. [CrossRef]
26. Wilson, W. The Quantum Theory of Radiation and Line Spectra. *Philos. Mag.* **1915**, *29*, 795–802. [CrossRef]
27. Yang, C.D.; Cheng, L.L. Optimal Guidance Law in Quantum Mechanics. *Ann. Phys.* **2013**, *338*, 167–185. [CrossRef]

28. Bhalla, R.S.; Kapoor, A.K.; Panigrahi, P.K. Energy Eigenvalues for a Class of One-Dimensional Potentials via Quantum Hamilton–Jacobi Formalism. *Mod. Phys. Lett. A* **1997**, *12*, 295–306. [CrossRef]

29. Bhalla, R.S.; Kapoor, A.K.; Panigrahi, P.K. Quantum Hamilton-Jacobi Formalism and the Bound State Spectra. *Am. J. Phys.* **1997**, *65*, 1187–1194. [CrossRef]

30. Arfken, G.B.; Weber, H.J. *Mathematical Methods for Physicists*, 6th ed.; Elsevier Academic Press: New York, NY, USA, 2005.

31. John, M.V. Modified De Broglie-Bohm Approach to Quantum Mechanics. *Found. Phys. Lett.* **2002**, *15*, 329–343. [CrossRef]

32. Yang, C.D. Modeling Quantum Harmonic Oscillator in Complex Domain. *Chaos Solitons Fractals* **2006**, *30*, 342–362. [CrossRef]

33. Hestenes, D. Consistency in the formulation of the Dirac, Pauli, and Schrödinger theories. *J. Math. Phys.* **1975**, *16*, 573–584.

34. Holland, P.R. Implications of Lorentz covariance for the guidance equation in two-slit quantum interference. *Phys. Rev. A* **2003**, *67*, 062105. [CrossRef]

35. Colijn, C.; Vrscay, E.R. Spin-dependent Bohm trajectories associated with an electronic transition in hydrogen. *J. Phys. A* **2003**, *36*, 4689–4702. [CrossRef]

36. Ohanian, H.C. What is spin? *Am. J. Phys.* **1986**, *54*, 500–505. [CrossRef]

37. Mita, K. Virtual probability current associated with the spin. *Am. J. Phys.* **2000**, *68*, 259–264. [CrossRef]

MDPI

Article
Gudder's Theorem and the Born Rule

Francisco De Zela

Departamento de Ciencias, Sección Física, Pontificia Universidad Católica del Perú, Apartado 1761, Lima, Peru; fdezela@pucp.edu.pe

Received: 23 December 2017; Accepted: 6 February 2018; Published: 2 March 2018

Abstract: We derive the Born probability rule from Gudder's theorem—a theorem that addresses orthogonally-additive functions. These functions are shown to be tightly connected to the functions that enter the definition of a signed measure. By imposing some additional requirements besides orthogonal additivity, the addressed functions are proved to be linear, so they can be given in terms of an inner product. By further restricting them to act on projectors, Gudder's functions are proved to act as probability measures obeying Born's rule. The procedure does not invoke any property that fully lies within the quantum framework, so Born's rule is shown to apply within both the classical and the quantum domains.

Keywords: Born probability rule; quantum-classical relationship; spinors in quantum and classical physics

PACS: 03.65.Ta; 03.65.Ud; 02.50.Ey; 42.50.Xa

1. Introduction

Originally, Born's probability rule was considered to be one of those salient features of quantum theory which make it markedly depart from a classical description of physical phenomena. Born's rule was complemented by another one, which is a prescription that establishes how a system changes when submitted to measurement: the so-called collapse rule. There have been some attempts to derive the Born rule from basic concepts of probability theory, thereby reducing the axiomatic basis of quantum mechanics. Notably, Gleason's theorem [1] claims to achieve such a reduction by deriving the Born rule from the properties of a probability measure. However, Gleason's theorem does not hold for two-dimensional quantum systems (i.e., for qubits). This is also the case with a prominent corollary of Gleason's theorem, the Bell–Kochen–Specker (BKS) theorem [2,3], which disproves the assumption that it is always possible to assign noncontextual values to observables prior to measurement. Thus, in the quantum framework, it is not possible to interpret measurement outcomes as revealing pre-existing values of the measured observables. However, such a fundamental claim does not include qubits. Moreover, Bell violations showing the impossibility of hidden-variable models require composite systems [2,4]. It is thus possible to construct a hidden-variable model for a single qubit [3,5]. This state of affairs has prompted some people to place qubits—and them alone—into a sort of limbo, as being half quantum and half classical objects [6,7]. Indeed, as pointed out in [8], it is widely believed that "a single qubit is not a truly quantum system". No matter how appealing the motivations for such a belief might seem, its untenability becomes clear when seen from the perspective of the quantum formalism alone: there is nothing in this formalism that distinguishes two-level systems from other systems of higher dimensionality. We should therefore simply admit that Gleason's approach does not meet its intended goal.

The inclusion of qubits was achieved in Busch's extension [9] of Gleason's theorem. Instead of the pairwise orthogonal projectors P_i entering Gleason's theorem, Busch addresses positive operator-valued measures (POVMs) E_n. However, the inclusion of qubits in Busch's approach was obtained at the cost of departing from our most intuitive notion of a measure. The mathematical tool that corresponds to our basic notion of a measure is a non-negative function m over a σ-algebra. This function is required to satisfy $m(A \cup B) = m(A) + m(B)$, whenever $A \cap B = \varnothing$. The last condition must hold because in case $A \cap B \neq \varnothing$, we should subtract $m(A \cap B)$ from $m(A) + m(B)$ in order to encompass our intuitive notion of a measure. A particular and important case is the "probability measure". In quantum mechanics, this measure is defined over the projection lattice $\mathcal{P}(\mathcal{H})$ of a Hilbert space $\mathcal{H}$, and it is thus consistent to require for $P_i, P_j \in \mathcal{P}(\mathcal{H})$ that $m(P_i + P_j) = m(P_i) + m(P_j)$, whenever $P_i P_j = 0$. On the other hand, it is rather unnatural to call v a measure if it is required to satisfy $v(E_n + E_m) = v(E_n) + v(E_m)$, even though $E_n E_m \neq 0$. However, this is the case in Busch's extension of Gleason's theorem, in which projectors are replaced by POVMs. As for the BKS theorem, Cabello [8] has similarly proved its validity in the case of qubits by replacing projective measurements with POVMs, while Aravind [10] extended Cabello's proof to arbitrary finite dimensions. The introduction of POVMs in the quantum formalism as a generalization of von Neumann's projection-valued measures has been required for various reasons, such as the quantum information approach to quantum mechanics, the employment of non-optimal devices that deliver unsharp measurement outcomes, the description of composite measurements, etc. However, none of these reasons bears any particular connection with two-state systems. It is thus unclear why the inclusion of qubits in the aforementioned theorems should require the replacement of projective measurements by POVMs.

Recently, we have presented an alternative derivation of the Born rule [11], starting from Gudder's theorem [12]—a theorem which is in a sense the reciprocal of Pythagoras's theorem. Such a derivation begins with two-dimensional systems and then extends to higher-dimensional ones, including both pure and mixed states. By observing that the Born rule involves only two states, its derivation can be generally reduced to the two-dimensional case, irrespective of the (finite or infinite) dimensionality of the addressed vector space. Moreover, the derivation blurs the distinction between quantum and classical measurements, so Born's rule is shown to apply beyond its original purely quantum domain. This opens the way for the construction of hidden-variable models of Bell violations produced by maximally entangled states [13].

Hall [14] recently criticized our derivation of the Born rule, arguing that a non-linear counterexample that shows why qubits are excluded from the scope of Gleason's theorem also applies in our approach. One of the purposes of the present work is to show that this is not so. The reason can be stated very simply and in advance: the assumptions underlying our approach imply that any function we deal with is a linear one. This was not explicitly shown in [11], but only implicitly, by deriving Born's linear expression. We present here an explicit demonstration of linearity, and moreover, go beyond the goals of our previous work. Indeed, Hall's criticisms represent a welcome opportunity to expand the scope of Ref. [11], as well as to clear up the physical content of the proposed extension of Gleason's theorem.

We should stress that we do not attempt to solve the so-called "measurement problem"; that is, we do not attempt to answer the question as to how measurements fit into the quantum formalism. Instead, we follow a similar approach as in Ref. [15] and take measurements as something fundamental that require a proper self-consistent description. Thus, we restrict ourselves to the probability rule, leaving aside the collapse rule and the question as to whether collapse is a physical process or just an updating of our system's knowledge. On the other hand, we do address the question about the placement of the Born rule with respect to the quantum–classical border. To this day, the latter remains a controversial issue [16–24], to which the present work intends to make a contribution.

This paper is organized as follows. In Section 2 we recall Gleason's theorem and in Section 3 we reproduce—for the sake of completeness—the essential points of Ref. [11]. At the same time, we extend somewhat the results presented in Ref. [11], by completely fixing the orthogonally additive function that we addressed there and that was left partially undefined in the cited work. We also address Hall's criticisms. In Section 4 we present an alternative derivation of the Born rule which bypasses the reduction to two-level systems that was used in Ref. [11], and generally applies to N-level systems, with $N \geq 2$. We close the paper by discussing our results.

2. Gleason's Theorem and Its Restriction to Dimensions Greater Than Two

Let us recall Gleason's theorem. It states that any probability measure over the lattice $\mathcal{P}(\mathcal{H})$ of orthogonal projectors $P_i \in \mathcal{P}(\mathcal{H})$ acting on a Hilbert space $\mathcal{H}$ has the form given by the Born rule [1]. The defining properties of a probability measure $m(P) : \mathcal{P}(\mathcal{H}) \to [0, 1]$ read as follows:

$$m(\mathbb{I}) = 1, \tag{1}$$

$$m\left(\sum_i P_i\right) = \sum_i m\left(P_i\right). \tag{2}$$

It is straightforward to show that $\sum_i P_i \in \mathcal{P}(\mathcal{H})$ implies that $P_i P_j = 0$, for $i \neq j$. Gleason proved that whenever $\dim \mathcal{H} \geq 3$, there exists a unique density operator ρ such that

$$m(P) = \mathrm{Tr}\left(\rho P\right), \quad \forall P \in \mathcal{P}(\mathcal{H}), \tag{3}$$

which is the Born rule.

The exclusion of qubits from the scope of Gleason's theorem may be traced back to the fact that assumptions ((1) and (2))—in particular (2)—are not strong enough to imply Equation (3) in the two-dimensional case. Indeed, Gleason's proof requires showing that m is continuous. This can be done only for $\dim \mathcal{H} \geq 3$. In the 2D case, there are discontinuous measures satisfying assumptions ((1) and (2)). While Gleason's proof is technically difficult (and for this reason the exclusion of the 2D case is not quite transparent), in the case of its prominent corollary, the BKS theorem, it is easier to understand why the latter does not hold in the 2D case. Indeed, an independent demonstration of the BKS theorem—i.e., not as a corollary of Gleason's—can be reduced to the task of coloring the surface of a unit hyper-sphere with two colors [7]. This is possible for two dimensions—viz., in the case of the unit circle—but not for higher dimensions.

There is yet another way to show that the 2D case must lie outside the domain of Gleason's theorem. We observe that measure $m(P)$ entering Born's rule (see Equation (3)) is not only continuous, but also linear. Hall [14] provided a non-linear measure m over the set of qubit-projectors which satisfies conditions ((1) and (2)), thereby proving that Gleason's theorem cannot hold for qubits. As for the derivation of the Born rule that we reported in [11], the conditions we impose on the addressed measures can be satisfied only by linear functions. This notwithstanding, Hall claimed to have provided a non-linear function satisfying said conditions [14]. Below, we will discuss what went wrong in Hall's reasoning.

3. Gudder's Theorem and the Born Rule for Two-Level Systems

Linearity is a central issue in the derivation of Born's rule from any chosen assumptions [9,15,25–28]. For instance, the derivation in Ref. [9]—which includes qubits—entails the demonstration that the measure $v(E)$ over POVMs is a positive linear functional that can be obtained from a density operator. As we have seen, Gleason's assumptions are instead too weak to enforce linearity in the case of qubits. In our approach, linearity is enforced by imposing upon the concept of a measure a series of requirements that reflect the most general experimental procedures. These requirements generally apply when submitting any system to measurement. As stressed in Ref. [11], our assumptions are not restricted to the quantum case, and therefore some classical measurements can also be encoded in terms of the Born rule. Said assumptions are strongly driven by physical considerations rather than by mathematical motivations.

Most measurement procedures in physics are essentially "counting" procedures. They consist of counting how many times a given unit—a measure—fits into the observable that is submitted to measurement. As already said, the primary standard mathematical tool that captures our basic notion of a measure is a non-negative function m over a σ-algebra. The restriction to be non-negative is a convenient one in some cases, such as integration theory. Instead, in physics it is often convenient to distinguish between, e.g., two sides (left and right), or to be able to add and subtract a given amount. Hence, a generalization of the original concept of measure is convenient, to what is called a signed measure μ. A signed measure is defined over a σ-algebra $\mathcal{A}_\sigma$, as $\mu : \mathcal{A}_\sigma \to \mathbb{R}$, with $\mu(\cup_n A_n) = \sum_n \mu(A_n)$, for any sequence $A_1, A_2, \ldots, A_n$ of pairwise disjoint sets in $\mathcal{A}_\sigma$. Besides these mathematical requirements, we can include some additional ones that reflect our dealing with physical measurements. First of all, we restrict ourselves to dealing with continuous functions f. This requirement captures our basic notion that infinitesimal variations of the observable being measured should lead to infinitesimal variations of the measurement result. Second, we restrict ourselves to dealing with functions f that are defined over an inner product vector space V. With these restrictions, what was initially a signed measure ends up being the subject matter of Gudder's theorem [12]. Indeed, Gudder's theorem deals with an inner product vector space V and a continuous function f that is orthogonally additive. The definition of such a function reads as follows:

Definition 1.

$$f : V \to \mathbb{R} \text{ is orthogonally additive if } \quad f(r + r') = f(r) + f(r') \quad \text{whenever} \quad r \cdot r' = 0. \tag{4}$$

Gudder proves that the following result holds true:

Theorem 1. *If $f : V \to \mathbb{R}$ is orthogonally additive and continuous, then it has the form*

$$f(r) = c(r \cdot r) + k \cdot r, \tag{5}$$

where $c \in \mathbb{R}$ and $k \in V$.

Our aim is to show how Born's rule arises from Gudder's theorem. To this end, we first focus on qubits. A qubit can be represented by a unit vector $|\phi\rangle \in \mathcal{H}_2$ of an equivalence class—a so-called "ray"—or alternatively, it can be represented by the corresponding projector

$$P_\phi \equiv |\phi\rangle\langle\phi| = \frac{1}{2} \left(\mathbb{I}_2 + \hat{\mathbf{n}}_\phi \cdot \sigma \right). \tag{6}$$

Here, $\mathbb{I}_2$ is the identity operator in $\mathcal{H}_2$ and the unit vector $\hat{\mathbf{n}}_\phi = \mathrm{Tr}\left(\sigma P_\phi\right)$, with σ standing for the triple of Pauli matrices. In general, for a non-normalized qubit $|\psi\rangle \in \mathcal{H}_2$, we can write

$$R_\psi \equiv |\psi\rangle\langle\psi| = \frac{1}{2} \sum_{\mu=0}^{3} r_\mu \sigma_\mu, \tag{7}$$

with $\sigma_0 \equiv \mathbb{I}_2$ and $r_\mu = \mathrm{Tr}\left(\sigma_\mu R_\psi\right)$. We see that $R_\psi = P_\psi$ whenever $\langle\psi|\psi\rangle = 1$. There is a one-to-one correspondence between operators R_ψ and vectors $r := (r_0, r_1, r_2, r_3) \equiv (r_0, \mathbf{r})$. The latter span a four-dimensional real vector space V_4 that can be made an inner product space by defining the Euclidean inner product

$$r \cdot r' = \sum_{\mu=0}^{3} r_\mu r'_\mu. \tag{8}$$

We now wish to define a measure f_ϕ that is associated to a particular qubit $|\phi\rangle \leftrightarrow r_\phi \equiv (1, \hat{\mathbf{n}}_\phi)$. In a sense, f_ϕ and $|\phi\rangle$ represent one and the same physical object that is mathematically encoded in two alternative ways [11]. To start with, f_ϕ must satisfy the following requirements.

(1) f_ϕ must satisfy the assumptions of Theorem 1.
(2) $f_\phi(r_\phi) = 1$, which corresponds to requiring that our unit of measure fits exactly one time into itself.
(3) $f_\phi(r_{\phi_\perp}) = 0$ for the vector $|\phi_\perp\rangle \leftrightarrow r_{\phi_\perp} \equiv (1, -\hat{\mathbf{n}}_\phi)$ that is orthogonal to $|\phi\rangle$.

On applying Gudder's theorem with $k = (k_0, \mathbf{k})$, we obtain

$$f_\phi\left[(1, \hat{\mathbf{n}}_\phi)\right] = 2c + k_0 + \hat{\mathbf{n}}_\phi \cdot \mathbf{k} = 1, \tag{9}$$

$$f_\phi\left[(1, -\hat{\mathbf{n}}_\phi)\right] = 2c + k_0 - \hat{\mathbf{n}}_\phi \cdot \mathbf{k} = 0. \tag{10}$$

From these equations, we get $2c + k_0 = 1/2$ and $\hat{\mathbf{n}}_\phi \cdot \mathbf{k} = 1/2$. Up to this point, we have been dealing with a function f_ϕ that is not necessarily identifiable with a probability measure. Let us further restrict f_ϕ to satisfy the following requirement:

(4) $f_\phi\left[(1, \hat{\mathbf{n}}_\psi)\right] \in [0,1]$ for any four-vector $(1, \hat{\mathbf{n}}_\psi) \leftrightarrow |\psi\rangle\langle\psi| = P_\psi$.

In such a case, $f_\phi\left[(1, \hat{\mathbf{n}}_\psi)\right] = 2c + k_0 + \hat{\mathbf{n}}_\psi \cdot \mathbf{k} = 1/2 + \hat{\mathbf{n}}_\psi \cdot \mathbf{k} \in [0,1]$; i.e.,

$$-\frac{1}{2} \le |\mathbf{k}| \cos\theta \le \frac{1}{2}, \tag{11}$$

where $\cos\theta = \hat{\mathbf{n}}_\psi \cdot \hat{\mathbf{k}}$ spans the interval $[-1,1]$ under variation of $\hat{\mathbf{n}}_\psi$. This implies that $|\mathbf{k}| = 1/2$, hence $\mathbf{k} = \hat{\mathbf{n}}_\phi/2$, and we can finally write

$$f_\phi\left[(1, \hat{\mathbf{n}}_\psi)\right] = \frac{1}{2}\left(1 + \hat{\mathbf{n}}_\phi \cdot \hat{\mathbf{n}}_\psi\right). \tag{12}$$

Using $P_\psi = |\psi\rangle\langle\psi| = \left(\mathbb{I}_2 + \hat{\mathbf{n}}_\psi \cdot \sigma\right)/2$ and similarly for $P_\phi = |\phi\rangle\langle\phi|$, we can write $f_\phi(P_\psi)$ in the standard form

$$f_\phi(P_\psi) = |\langle\phi|\psi\rangle|^2 = \mathrm{Tr}\left(P_\phi P_\psi\right). \tag{13}$$

The measure f_ϕ we have obtained under the above requirements can be consistently interpreted as a probability measure. We have put our requirements on a function f_ϕ that applies to vectors $r \in V_4$ in general. It is just in order to fix some of the parameters that define f_ϕ (i.e., c and $k = (k_0, \mathbf{k})$) that we conveniently applied f_ϕ to some particular vectors $(1, \hat{\mathbf{n}}) \in V_4$. These vectors belong to V_4 in spite of carrying only two independent parameters—the ones fixing $\hat{\mathbf{n}}$. Now, as for the function f_ϕ, it has not been completely fixed. Though we know its action on vectors of the form $(1, \hat{\mathbf{n}})$ (see Equations (9) and (10)), we do not know its action on more general vectors $r \in V_4$. This is because we have fixed only $\mathbf{k} = \hat{\mathbf{n}}_\phi/2$, while c and k_0 remain yet undetermined. In order to fix them, we can consider the vector $(-1, \hat{\mathbf{n}}_\phi)$, which is orthogonal to $|\phi\rangle \leftrightarrow r_\phi \equiv (1, \hat{\mathbf{n}}_\phi)$. Thus, we must consistently require that

$$3a)\ f_\phi\left[(-1, \hat{\mathbf{n}}_\phi)\right] = 2c - k_0 + \hat{\mathbf{n}}_\phi \cdot \mathbf{k} = 2c - k_0 + \frac{1}{2} = 0. \tag{14}$$

On account of the above equation and $2c + k_0 = 1/2$, we get $c = 0$ and $k_0 = 1/2$. Hence, $k = r_\phi/2$ and Theorem 1 establishes that f_ϕ is a linear function given by $f_\phi(r) = k \cdot r$; i.e.,

$$f_\phi[(r_0, \mathbf{r})] = \frac{1}{2}\left(r_0 + \hat{\mathbf{n}}_\phi \cdot \mathbf{r}\right). \tag{15}$$

On view of $(r_0, \mathbf{r}) \leftrightarrow R_\psi \equiv \rho_\psi = \sum_\mu r_\mu \sigma_\mu/2$ (see Equation (7)), and $(1, \hat{\mathbf{n}}_\phi) \leftrightarrow P_\phi \equiv \rho_\phi = \left(\mathbb{I}_2 + \hat{\mathbf{n}}_\phi \cdot \sigma\right)/2$ (see Equation (6)), we can also write

$$f_\phi[(r_0, \mathbf{r})] = \mathrm{Tr}(\rho_\phi^\dagger \rho_\psi). \tag{16}$$

In summary, under the above assumptions, $f_\phi(r)$ has reduced to be a scalar product. It can be specified either in vector space V_4, where it is given by the Euclidean scalar product, or in the space of linear operators acting on $\mathcal{H}_2$, where it is given by the Hilbert–Schmidt inner product $\mathrm{Tr}(A^\dagger B)$. Of course, $f_\phi(r)$ can be negative for some $r \in V_4$. However, if we restrict ourselves to applying $f_\phi(r)$ on vectors $(1, \hat{\mathbf{n}}_\psi) \in V_4$, then $f_\phi[(1, \hat{\mathbf{n}}_\psi)] \in [0,1]$, and in this case we may use f_ϕ as a probability measure. It is up to us to decide which mathematical tools we employ in order to describe our experimental observations. The probability measure f_ϕ is just one of these tools. As discussed in [11], it is not exclusively connected to quantum phenomena.

Let us now briefly refer to Hall's criticisms [14] of our derivation of Born's rule. Hall claims that our defining conditions for a measure f_ϕ are satisfied by the following non-linear measure:

$$f_\phi(P_\psi) = \frac{1}{2}\left[1 + f(\hat{\mathbf{n}}_\phi \cdot \hat{\mathbf{n}}_\psi)\right]. \tag{17}$$

Here, $f(x)$ "is any non-linear function mapping the interval $[-1,1]$ into itself, with $f(-x) = -f(x)$ and $f(1) = 1$" [14]. The above f_ϕ can be proved to satisfy Gleason's assumptions ((1) and (2)) in the 2D case, thereby showing that Gleason's theorem does not hold for qubits. If $f(x)$ is also required to be continuous, then f_ϕ should allegedly satisfy our defining conditions [14]. However, our function f_ϕ maps vectors in V_4 to the reals. For instance, these vectors may be of the form $(\pm 1, \hat{\mathbf{n}}_\psi)$. On the other hand, the subject of the above definition, Equation (17), is a function whose domain is not V_4. Instead of Hall's notation, $f_\phi(P_\psi)$, one should more properly write $f_\phi(\hat{\mathbf{n}}_\psi)$ on the lhs of Equation (17). The domain of Hall's f_ϕ is thus the unit sphere. In particular, one cannot tell the results of applying this f_ϕ to vectors such as $(1, \hat{\mathbf{n}}_\psi)$ and $(-1, \hat{\mathbf{n}}_\psi)$. Hence, one cannot claim that this $f_\phi(P_\psi)$ satisfies, for example, the requirement given by Equation (14): $f_\phi\left[(-1, \hat{\mathbf{n}}_\phi)\right] = 0$.

One can try to circumvent Hall's technical flaw and still seek to object to our derivation of Born's rule by arguing that qubits should not be treated as belonging to V_4. Such a claim connects with the belief that qubits are bijectively mapped to the points on the surface of the unit (Bloch/Poincaré) sphere, so that any given qubit $|\psi\rangle$ may be represented by some unit vector $\hat{\mathbf{n}}_\psi$. This is wrong. Qubits (viz., spinors) span $V_4 \sim \mathbb{C}^2 \ni |\psi\rangle = \alpha|\uparrow\rangle + \beta|\downarrow\rangle$, under variation of the complex-valued coefficients α and β. In order to restrict spinors $|\psi\rangle$ so as to span only the unit sphere $S^2 := \left\{\hat{\mathbf{n}} \in \mathbb{R}^3 : |\hat{\mathbf{n}}| = 1\right\} \subset \mathbb{R}^3$, we need to normalize $|\psi\rangle$ and discard a global phase. This amounts to neglecting some information that we deem unimportant, whatever the reason. However, under different circumstances, this information may turn out to be physically meaningful; see our closing remarks below, Section 5. An exhaustive description of qubits should therefore be given by the elements of $\mathbb{C}^2 \sim V_4$.

The generalization of the above results to higher dimensional vector spaces and to mixed states is straightforward, and has been discussed in Ref. [11]. The generalization is based on the observation that two-dimensional Hilbert spaces are in fact general enough for dealing with the Born rule. Indeed, this rule involves only two states and therefore effectively limits itself—in each concrete case—to dealing with a two-dimensional subspace of the addressed vector space. This also holds in the case of infinite-dimensional spaces with continuous basis vectors $|\phi(\alpha)\rangle$, which may be thought of as eigenvectors of some observable with a continuous spectrum given by α. In such a case, one replaces the probability $f_\phi(P_\psi)$ in Born's formula (13) by $df_{\phi(\alpha)}(P_\psi) = |\langle\phi(\alpha)|\psi\rangle|^2 d\alpha$, corresponding to measurement results between α and $\alpha + d\alpha$. Although this procedure leads to our intended goal, it is instructive to follow an alternative approach, in which we apply algebraic tools similar to those related to the Pauli algebra. This puts the qubit case on the same footing as the higher-dimensional ones. We present this approach next, restricted to systems of arbitrary finite dimension.

4. Gudder's Theorem and the Born Rule for *N*-Level Systems

Let us first recall that the Pauli matrices are generators of the SU(2) group. Together with the 2×2 unit matrix, they constitute an orthonormal basis, in terms of which we can express any operator acting on the two-dimensional Hilbert space $\mathcal{H}_2$. When dealing with higher dimensional spaces $\mathcal{H}_N$, we can resort to the $N^2 - 1$ generators $G_i = G_i^\dagger$ of the SU(N) group. These can be chosen so as to satisfy

$$\operatorname{Tr} G_i = 0, \quad \operatorname{Tr}(G_i G_j) = N\delta_{ij}. \tag{18}$$

Notice that our choice of normalization is best suited to our present purposes and differs from the most commonly employed one, namely $\text{Tr}(G_i G_j) = 2\delta_{ij}$ [29–32]. Any operator $\rho = \rho^\dagger$ with $\text{Tr}\,\rho \equiv \sqrt{N} r_0$ can be expressed as

$$\rho_r = \frac{1}{\sqrt{N}}\left(r_0 \mathbb{I}_N + \sum_{k=1}^{N^2-1} r_k G_k\right), \tag{19}$$

where $r_k \in \mathbb{R}$, for $k = 0, \ldots, N^2 - 1$. This establishes a one-to-one correspondence between Hermitian operators ρ acting on $\mathcal{H}_N$ and vectors $r \in V_N$. Let us now choose one of these vectors, $r_\phi = (r_0, \ldots, r_d) \in V_N$, where $d = N^2 - 1$. It corresponds to a fixed state ρ_ϕ, a Hermitian operator that acts on $\mathcal{H}_N$. We can represent the state ρ_ϕ in an alternative way, namely by means of Gudder's measure f_ϕ, the one that is the subject matter of Gudder's Theorem 1. To begin with, we consider a vector $r_\perp$ orthogonal to r_ϕ (i.e., $r_\phi \cdot r_\perp = 0$), and require that our measure yields a null result in this case: $f_\phi(r_\perp) = 0$. The same requirement holds for vector $-r_\perp$, so that on view of Gudder's theorem we have:

$$f_\phi(r_\perp) = c_\phi r_\perp \cdot r_\perp + k_\phi \cdot r_\perp = 0, \tag{20}$$

$$f_\phi(-r_\perp) = c_\phi r_\perp \cdot r_\perp - k_\phi \cdot r_\perp = 0. \tag{21}$$

The above requirements imply that $c_\phi = 0$. Thus, Gudder's measure f_ϕ reads $f_\phi(r) = k_\phi \cdot r$ in our case, with $k_\phi \in V_N$ yet to be determined. With r_ϕ and $d = N^2 - 1$ additional vectors $s_{(1)}, \ldots, s_{(d)}$, we can conform an orthogonal basis, in terms of which we can write $k_\phi = \lambda r_\phi + \sum_{j=1}^{d} \lambda_j s_{(j)}$. For the same reasons as before, we require that $f_\phi(s_{(j)}) = k_\phi \cdot s_{(j)} = 0$ for $j = 1, \ldots, d$. This leads us to conclude that k_ϕ is parallel to r_ϕ; i.e., $k_\phi = \lambda r_\phi$. If we finally require that $f_\phi(r_\phi) = 1$, we end up with

$$f_\phi(r) = \frac{1}{r_\phi \cdot r_\phi} r_\phi \cdot r. \tag{22}$$

By choosing the normalization $r_\phi \cdot r_\phi = 1$, we have $f_\phi(r) = r_\phi \cdot r$. The normalization in Equation (19) has been chosen so as to render

$$\text{Tr}(\rho_r \rho_s) = r \cdot s. \tag{23}$$

This allows us to write

$$f_\phi(r) = r_\phi \cdot r = \text{Tr}(\rho_\phi \rho_r). \tag{24}$$

It is a matter of convention which normalization we use; e.g., that of Equations (19) and (22), or else that of Equations (6) and (12). The Born rule is contained in Equation (24) when we restrict ourselves to suitably normalized vectors and operators. In that case, Gudder's measure may be used as a probability measure. The general case corresponds instead to an inner product, which can be seen as a signed measure.

5. Closing Remarks and Discussion

According to Bohr, all quantum measurements require the involvement of a classical device. This assertion implies the unavoidable existence of two different domains—the classical and the quantal. That is, the quantum domain cannot be extended to embrace all physical phenomena, because these phenomena would include measurements themselves. Moreover, if we explicitly avoid dealing with the physical process that takes place during a measurement—that is, with possible changes suffered by a system when submitted to measurement—and focus on the quantification of the outcomes, then we cannot expect that this quantification has peculiar features that are exclusively ascribable to the quantum or to the classical domain. In other words, the Born rule by itself should equally well fit into a quantum and into a classical framework. The derivation of the Born rule presented here is in accordance with such a view. There is nothing in the framework we have used that can be identified as purely quantal. In particular, spinors—or their corresponding density matrices—are an appropriate and useful tool in both the quantum framework (e.g., spin-1/2 particles) and the classical framework (e.g., polarized light beams).

In order to obtain the Born rule, we drew upon Gudder's theorem—a result that is tightly connected with a signed measure. By adding some requirements to the orthogonally-additive functions that are the subject matter of Gudder's theorem, we got a twofold extension of Gleason's theorem in which, first, qubits are included within the scope of the theorem and, second, Born's probability rule arises as a special case of an inner product. Qubits may be understood as spanning a four-dimensional real vector space V_4 whose elements are of the form $(r_0, \mathbf{r})$. The function f in Gudder's theorem acts on this space, and is assumed to be continuous and orthogonally additive. When dealing with vectors of the particular form $(1, \hat{\mathbf{n}})$, we impose some additional requirements on f. These requirements let us interpret f as a probability measure f_ϕ, which is defined in terms of some fixed state $(1, \hat{\mathbf{n}}_\phi)$. When f_ϕ acts on more general vectors $(r_0, \mathbf{r})$, then it acts as an inner product. As pointed out in Ref. [11], having discussed the two-dimensional Hilbert space, we have essentially discussed all higher-dimensional Hilbert spaces, at least with respect to Born's rule. It is worthwhile to stress that the key requirements leading to the linearity of f_ϕ (i.e., $f_\phi(r) = k_\phi \cdot r$) are just two: $f_\phi(r_\perp) = 0$ and $f_\phi(-r_\perp) = 0$, cf. Equations (20) and (21). From them, it follows that $c = 0$ in Theorem 1. Hence, as a consequence of these assumptions, f_ϕ turns out to be an odd function: $f_\phi(-r) = -f_\phi(r)$. Reciprocally, if f_ϕ is assumed to be odd, then it must be linear [12].

Concerning dimensionality, we should emphasize why we have dealt with V_4 in the case of qubits, instead of dealing with a space of lower dimensionality. Qubits are usually defined as normalized vectors in a two-dimensional Hilbert space, or equivalently, as projectors (i.e., density operators acting on this space). They can thus be represented as points on the $2D$ surface of a unit sphere that is embedded in $3D$ space. There are many ways in which one can embed a $2D$ surface in a higher-dimensional space. One can then ask about the physical motivation for dealing with V_4. Why do we not stay dealing with a $2D$ sphere? The physical motivation is given by mixed states in the case of spin-1/2 particles and by partially polarized light in the optical case. In these cases, we must deal with the whole Bloch ball and with the whole Poincaré ball, respectively, and not only with their surfaces. This is because the first component of a Poincaré or a Bloch vector $r \in V_4$ generally carries some physical information. For example, the intensity of polarized light is encoded in this first component. Although it might occur that we are not interested in knowing absolute but only relative intensity values and we consequently normalize our vectors, our formalism should nonetheless provide us with the option of accessing all the physical information that is connected with the phenomenon it is supposed to describe. This brings us outside the unit ball, and so we have to consider balls of arbitrary radii—the union of which makes up V_4. In the case of spin-1/2 particles, we naturally unit-normalize the density operator due to its interpretation in terms of probability. In that case, we usually do not need to go beyond the unit sphere. However, we could find it useful to connect probability with the actual number of particles we expect to detect in a given experiment. This could happen because of practical reasons, for example in order to avoid saturation of some detectors. In cases like this, we again need to go beyond the unit sphere in V_4. As an example of current theoretical interest, we may mention the study of qubits evolving according to quantum maps that are not completely positive, and therefore generally map the unit ball onto a set that is not contained in this ball [33]. The point in question seems to have been better appreciated by the classical community than by the quantum community, at least in the case of classical and quantum optics. Indeed, in classical optics one routinely uses either the Jones or the Mueller formalism. The latter deals with vectors in V_4, and perhaps no one would object that all four components of Mueller vectors have physical meaning. Some researchers even think that the Mueller formalism is more general and better suited than the Jones formalism to address physically-motivated inquiries [34]. Our approach acknowledges the fact that by dealing with $2D$ spinors some portion of physical information has been discarded. To take full account of this information, a $4D$ formalism is required, with the corresponding generalization in the SU(N) case.

Finally, we should emphasize that our goals substantially differ from Gleason's. Indeed, we are not interested in showing that the structure of the Hilbert space naturally arises as the scenario in which quantum mechanics should be formulated. We have instead assumed that, say, qubits can be represented by density matrices in a Hilbert space, or else by four-dimensional vectors of a linear space. Our aim was to expose the fundamental underlying assumptions leading to a probability rule that has the structure of Born's rule. By so doing, we can see the extent to which these assumptions lie in the quantum or in the classical domain.

Acknowledgments: This work was partially supported by DGI-PUCP (Grant-Nr. 441), which also covered the costs to publish in open access.

Conflicts of Interest: The author declares no conflict of interest.

References

1. Gleason, A.M. Measures on the closed subspaces of a Hilbert space. *J. Math. Mech.* **1957**, *6*, 885–893.
2. Bell, J.S. On the Problem of Hidden Variables in Quantum Mechanics. *Rev. Mod. Phys.* **1966**, *38*, 447.
3. Kochen, S.; Specker, E.P. The Problem of Hidden Variables in Quantum Mechanics. *J. Math. Mech.* **1967**, *17*, 59–87.
4. Clauser, J.F.; Horne, M.A.; Shimony, A.; Holt, A.R. Proposed Experiment to Test Local Hidden-Variable Theories. *Phys. Rev. Lett.* **1969**, *23*, 880; Erratum in **1970**, *24*, 549.
5. Clauser, J.F. Von Neumann's Informal Hidden-Variable Argument. *Am. J. Phys.* **1971**, *39*, 1095–1096.
6. Van Enk, S.J. Quantum and Classical Game Strategies. *Phys. Rev. Lett.* **2000**, *84*, 789.
7. Redhead, M. *Incompleteness, Nonlocality, Realism*; Clarendon: Oxford, UK, 1987.
8. Cabello, A. Kochen-Specker Theorem for a Single Qubit using Positive Operator-Valued Measures. *Phys. Rev. Lett.* **2003**, *90*, 190401.
9. Busch, P. Quantum States and Generalized Observables: A Simple Proof of Gleason's Theorem. *Phys. Rev. Lett.* **2003**, *91*, 120403.
10. Aravind, P.K. Generalized Kochen-Specker theorem. *Phys. Rev. A* **2003**, *68*, 052104.
11. De Zela, F. Gleason-Type Theorem for Projective Measurements, Including Qubits: The Born Rule Beyond Quantum Physics. *Found. Phys.* **2016**, *46*, 1293–1306.
12. Gudder, S.P. *Stochastic Methods in Quantum Mechanics*; North-Holland: New York, NY, USA, 1979.
13. De Zela, F. Beyond Bell's theorem: Realism and locality without Bell-type correlations. *Sci. Rep.* **2017**, *7*, 14570.
14. Hall, M.J.W. Comment on "Gleason-Type Theorem for Projective Measurements, Including Qubits" by F. De Zela. *arXiv* **2016**, arXiv:1611.00613v2.
15. Caves, C.M.; Fuchs, C.A.; Manne, K.K.; Rennes, J.M. Gleason-Type Derivations of the Quantum Probability Rule for Generalized Measurements. *Found. Phys.* **2004**, *34*, 193–209.
16. Abouraddy, F.A.; Yarnall, T.; Saleh, A. BE.; Teich, C.M. Violation of Bell's inequality with continuous spatial variables. *Phys. Rev. A* **2007**, *75*, 052114.
17. Borges, S.C.V.; Hor-Meyll, M.; Huguenin, O.J.A.; Khoury, Z.A. Bell-like inequality for the spin-orbit separability of a laser beam. *Phys. Rev. A* **2010**, *82*, 033833.
18. Chen, H.; Peng, T.; Karmakar, S.; Shih, Y. Simulation of Bell states with incoherent thermal light. *New J. Phys.* **2011**, *13*, 083018.
19. Kagalwala, H.K.; Di Giuseppe, G.; Abouraddy, F.A.; Saleh, A.B.E. Bell's measure in classical optical coherence. *Nat. Photonics* **2013**, *7*, 72–78.
20. Qian, X.-.F.; Little, B.; Howell, J.C.; Eberly, J.H. Shifting the quantum-classical boundary: theory and experiment for statistically classical optical fields. *Optica* **2015**, *2*, 611–615.
21. Eberly, H.J. Shimony–Wolf states and hidden coherences in classical light. *Contemp. Phys.* **2015**, *56*, 407–416.
22. Eberly, H.J. Correlation, coherence and context. *Laser Phys.* **2016**, *26*, 084004.
23. Sandeau, N.; Akhouayri, H.; Matzkin, A.; Durt, T. Experimental violation of Tsirelson's bound by Maxwell fields. *Phys. Rev. A* **2016**, *93*, 053829.

24. Eberly, J.H.; Qian, X.-F.; Vamivakas, A.N. Polarization coherence theorem. *Optica* **2017**, *4*, 1113–1114.

25. Barnett, M.S; Cresser, D.J; Jeffers, J.; Pegg, D.T. Quantum probability rule: a generalization of the theorems of Gleason and Busch. *New J. Phys.* **2014**, *16*, 043025.

26. Marzlin, K.-P.; Landry, T. On the connection between the theorems of Gleason and of Kochen and Specker. *Can. J. Phys.* **2015**, *93*, 1446–1452.

27. Granström, H. Some remarks on the theorems of Gleason and Kochen-Specker. *arXiv* **2007**, arXiv:0612103v2.

28. Shrapnel, S.; Costa, F.; Milburn, G. Updating the Born rule. *arXiv* **2017**, arXiv:1702.01845.

29. Hioe, F.T.; Eberly, H.J. *N*-Level Coherence Vector and Higher Conservation Laws in Quantum Optics and Quantum Mechanics. *Phys. Rev. Lett.* **1981**, *47*, 838.

30. Jakóbczyk, L.; Siennicki, M. Geometry of Bloch vectors in two-qubit system. *Phys. Lett. A* **2001**, *286*, 383–390.

31. Kimura, G. The Bloch vector for *N*-level systems. *Phys. Lett. A* **2003**, *314*, 339–349.

32. Byrd, S.M.; Bishop, C.A.; Ou, Y.-C. General open-system quantum evolution in terms of affine maps of the polarization vector. *Phys. Rev. A* **2011**, *83*, 012301.

33. Bernardes, N.K.; Cuevas, A.; Orieux, A.; Monken, C.H.; Mataloni, P.; Sciarrino, F.; Santos, M.F. Experimental observation of weak non-Markovianity. *Sci. Rep.* **2015**, *5*, 17520.

34. Simon, B.N.; Simon, S.; Gori, F.; Santarsiero, M.; Borghi, R.; Mukunda, N.; Simon, R. Nonquantum Entanglement Resolves a Basic Issue in Polarization Optics. *Phys. Rev. Lett.* **2010**, *104*, 023901.

entropy

MDPI

Article

Uncertainty Relation Based on Wigner–Yanase–Dyson Skew Information with Quantum Memory

Jun Li [1] and Shao-Ming Fei [1,2,*]

[1]   School of Mathematical Sciences, Capital Normal University, Beijing 100048, China; lijunnl123@163.com
[2]   Max-Planck-Institute for Mathematics in the Sciences, Leipzig 04103, Germany
*   Correspondence: feishm@cnu.edu.com

Received: 2 January 2018; Accepted: 15 February 2018; Published: 20 February 2018

Abstract: We present uncertainty relations based on Wigner–Yanase–Dyson skew information with quantum memory. Uncertainty inequalities both in product and summation forms are derived. It is shown that the lower bounds contain two terms: one characterizes the degree of compatibility of two measurements, and the other is the quantum correlation between the measured system and the quantum memory. Detailed examples are given for product, separable and entangled states.

Keywords: uncertainty relation; Wigner–Yanase–Dyson skew information; quantum memory

1. Introduction

The uncertainty principle is an essential feature of quantum mechanics, characterizing the experimental measurement incompatibility of non-commuting quantum mechanical observables in the preparation of quantum states. Heisenberg first introduced variance-based uncertainty [1]. Later, Robertson [2] proposed the well-known formula of the uncertainty relation, $V(\rho, R)V(\rho, S) \geq \frac{1}{4}|Tr\rho[R,S]|^2$, for arbitrary observables R and S, where $[R,S] = RS - SR$ and $V(\rho, R)$ is the standard deviation of R. Schrödinger gave a further improved uncertainty relation [3]:

$$V(\rho, R)V(\rho, S) \geq \frac{1}{4}|\langle[R,S]\rangle|^2 + |\frac{1}{2}\langle\{R,S\}\rangle - \langle R\rangle\langle S\rangle|^2$$

where $\langle R\rangle = Tr(\rho R)$, and $\{R,S\} = RS + SR$ is the anti-commutator. Since then many kinds of uncertainty relations have been presented [4–11]. In addition to the uncertainty of the standard deviation, entropy can be used to quantify uncertainties [12]. The first entropic uncertainty relation was given by Deutsch [13] and was then improved by Maassen and Uffink [14]:

$$H(R) + H(S) \geq \log_2 \frac{1}{c}$$

where $R = \{|u_j\rangle\}$, and $S = \{|v_k\rangle\}$ are two orthonormal bases on d-dimensional Hilbert space H, and $H(R) = -\Sigma_j p_j \log p_j$ ($H(S) = -\Sigma_k q_k \log q_k$) is the Shannon entropy of the probability distribution $p_j = \langle u_j|\rho|u_j\rangle$ ($q_k = \langle v_k|\rho|v_k\rangle$) for state ρ of H. The number c is the largest overlap among all $c_{jk} = |\langle u_j|v_k\rangle|^2$ between the projective measurements R and S. Berta et al. [15] bridged the gap between cryptographic scenarios and the uncertainty principle and derived this landmark uncertainty relation for measurements R and S in the presence of quantum memory B:

$$H(R|B) + H(R|B) \geq \log_2 \frac{1}{c} + H(A|B)$$

where $H(R|B) = H(\rho_{RB}) - H(\rho_B)$ is the conditional entropy with $\rho_{RB} = \Sigma_j(|u_j\rangle\langle u_j| \otimes I)\rho_{AB}(|u_j\rangle\langle u_j| \otimes I)$ (similarly for $H(S|B)$), and d is the dimension of the subsystem A. The term

$H(A|B) = H(\rho_{AB}) - H(\rho_B)$ appearing on the right-hand side is related to the entanglement between the measured particle A and the quantum memory B. The bound of Berta et al. has been further improved [16–18]. Moreover, there are also some uncertainty relations given by the generalized entropies, such as the Rényi entropy [19–21] and the Tsallis entropy [22–24], and even more general entropies such as the (h, Φ) entropies [25]. These uncertainty relations not only manifest the physical implications of the quantum world but also play roles in entanglement detection [26,27], quantum spin squeezing [28,29] and quantum metrology [30,31].

In [32], an uncertainty relation based on Wigner–Yanase skew information $I(\rho, H)$ has been obtained with quantum memory, where $I(\rho, H) = \frac{1}{2}Tr[(i[\sqrt{\rho}, H])^2] = Tr(\rho H^2) - Tr(\sqrt{\rho}H\sqrt{\rho}H)$ quantifies the degree of non-commutativity between a quantum state ρ and an observable H, which is reduced to the variance $V(\rho, H)$ when ρ is a pure state. In fact, the Wigner–Yanase skew information $I(\rho, H)$ is generalized to Wigner–Yanase–Dyson skew information $I_\alpha(\rho, H)$, $\alpha \in [0, 1]$ (see [33]):

$$
\begin{aligned}
I_\alpha(\rho, H) &= \tfrac{1}{2}Tr[(i[\rho^\alpha, H])(i[\rho^{1-\alpha}, H])] \\
&= Tr(\rho H^2) - Tr(\rho^\alpha H \rho^{1-\alpha} H) \qquad \alpha \in [0, 1]
\end{aligned}
\tag{1}
$$

Here the Wigner–Yanase–Dyson skew information $I_\alpha(\rho, H)$ reduces to the Wigner–Yanase skew information $I(\rho, H)$ when $\alpha = \frac{1}{2}$. The Wigner–Yanase–Dyson skew information $I_\alpha(\rho, H)$ reduces to the standard deviation $V(\rho, H)$ when ρ is a pure state.

The convexity of $I_\alpha(\rho, H)$ with respect to ρ has been proven by Lieb in [34]. In [35], Kenjiro introduced another quantity:

$$
\begin{aligned}
J_\alpha(\rho, H) &= \tfrac{1}{2}Tr[(\{\rho^\alpha, H_0\})(\{\rho^{1-\alpha}, H_0\})] \\
&= Tr(\rho H_0^2) + Tr(\rho^\alpha H_0 \rho^{1-\alpha} H_0) \qquad \alpha \in [0, 1]
\end{aligned}
\tag{2}
$$

where $H_0 = H - Tr(\rho H)I$ with I being the identity operator.

For a quantum state ρ and observables R, S and $0 \le \alpha \le 1$, the following inequality holds [35]:

$$
U_\alpha(\rho, R)U_\alpha(\rho, S) \ge \alpha(1 - \alpha)|Tr\rho[R, S]|^2
\tag{3}
$$

where $U_\alpha(\rho, R) = \sqrt{I_\alpha(\rho, R)J_\alpha(\rho, R)}$ can be regarded as a kind of measure for quantum uncertainty, in the sense given by [35]. For a pure state, a standard deviation-based relation is recovered from Equation (3). When $\alpha = \frac{1}{2}$, it is reduced to the result of [36].

Inspired by the works [32,35], in this paper, we study the uncertainty relations based on Wigner–Yanase–Dyson skew information in the presence of quantum memory, which generalize the results in [32] to the case of Wigner–Yanase–Dyson skew information, and the results in [35], which generalize to the case with the presence of quantum memory. We present uncertainty inequalities both in product and summation forms, and show that the lower bounds contain two terms: one concerns the compatibility of two measurement observables, and the other concerns the quantum correlations between the measured system and the quantum memory. We compare the lower bounds for product, separable and entangled states by detailed examples.

2. Results

Let $\phi_k = |\phi_k\rangle\langle\phi_k|$ and $\psi_k = |\psi_k\rangle\langle\psi_k|$ be the rank 1 spectral projectors of two non-degenerate observables R and S with the eigenvectors $|\phi_k\rangle$ and $|\psi_k\rangle$, respectively. Similarly to [32], we define $UN_\alpha(\rho, \phi) = \sum_k U_\alpha(\rho, \phi_k) = \sum_k \sqrt{I_\alpha(\rho, \phi_k)J_\alpha(\rho, \phi_k)}$ as the uncertainty of ρ associated to the projective measurement $\{\phi_k\}$, and $U_\alpha(\rho, \psi)$ to $\{\psi_k\}$.

Let ρ_{AB} be a bipartite state on $H_A \otimes H_B$, where H_A and H_B denote the Hilbert space of subsystems A and B, respectively. Let V be any orthogonal basis space on H_A and $|\phi_k\rangle$ be an orthogonal basis of H_A. We define a quantum correlation of ρ_{AB} as

$$\tilde{D}_\alpha(\rho_{AB}) = \min_V \sum_k [I_\alpha(\rho_{AB}, \phi_k \otimes I_B) - I_\alpha(\rho_A, \phi_k)] \tag{4}$$

where the minimum is taken over all the orthogonal bases on H_A, $\rho_A = Tr_B \rho_{AB}$.

For any bipartite state ρ_{AB} and any observable X_A on H_A, we have $I_\alpha(\rho_{AB}, X_A \otimes I_B) \geq I_\alpha(\rho_A, X_A)$, which follows from Corollary 1.3 in [34] and Lemma 2 in [37]. Therefore, $\tilde{D}_\alpha(\rho_{AB}) \geq 0$. Furthermore, $\tilde{D}_\alpha(\rho_{AB}) = 0$ when ρ_{AB} is a classical quantum correlated state, which follows from the proof in Theorem 1 of [38]. $\tilde{D}_\alpha(\rho_{AB})$ has a measurement on subsystem A, which gives an explicit physical meaning: it is the minimal difference of incompatibility of the projective measurement on the bipartite state ρ_{AB} and on the local reduced state ρ_A. $\tilde{D}_\alpha(\rho_{AB})$ quantifies the quantum correlations between the subsystems A and B. We have the following.

Theorem 1. *Let ρ_{AB} be a bipartite quantum state on $H_A \otimes H_B$ and $\{\phi_k\}$ and $\{\psi_k\}$ be two sets of rank 1 projective measurements on H_A. Then*

$$UN_\alpha(\rho_{AB}, \phi \otimes I)UN_\alpha(\rho_{AB}, \psi \otimes I) \geq \sum_k L^2_{\alpha,\rho_A}(\phi_k, \psi_k) + \tilde{D}^2_\alpha(\rho_{AB}) \tag{5}$$

where $L_{\alpha,\rho_A}(\phi_k, \psi_k) = \alpha(1-\alpha)\dfrac{|Tr\rho_A[\phi_k,\psi_k]|^2}{\sqrt{J_\alpha(\rho_A,\phi_k)\cdot J_\alpha(\rho_A,\psi_k)}}$.

Proof of Theorem 1. By definition, we have

$$
\begin{aligned}
&UN_\alpha(\rho_{AB}, \phi \otimes I)UN_\alpha(\rho_{AB}, \psi \otimes I) \\
&= \sum_k \sqrt{I_\alpha(\rho_{AB}, \phi_k \otimes I) \cdot J_\alpha(\rho_{AB}, \phi_k \otimes I)} \cdot \sum_k \sqrt{I_\alpha(\rho_{AB}, \psi_k \otimes I) \cdot J_\alpha(\rho_{AB}, \psi_k \otimes I)} \\
&\geq \sum_k I_\alpha(\rho_{AB}, \phi_k \otimes I) \cdot \sum_k I_\alpha(\rho_{AB}, \psi_k \otimes I) \\
&= [\sum_k (I_\alpha(\rho_{AB}, \phi_k \otimes I) - I_\alpha(\rho_A, \phi_k)) + \sum_k I_\alpha(\rho_A, \phi_k)] \\
&\quad \cdot [\sum_k (I_\alpha(\rho_{AB}, \psi_k \otimes I) - I_\alpha(\rho_A, \psi_k)) + \sum_k I_\alpha(\rho_A, \psi_k)] \\
&\geq [\tilde{D}_\alpha(\rho_{AB}) + \sum_k I_\alpha(\rho_A, \phi_k)] \cdot [\tilde{D}_\alpha(\rho_{AB}) + \sum_k I_\alpha(\rho_A, \psi_k)] \\
&\geq \tilde{D}^2_\alpha(\rho_{AB}) + \sum_k I_\alpha(\rho_A, \phi_k) I_\alpha(\rho_A, \psi_k) \\
&\geq \tilde{D}^2_\alpha(\rho_{AB}) + \sum_k \frac{\alpha^2(1-\alpha)^2 |Tr\rho_A[\phi_k,\psi_k]|^4}{J_\alpha(\rho_A,\phi_k)J_\alpha(\rho_A,\psi_k)} \\
&\triangleq \tilde{D}^2_\alpha(\rho_{AB}) + \sum_k L^2_{\alpha,\rho_A}(\phi_k, \psi_k)
\end{aligned}
\tag{6}
$$

where the first inequality is due to $J_\alpha(\rho, H) \geq I_\alpha(\rho, H)$ [35], and the last inequality follows from Equation (3). $\square$

Theorem 1 gives a product form of the uncertainty relation. Comparing the results (Equation (3)) without quantum memory with those (Equation (5)) with quantum memory, one finds that if the observables A and B satisfy $[A, B] = 0$, the bound is trivial in Equation (3), while in Equation (5), even if the projective measurements ϕ_k and ψ_k satisfy $[\phi_k, \psi_k] = 0$, that is, $L_{\alpha,\rho_A}(\phi_k, \psi_k) = 0$, $\tilde{D}_\alpha(\rho_{AB})$ may still not be trivial because of correlations between the system and the quantum memory.

Corresponding to the product form of the uncertainty relation, we can also derive the sum form of the uncertainty relation:

Theorem 2. *Let ρ_{AB} be a quantum state on $H_A \otimes H_B$ and $\{\phi_k\}$ and $\{\psi_k\}$ be two sets of rank 1 projective measurements on H_A. Then*

$$UN_\alpha(\rho_{AB}, \phi \otimes I) + UN_\alpha(\rho_{AB}, \psi \otimes I) \geq 2 \sum_k L_{\alpha,\rho_A}(\phi_k, \psi_k) + 2\tilde{D}_\alpha(\rho_{AB}) \tag{7}$$

Proof of Theorem 2. By definition and taking into account the fact that $J_\alpha(\rho, H) \geq I_\alpha(\rho, H)$ [35], we have

$$UN_\alpha(\rho_{AB}, \phi \otimes I) + UN_\alpha(\rho_{AB}, \psi \otimes I)$$
$$= \sum_k \sqrt{I_\alpha(\rho_{AB}, \phi_k \otimes I) \cdot J_\alpha(\rho_{AB}, \phi_k \otimes I)} + \sum_k \sqrt{I_\alpha(\rho_{AB}, \psi_k \otimes I) \cdot J_\alpha(\rho_{AB}, \psi_k \otimes I)}$$
$$\geq \sum_k I_\alpha(\rho_{AB}, \phi_k \otimes I) + \sum_k I_\alpha(\rho_{AB}, \psi_k \otimes I)$$

While

$$\sum_k I_\alpha(\rho_{AB}, \phi_k \otimes I) + \sum_k I_\alpha(\rho_{AB}, \psi_k \otimes I)$$
$$= \sum_k I_\alpha(\rho_A, \phi_k) + \sum_k I_\alpha(\rho_A, \psi_k) + \sum_k [I_\alpha(\rho_{AB}, \phi_k \otimes I) - I_\alpha(\rho_A, \phi_k)]$$
$$+ \sum_k [I_\alpha(\rho_{AB}, \psi_k \otimes I) - I_\alpha(\rho_A, \psi_k)]$$
$$\geq \sum_k I_\alpha(\rho_A, \phi_k) + \sum_k I_\alpha(\rho_A, \psi_k) + 2\tilde{D}_\alpha(\rho_{AB})$$

where the inequality follows from Equation (4). By using the inequality $a + b \geq 2\sqrt{ab}$ for positive $a = I_\alpha(\rho_A, \phi_k)$ and $b = I_\alpha(\rho_A, \psi_k)$, we further obtain

$$UN_\alpha(\rho_{AB}, \phi \otimes I) + UN_\alpha(\rho_{AB}, \psi \otimes I)$$
$$\geq 2 \sum_k \sqrt{I_\alpha(\rho_A, \phi_k) \cdot I_\alpha(\rho_A, \psi_k)} + 2\tilde{D}_\alpha(\rho_{AB})$$
$$\geq 2 \sum_k \alpha(1 - \alpha) \frac{|Tr\rho_A[\phi_k, \psi_k]|^2}{\sqrt{J_\alpha(\rho_A, \phi_k) \cdot J_\alpha(\rho_A, \psi_k)}} + 2\tilde{D}_\alpha(\rho_{AB}) \tag{8}$$
$$\triangleq 2 \sum_k L_{\alpha,\rho_A}(\phi_k, \psi_k) + 2\tilde{D}_\alpha(\rho_{AB})$$

where the second inequality follows from Equation (3). □

We note that Equation (7) reduces to an inequality that agrees with the result of [32] when $\alpha = \frac{1}{2}$. Theorem 2 is a generalization of the theorem in [32].

From Theorems 1 and 2, we obtain uncertainty relations in the form of the product and sum of skew information, which are different from the uncertainty of [39], which only deals with the single partite state. However, we treat the bipartite case with quantum memory B. It is shown that the lower bound contains two terms: one is the quantum correlation $\tilde{D}_\alpha(\rho_{AB})$, and the other is $\sum_k L_{\alpha,\rho_A}(\phi_k, \psi_k)$, which characterizes the degree of compatibility of the two measurements, just as for the meaning of $\log_2 \frac{1}{c}$ in the entropy uncertainty relation [15].

Example 1. *We consider the 2-qubit Werner state $\rho = \frac{2-p}{6}I + \frac{2p-1}{6}V$, where $p \in [-1, 1]$ and $V = \sum_{kl} |kl\rangle\langle lk|$.*

Let the Pauli matrices σ_x and σ_z be the two observables and $\{|\psi_k\rangle\}$ and $\{|\varphi_k\rangle\}$ be the eigenvectors of σ_x and σ_z, respectively, which satisfy $|\langle\psi_i|\varphi_j\rangle|^2 = \frac{1}{2}$, $i, j = 1, 2$. For all k, we have $Tr\rho_A[\psi_k, \varphi_k] = 0$, that is, $L_{\alpha,\rho_A}(\psi_k, \varphi_k) = 0$. The values of the left- and right-hand sides of Equation (5) are given by

$$4(\frac{2-p}{12} - \frac{(3-3p)^\alpha(1+p)^{1-\alpha}+(1+p)^\alpha(3-3p)^{1-\alpha}}{24})$$
$$\times (\frac{4+p}{12} + \frac{(3-3p)^\alpha(1+p)^{1-\alpha}+(1+p)^\alpha(3-3p)^{1-\alpha}}{24})$$

and

$$(\frac{2-p}{6} - \frac{(3-3p)^\alpha(1+p)^{1-\alpha}+(1+p)^\alpha(3-3p)^{1-\alpha}}{12})^2$$

respectively; see Figure 1a for the uncertainty relations with different values of α.

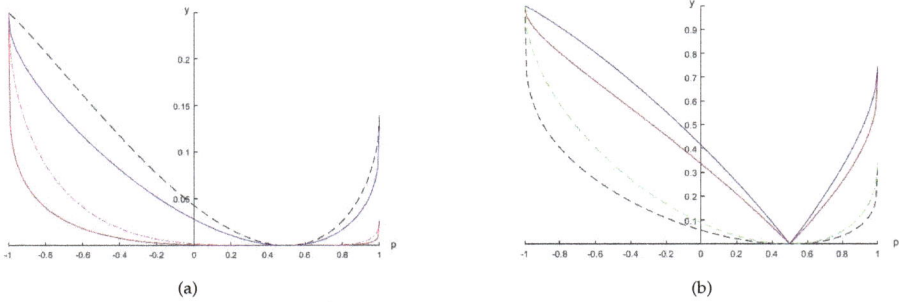

Figure 1. The y-axis shows the uncertainty and its lower bounds. (**a**) Blue (red) solid line for the value of the left (right)-hand side of Equation (5) with $\alpha = 0.2$; black dotted (red dot-dashed) line represents the value of the left (right)-hand side of Equation (5) with $\alpha = 0.5$. (**b**) Red solid (black dotted) line represents the value of the left (right)-hand side of Equation (7) with $\alpha = 0.2$; blue solid (green dotted) line represents the value of the left (right)-hand side of Equation (7) with $\alpha = 0.5$, which corresponds to Figure 1 in [32].

Similarly, we can obtain the values of the left- and right-hand sides of Equation (7):

$$4\sqrt{(\frac{2-p}{12} - \frac{(3-3p)^\alpha(1+p)^{1-\alpha}+(1+p)^\alpha(3-3p)^{1-\alpha}}{24})}$$
$$\times \sqrt{(\frac{4+p}{12} + \frac{(3-3p)^\alpha(1+p)^{1-\alpha}+(1+p)^\alpha(3-3p)^{1-\alpha}}{24})}$$

and

$$\frac{2-p}{3} - \frac{(3-3p)^\alpha(1+p)^{1-\alpha}+(1+p)^\alpha(3-3p)^{1-\alpha}}{6}$$

respectively; see Figure 1b.

Here we see explicitly that, just as for the Shannon entropy, Rényi entropy, Tsallis entropy, (h, Φ) entropies and Wigner–Yanase skew information, the Wigner–Yanase–Dyson skew information characterizes a special kind of information of a system or measurement outcomes, which needs to satisfy certain restrictions for given measurements and correlations between the system and the memory. Different α parameter values give rise to different kinds of information. From Figure 1, we see that for a given state and measurements, the differences between the left- and right-hand sides of the inequalities given by Equation (5) or (7) vary with the parameter α. Moreover, the degree of compatibility of the two measurements, $L_{\alpha,\rho_A}(\phi_k, \psi_k)$, vanishes for $\alpha = 0$ or 1, which is a fact in accordance with Equation (3), the case without quantum memory. For $p = 1/2$, the state ρ is maximally mixed. In this case, both sides of the inequalities given by Equations (5) and (7) vanish for any α.

Example 2. *Consider a separable bipartite state, $\rho^{AB} = \frac{1}{2}[|+\rangle\langle+| \otimes |0\rangle\langle0| + |-\rangle\langle-| \otimes |1\rangle\langle1|]$, where $|+\rangle = \frac{1}{\sqrt{2}}(|0\rangle + |1\rangle)$, $|-\rangle = \frac{1}{\sqrt{2}}(|0\rangle - |1\rangle)$.*

We still choose σ_x and σ_z to be the two observables. By calculation we obtain the following: For product states $|+\rangle\langle+| \otimes |0\rangle\langle0|$ and $|-\rangle\langle-| \otimes |1\rangle\langle1|$, both the left- and right-hand sides of Equation (5) are zero, and the right-hand side of Equation (7) is zero. For the separable bipartite state ρ^{AB}, the left- and right-hand sides of Equation (5) are $\frac{1}{2}$ and 0, respectively. Both the left- and right-hand sides of Equation (7) are zero.

Example 3. *For the Werner state $\rho_w^{AB} = (1-p)\frac{I}{4} + p|\varphi\rangle\langle\varphi|$, where $|\varphi\rangle = \frac{1}{\sqrt{2}}(|00\rangle + |11\rangle)$ is the Bell state, $p \in [0,1]$, and the state is separable when $p \leq \frac{1}{3}$.*

We have the values of the left- and right-hand sides of Equation (5), respectively:

$$4\left(\frac{1+p}{8} - \frac{(1-p)^\alpha(1+3p)^{1-\alpha} + (1-p)^{1-\alpha}(1+3p)^\alpha}{16}\right)$$
$$\times \left(\frac{3-p}{8} + \frac{(1-p)^\alpha(1+3p)^{1-\alpha} + (1-p)^{1-\alpha}(1+3p)^\alpha}{16}\right)$$

and

$$4\left(\frac{1+p}{8} - \frac{(1-p)^\alpha(1+3p)^{1-\alpha} + (1-p)^{1-\alpha}(1+3p)^\alpha}{16}\right)^2$$

See Figure 2a for a comparison with different values of α.

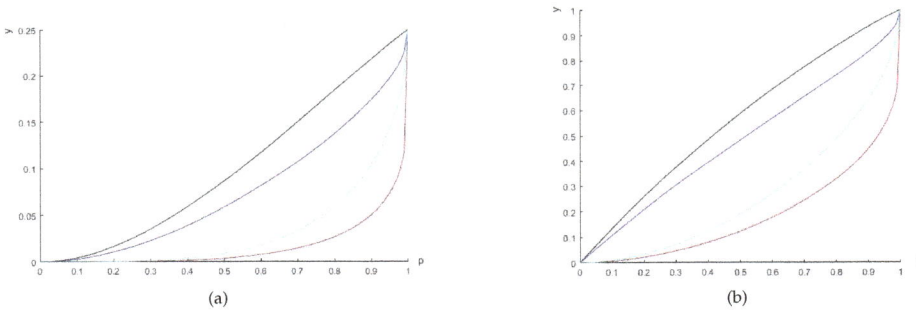

(a)

(b)

Figure 2. The y-axis shows the uncertainty and the lower bounds. (**a**) Blue (red) solid line is the value of the left (right)-hand side of Equation (5) for $\alpha = 0.2$; black (blue-green) solid line represents the value of the left (right)-hand side of Equation (5) for $\alpha = 0.5$. (**b**) Blue (red) solid line represents value of the left (right)-hand side of Equation (7) for $\alpha = 0.2$; black (blue-green) solid line represents the value of the left (right)-hand side of Equation (7) for $\alpha = 0.5$.

We can also obtain the values of the left- and right-hand sides of Equation (7):

$$4\sqrt{\frac{1+p}{8} - \frac{(1-p)^\alpha(1+3p)^{1-\alpha} + (1-p)^{1-\alpha}(1+3p)^\alpha}{16}}$$
$$\times \sqrt{\frac{3-p}{8} + \frac{(1-p)^\alpha(1+3p)^{1-\alpha} + (1-p)^{1-\alpha}(1+3p)^\alpha}{16}}$$

and

$$\frac{1+p}{2} - \frac{(1-p)^\alpha(1+3p)^{1-\alpha} + (1-p)^{1-\alpha}(1+3p)^\alpha}{4}$$

respectively; see Figure 2b.

Moreover, when ρ_w^{AB} is separable, namely, $p \leq \frac{1}{3}$, the differences between the left- and right-hand sides of the inequalities are smaller than those of the entangled states. Figure 3 shows the differences for different values of p.

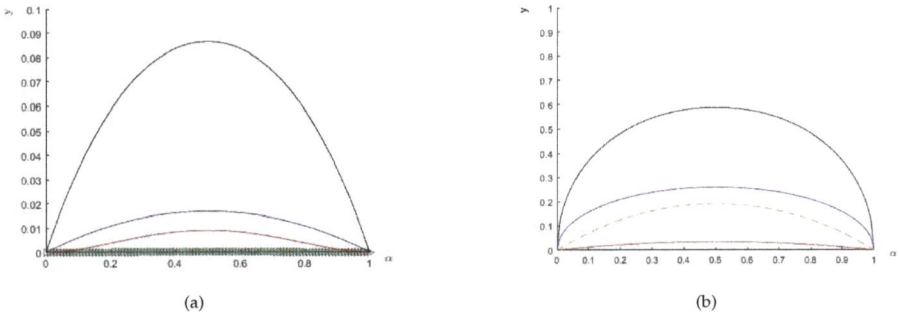

(a)   (b)

Figure 3. The *y*-axis shows the uncertainty and its lower bound; (**a**) $p = 0.2$ (ρ_w^{AB} is a separable state): blue solid line represents the value of the left-hand side of Equation (5), and the line (very near the *x*-axis) marked by triangles represents the corresponding lower bound; $p = 0.5$ (ρ_w^{AB} is an entangled state): the black (red) solid line represents the value of the left (right)-hand side of Equation (5). (**b**) Blue (red) solid line represents the value of the left (right)-hand side of Equation (7) for $p = 0.2$; black solid (red dashed) line represents the value of the left (right)-hand side of Equation (7) for $p = 0.5$.

3. Conclusions

We have investigated the uncertainty relations both in product and summation forms in terms of the Wigner–Yanase–Dyson skew information with quantum memory. It has been shown that the lower bounds contain two terms: one is the quantum correlation $\tilde{D}_\alpha(\rho_{AB})$, and the other is $\sum_k L_{\alpha,\rho_A}(\phi_k, \psi_k)$, which characterizes the degree of compatibility of the two measurements. By detailed examples, we have compared the lower bounds for product, separable and entangled states.

Acknowledgments: This work is supported by the NSF of China under Grant No. 11675113.

Author Contributions: Shao-Ming Fei guided the research; Jun Li finished the calculations and wrote the paper. Both authors have read and approved the final manuscript.

Conflicts of Interest: The authors declare no conflict of interest.

References

1. Heisenberg, W. Über den anschaulichen Inhalt der quantentheoretischen Kinematik und Mechanik. *Z. Phys.* **1927**, *43*, 172–198. (In German)
2. Robertson, H.P. The Uncertainty Principle. *Science* **1929**, *73*, 653.
3. Schrödinger, E. Zum Heisenbergschen Unschärfeprinzip. *Sitzungsberichte Akad. Berl.* **1930**, *19*, 296–303. (In German)
4. Busch, P.; Lahti, P.; Werner, R.F. Proof of Heisenberg's Error-Disturbance Relation. *Phys. Rev. Lett.* **2013**, *111*, 160405.
5. Busch, P.; Lahti, P.; Werner, R.F. Heisenberg uncertainty for qubit measurements. *Phys. Rev. A* **2014**, *89*, 012129.
6. Sulyok, G.; Sponar, S.; Demirel, B.; Buscemi, F.; Hall, M.J.W.; Ozawa, M.; Hasegawa, Y. Experimental test of entropic noise-disturbance uncertainty relations for spin-1/2 measurements. *Phys. Rev. Lett.* **2015**, *115*, 030401.

7. Ma, W.; Ma, Z.; Wang, H.; Liu, Y.; Chen, Z.; Kong, F.; Li, Z.; Shi, M.; Shi, F.; Fei, S.-M.; et al. Experimental demonstration of Heisenberg's measurement uncertainty relation based on statistical distances. *Phys. Rev. Lett.* **2016**, *116*, 160405.

8. Puchała, Z.; Rudnicki, Ł.; Zyczkowski, K.J. Majorization entropic uncertainty relations. *Phys. A Math. Theor.* **2013**, *46*, 272002.

9. Friedland, S.; Gheorghiu, V.; Gour, G. Universal Uncertainty Relations. *Phys. Rev. Lett.* **2013**, *111*, 230401.

10. Maccone, L.; Pati, A.K. Stronger Uncertainty Relations for All Incompatible Observables. *Phys. Rev. Lett.* **2014**, *113*, 260401.

11. Ma, W.C.; Chen, B.; Liu, Y.; Wang, M.Q.; Ye, X.Y.; Kong, F.; Shi, F.Z.; Fei, S.M.; Du, J.F. Experimental Demonstration of Uncertainty Relations for the Triple Components of Angular Momentum. *Phys. Rev. Lett.* **2014**, *118*, 180402.

12. Coles, P.J.; Berta, M.; Tomamichel, M.; Wehner, S. Entropic uncertainty relations and their applications. *Rev. Mod. Phys.* **2017**, *89*, 015002.

13. Deutsch, D. Uncertainty in Quantum Measurements. *Phys. Rev. Lett.* **1983**, *50*, 631.

14. Maassen, H.; Uffink, J.B.M. Generalized entropic uncertainty relations. *Phys. Rev. Lett.* **1988**, *60*, 1103.

15. Berta, M.; Christandl, M.; Colbeck, R.; Renes, J.; Renner, R. The uncertainty principle in the presence of quantum memory. *Nat. Phys.* **2010**, *6*, 659–662.

16. Coles, P.J.; Piani, M. Improved entropic uncertainty relations and information exclusion relations. *Phys. Rev. A* **2014**, *89*, 022112.

17. Rudnicki, Ł.; Puchała, Z.; Życzkowski, K. Strong majorization entropic uncertainty relations. *Phys. Rev. A* **2014**, *89*, 052115.

18. Xiao, Y.; Jing, N.; Fei, S.M.; Li-Jost, X.J. Improved uncertainty relation in the presence of quantum memory. *Phys. A Math. Theor.* **2016**, *49*, 49LT01.

19. Romera, E.; de los Santos, F. Fractional revivals through Rényi uncertainty relations. *Phys. Rev. A* **2008**, *78*, 013837.

20. Bialynicki-Birula, I. Formulation of uncertainty relations in terms of the Rényi entropies. *Phys. Rev. A* **2012**, *74*, 052101.

21. Zhang, J.; Zhang, Y.; Yu, C.S. Rényi entropy uncertainty relation for successive projective measurements. *Quantum Inf. Process.* **2015**, *14*, 2239–2253.

22. Rastegin, A.E. Uncertainty and certainty relations for complementary qubit observables in terms of Tsallis' entropies. *Quantum Inf. Process.* **2013**, *12*, 2947–2963.

23. Rastegin, A.E. Uncertainty and Certainty Relations for Successive Projective Measurements of a Qubit in Terms of Tsallis' Entropies. *Commun. Theor. Phys.* **2015**, *63*, 687–694.

24. Kurzyk, D.; Pawela, L.; Puchała, Z. Conditional entropic uncertainty relations for Tsallis entropies. *arXiv* **2017**, arXiv:1707.09278.

25. Zozor, S.; Bosyk, G.M.; Portesi, M. General entropy-like uncertainty relations in finite dimensions. *J. Phys. A Math. Theor.* **2014**, *47*, 495302.

26. Prevedel, R.; Hamel, D.R.; Colbeck, R.; Fisher, K.; Resch, K.J. Experimental investigation of the uncertainty principle in the presence of quantum memory and its application to witnessing entanglement. *Nat. Phys.* **2011**, *7*, 757–761.

27. Hofmann, H.F.; Takeuchi, S. Violation of local uncertainty relations as a signature of entanglement. *Phys. Rev. A* **2003**, *68*, 032103.

28. Walls, D.F.; Zoller, P. Reduced Quantum Fluctuations in Resonance Fluorescence. *Phys. Rev. Lett.* **1981**, *47*, 709–711.

29. Ma, J.; Wang, X.; Sun, C.P.; Nori, F. Quantum spin squeezing. *Phys. Rep.* **2011**, *509*, 89–165.

30. Giovannetti, V.; Lloyd, S.; Maccone, L. Quantum-Enhanced Measurements: Beating the Standard Quantum Limit. *Science* **2004**, *306*, 1330–1336.

31. Giovannetti, V.; Lloyd, S.; Maccone, L. Advances in quantum metrology. *Nat. Photonics* **2011**, *5*, 222–229.

32. Ma, Z.H.; Chen, Z.H.; Fei, S.M. Uncertainty relations based on skew information with quantum memory. *Sci. China Phys. Mech. Astron.* **2017**, *60*, 010321.

33. Wigner, E.P.; Yanase, M.M. Information contents of distributions. *Proc. Natl. Acad. Sci. USA* **1963**, *49*, 910–918.

34. Lieb, E.H. Convex trace functions and the Wigner-Yanase-Dyson conjecture. *Adv. Math.* **1973**, *11*, 267–288.

35. Yanagi, K. Uncertainty relation on Wigner–Yanase–Dyson skew information. *J. Math. Anal. Appl.* **2010**, *365*, 12–18.
36. Luo, S. Heisenberg uncertainty relation for mixed states. *Phys. Rev. A* **2005**, *72*, 042110.
37. Luo, S. Notes on Superadditivity of Wigner–Yanase–Dyson Information. *J. Stat. Phys.* **2007**, *128*, 1177–1188.
38. Luo, S.; Fu, S.; Oh, C.H. Quantifying correlations via the Wigner-Yanase skew information. *Phys. Rev. A* **2012**, *85*, 32117.
39. Chen, B.; Fei, S.M. Sum uncertainty relations for arbitrary N incompatible observables. *Sci. Rep.* **2015**, *5*, 14238.

entropy

MDPI

Review

Uncertainty Relations for Coarse–Grained Measurements: An Overview

Fabricio Toscano [1,*], Daniel S. Tasca [2], Łukasz Rudnicki [3,4] and Stephen P. Walborn [1]

[1]   Instituto de Física, Universidade Federal do Rio de Janeiro, Caixa Postal 68528, Rio de Janeiro 21941-972, Brazil; swalborn@if.ufrj.br

[2]   Instituto de Física, Universidade Federal Fluminense, Niteroi 24210-346, Brazil; dan.tasca@gmail.com

[3]   Max-Planck-Institut für die Physik des Lichts, Staudtstraße 2, Erlangen 91058, Germany; rudnicki@cft.edu.pl

[4]   Center for Theoretical Physics, Polish Academy of Sciences, Aleja Lotników 32/46, Warsaw 02-668, Poland

*   Correspondence: toscano@if.ufrj.br; Tel.: +55-21-3938-7477

Received: 30 April 2018; Accepted: 6 June 2018; Published: 10 June 2018

Abstract: Uncertainty relations involving incompatible observables are one of the cornerstones of quantum mechanics. Aside from their fundamental significance, they play an important role in practical applications, such as detection of quantum correlations and security requirements in quantum cryptography. In continuous variable systems, the spectra of the relevant observables form a continuum and this necessitates the coarse graining of measurements. However, these coarse-grained observables do not necessarily obey the same uncertainty relations as the original ones, a fact that can lead to false results when considering applications. That is, one cannot naively replace the original observables in the uncertainty relation for the coarse-grained observables and expect consistent results. As such, several uncertainty relations that are specifically designed for coarse-grained observables have been developed. In recognition of the 90th anniversary of the seminal Heisenberg uncertainty relation, celebrated last year, and all the subsequent work since then, here we give a review of the state of the art of coarse-grained uncertainty relations in continuous variable quantum systems, as well as their applications to fundamental quantum physics and quantum information tasks. Our review is meant to be balanced in its content, since both theoretical considerations and experimental perspectives are put on an equal footing.

Keywords: quantum uncertainty; quantum foundations; quantum information; continuous variables

---

1. Introduction

The physics of classical waves distinguishes itself from that of a classical point particle in several ways. Waves are spread-out packets of energy moving through a medium, while a particle is localized and follows a well-defined trajectory. It was thus most surprising when it was discovered in the early 20th century that quantum objects, such as electrons and atoms, could exhibit behavior that at times was best described according to wave mechanics. Moreover, it was shown that either wave or particle behavior could be observed depending almost entirely upon how an observer chooses to measure the system. This complementarity of wave and particle behavior played a key role in the early debates concerning the validity of quantum theory [1], and has been linked to several interesting and fundamental phenomena of quantum physics [2–5]. Though several complementarity relations have been cast in quantitative forms [6,7], perhaps complementarity is most frequently observed in terms of quantum uncertainty relations. In words, uncertainty relations establish the fact that the intrinsic uncertainties associated to measurement outcomes of two incompatible observations of a quantum system can never both be arbitrarily small. We note that this type of behavior appears in classical wave mechanics, for example in the form of time-bandwidth uncertainty relations, which are quite important in communications and signal processing [8]. In contrast, there is no aspect of a classical physics that prohibits us from measuring all of the relevant properties of a classical point particle, at least in principle.

In addition to quantum fundamentals, quantum uncertainty relations play an important role in several interesting tasks associated to quantum information protocols, such as the detection of quantum correlations and the security of quantum cryptography [9]. In this paper, we focus on continuous variable (CV) quantum systems [10,11]. Though many interesting results have been found for discrete systems, they are outside the scope of this manuscript. We refer the interested reader to Reference [9], being a comprehensive unification and extension of two older reviews on entropic uncertainty relations, more focused on the physical [12] and information-theoretic [13] side respectively. However, since the coarse-grained scenario situates itself somehow in-between the discrete and continuous description, we make a short introduction to discrete entropic uncertainty relations before discussing their coarse-grained relatives.

In CV systems, one encounters a fundamental problem when performing measurements. That is, the eigenspectra of the corresponding observables are infinite dimensional, and can be continuous or discrete. Since any measurement device registers measurement outcomes with a finite precision and within a finite range of values, the experimental assessment of CV observables can be quite different from theory. Of course, one can consider a truncation of the relevant Hilbert space [14], as well as some type of binning or coarse graining of the measurement outcomes. This is similar to the idea of coarse graining that was discussed by Gibbs [15] and used by Paul and Tanya Ehrenfest [16,17] in the early 20th century to account for imprecise knowledge of dynamical variables in statistical mechanics [18]. Coarse graining has also appeared in the quantum mechanical context as an attempt to describe the quantum-to-classical transition, where the idea is that measurement imprecision could be responsible for the disappearance of quantum properties [19–23]. Though this is quite an intuitive notion, it was recently shown that one can always find an uncertainty relation that is satisfied non-trivially for any amount of coarse graining [24]. That is, quantum mechanical uncertainty is always present in this type of "classical" limit. This motivates the formulation of coarse-grained uncertainty relations.

In addition to the necessity of coarse graining, there could be practical advantages: for tasks such as entanglement detection, it might be interesting to perform as few measurements as possible, advocating the use of coarse-grained measurements. However, improper handling of coarse graining can result in false detections of entanglement [25,26], pseudo-violation of Bell's inequalities or the Tsirelson bound [27,28], and sacrifice security in quantum key distribution [29], for example. Thus, the proper formulation and application of uncertainty relations for coarse-grained observables is both interesting and necessary.

In the present contribution we review the current state of the art of uncertainty relations (URs) for coarse-grained observables in continuous-variable quantum systems. In Section 2 we review the concept of uncertainty of continuous variable (CV) quantum systems in more depth and introduce several prominent URs. In Section 3 we discuss the utility of CV URs in quantum physics and quantum information, in particular for identifying non-classical states and quantum correlations. Section 4 presents the problem of coarse graining of CVs in detail, and two coarse-graining models are provided. The current status of URs for these coarse-graining models is reviewed in Section 5, where we present a series of coarse-grained URs previously reported in the literature [12,24,30–32]. In addition, we extend the validity of some of these URs to general linear combinations of canonical observables. Section 6 is devoted to the experimental investigation and application of coarse-grained URs in quantum physics and quantum information. Concluding remarks are provided in Section 7.

2. Uncertainty Relations

The history of uncertainty relations traces back to the early days of the formalization of quantum theory and begins with the celebrated work by Heisenberg in 1927 [33] (see [1] for an English version).

The work discussed what later became known as Heisenberg's uncertainty principle. The first mathematical formulation for this principle, in [33], essentially reads:

$$\Delta x \Delta p \gtrsim h \qquad (1)$$

where Δx and Δp are the uncertainties of the position and linear momentum of a particle, respectively, and h is the Planck constant. Although the existence of such a principle is ultimately due to the non-commutativity of the position and momentum observables, it took almost 80 years for all the physical meanings, scope and validity of this principle to be elucidated [34]. Distinct physical meanings emerge from different definitions for "uncertainty" of position or momentum, and in each case a proper multiplicative constant makes the lower bound sharp. All of these inequalities are known by the generic name of *Uncertainty Relations*, from the beginning of this review referred to as URs. Even though the inception of the URs was made in the context of position and momentum of a particle, their existence can be extended to the "uncertainties" associated with any pair of non-commuting observables in discrete or continuous variable quantum systems. Thus, generically we can define the URs as inequalities that stem from the fact that the measured quantities involved are associated to non-commuting observables.

Nowadays, we can say that it is clear that there are three conceptually distinct types of URs [34]: *(i)* URs associated with the statistics of the measurement results of non-commuting observables after preparing the system repeatedly in the same quantum state, or *statistical* URs for short, *(ii)* the *error-disturbance* URs, also known as *noise-disturbance* URs, for the relation of the imprecision in the measurement of one observable and the corresponding disturbance in the other, and, *(iii)* the *joint measurement* URs associated with the precision of the joint measurements of non-commuting observables. The *error-disturbance* URs has two main contributions: one in References [35–37] that present state-dependent *error-disturbance* URs and the other in References [38–40] that argue for a state-independent characterisation of the overall performance of measuring devices as a measure of uncertainty that satisfies an UR of the form given in Equation (1). There was a certain controversy involving these two contributions and we recommend the work [41] that discusses the limitations of the state-dependent error-disturbance URs. The development of *joint measurement* URs has an early contribution in Reference [42] and further developments were given in References [38,43–48].

The *statistical* URs are also referred to in the literature as *preparation* URs. This is because it is impossible to prepare a quantum system in a state for which two non-commuting observables have sharply defined values. However, here we prefer to call them *statistical* URs, as they express the limits to the amount of information that can be obtained about incompatible observables of a quantum system when it is repeatedly measured after being prepared in the same initial state in each round of the measurement process. We emphasize that there is not any attempt to measure the two non-commuting observables simultaneously. In each round of the measurement process only one observable is measured, the choice of which could be made randomly. In this sense the "uncertainties" contained in the *statistical* URs are of the statistical type: the more certain the sequence of outcomes of one observable is in a given state, then the more uncertain is the sequence of outcomes of the other non-commuting observable(s) considered.

This review focuses on *statistical* URs that are valid for coarse-grained measurements in continuous variable quantum systems, although a similar approach can be made for the other two types of URs mentioned above. There are two types of quantum mechanical degrees of freedom: the ones that can be described by a Hilbert space of quantum states with finite dimension and the others in which it has infinite dimension. In particular, we are interested only in continuous variable (CV) systems where the Hilbert space, $\mathcal{H}$, of pure states, $|\psi\rangle$, has an infinite dimension. The CV systems that we consider consist of a finite set of n bosonic modes, sometimes called "qumodes" [10], so that $\mathcal{H} := \mathcal{H}_1 \otimes \ldots \otimes \mathcal{H}_n$. Each mode is described by a pair of canonically conjugate operators, $\hat{x}_j$ and $\hat{p}_j$, such that

$$[\hat{x}_j, \hat{p}_k] = i\hbar \hat{1} \delta_{jk}. \qquad (2)$$

Alternatively, each mode can be described by a pair of ladder operators, $\hat{a}_j := (1/\sqrt{2\hbar})(\hat{x}_j + i\hat{p}_j)$ and $\hat{a}_j^\dagger := (1/\sqrt{2\hbar})(\hat{x}_j - i\hat{p}_j)$, with $[\hat{a}_j, \hat{a}_k^\dagger] = \hat{1}\delta_{jk}$. Therefore, the separable Hilbert space of each mode, $\mathcal{H}_j$, has a enumerable basis $\{|n_j\rangle\}_{n_j=1,...,\infty}$ consisting of eigenstates of the number operator, *viz.* $\hat{n}_j|n_j\rangle = n_j|n_j\rangle$), evidencing the infinite dimensionality of the Hilbert space of the quantum states. In the case of mixed states we use density operators represented by greek letters with a hat, i.e., $\hat{\rho}$, $\hat{\sigma}$ etc.

Important examples of CV systems are the motional degrees of freedom of atoms, ions and molecules, where $\hat{x}_j$ and $\hat{p}_j$ are the components of the position and linear momentum of the particles (in this case $\hbar$ in Equation (2) is the usual reduced Planck constant, i.e., $\hbar = h/2\pi$); the quadrature modes of the quantized electromagnetic field where $\hat{x}_j$ and $\hat{p}_j$ are canonically conjugate quadratures (in this case $\hbar$ in Equation (2) is just $\hbar = 1$) [10]; and the transverse spatial degrees of freedom of single photons propagating in the paraxial approximation (in this case $\hbar$ in Equation (2) is $\hbar = \lambda/2\pi$ where λ is the photon's wave length [49]).

In what follows, we summarize the principal *statistical* URs in CV systems that have been generalised to coarse-grained measurements. The corresponding coarse-grained URs will be presented in Section 5.

2.1. Heisenberg (or Variance) Uncertainty Relation

Let us consider two operators:

$$\hat{u} := \mathbf{d}^T \hat{\mathbf{x}} = \mathbf{a}^T \hat{\mathbf{q}} + \mathbf{a}'^T \hat{\mathbf{p}} \quad \text{and} \quad \hat{v} := \mathbf{d}'^T \hat{\mathbf{x}} = \mathbf{b}^T \hat{\mathbf{q}} + \mathbf{b}'^T \hat{\mathbf{p}}, \tag{3}$$

where T means transposition and we define the $2n$-dimensional vector of operators,

$$\hat{\mathbf{x}} := (\hat{\mathbf{q}}, \hat{\mathbf{p}})^T = (\hat{x}_1, \dots, \hat{x}_n, \hat{p}_1, \dots, \hat{p}_n)^T, \tag{4}$$

as well as the arbitrary real vectors,

$$\mathbf{d} = (\mathbf{a}, \mathbf{a}')^T = (a_1, \dots, a_n, a_1', \dots, a_n')^T \quad \text{and} \quad \mathbf{d}' = (\mathbf{b}, \mathbf{b}')^T = (b_1, \dots, b_n, b_1', \dots, b_n')^T. \tag{5}$$

The commutation relation of $\hat{u}$ and $\hat{v}$ is

$$[\hat{u}, \hat{v}] = i\hbar \mathbf{d}^T \mathbf{J} \mathbf{d}' \hat{1} =: i\hbar\gamma\hat{1}, \tag{6}$$

where $\mathbf{J}$ is the $2n \times 2n$-dimensional matrix of the symplectic norm [50]:

$$\mathbf{J} = \begin{pmatrix} \mathbf{O} & \mathbf{I} \\ -\mathbf{I} & \mathbf{O} \end{pmatrix}, \tag{7}$$

and the $n \times n$ matrices in the blocks are the identity matrix $\mathbf{I}$ and the null matrix $\mathbf{O}$. In this review, matrices of an arbitrary shape not treated as quantum-mechanical operators are denoted in bold and without a hat.

The parameter γ in definition Equation (6) is a scalar that in some sense quantifies the non-commutativity of $\hat{u}$ and $\hat{v}$. Commutation relations such as Equation (6) are called Canonical Commutation Relations (CCR) (sometimes the name CCR is used in the case when $\gamma = 1$, however, as $\hbar\gamma$ can be interpreted as an effective Planck constant, so the name CCR here is well justified). However, a CCR between two operators $\hat{u}$ and $\hat{v}$ does not guarantee that they are necessarily Canonically Conjugate Operators (CCOs). For this to be true we additionally need that the eigenvectors of $\hat{u}$ and $\hat{v}$ must be connected by a Fourier Transform. In such a case we call $\hat{u}$ and $\hat{v}$ CCOs (also note that when two operators like the ones defined in Equation (3) have their eigenstates connected by a Fourier Transform, they necessary satisfy a commutation relation like in Equation (6), as can be easily shown. However the converse is not true. Take for example the single mode operators $\hat{u} = \hat{x}$ and $\hat{v} = \hat{x} + \hat{p}$, which satisfy $[\hat{u}, \hat{v}] = [\hat{x}, \hat{p}] = i\hbar$ but are not a Fourier pair).

Every pair of operators, $\hat{u}$ and $\hat{v}$, that obey a CCR also satisfies the *statistical* UR:

$$\sigma_{P_u}^2 \sigma_{P_v}^2 \geq \frac{\hbar^2}{4}\gamma^2, \tag{8}$$

where

$$\sigma_{P_u}^2 := \langle \hat{u}^2 \rangle - \langle \hat{u} \rangle^2, \quad \text{and} \quad \sigma_{P_v}^2 := \langle \hat{v}^2 \rangle - \langle \hat{v} \rangle^2, \tag{9}$$

are the variances of the marginal probability distribution functions (pdf):

$$P_u(u) = \langle |u\rangle\langle u| \rangle, \quad \text{and} \quad P_v(v) = \langle |v\rangle\langle v| \rangle, \tag{10}$$

where we have defined

$$\langle \ldots \rangle := \mathrm{Tr}(\ldots \hat{\rho}), \tag{11}$$

with $\hat{\rho}$ being an arbitrary $n-$mode quantum state. We call the UR in Equation (8) the *Heisenberg* UR, or variance-product UR. For one mode CCOs, such as $\hat{u} = \hat{x}$ and $\hat{v} = \hat{p}$ (therefore $\gamma = 1$), the *Heisenberg* UR in Equation (8) was first proved by Kennard in 1927 [51], inspired by the inequality in Equation (1) of Heisenberg's seminal paper of the same year [33]. Later, it was also proved by Weyl in 1928 [52]. In 1929 Robertson [53] extended the *Heisenberg* UR for any pair of Hermitian operators $\hat{A}$ and $\hat{B}$:

$$\sigma_{P_A}^2 \sigma_{P_B}^2 \geq \frac{1}{4} \left| \langle [\hat{A}, \hat{B}] \rangle \right|^2. \tag{12}$$

This result extends the *Heisenberg* UR in Equation (8) to $\hat{u}$ and $\hat{v}$ that are not CCOs.

For every variance-product UR in Equation (12) there is an associated linear UR:

$$\sigma_{P_A}^2 + \sigma_{P_B}^2 \geq \left| \langle [\hat{A}, \hat{B}] \rangle \right|. \tag{13}$$

In fact, this UR is a consequence of Equation (12) and the trivial inequality $(\sigma_{P_A} - \sigma_{P_B})^2 \geq 0$, so that

$$\sigma_{P_A}^2 + \sigma_{P_B}^2 \geq 2\sigma_{P_A}\sigma_{P_B} \geq \left| \langle [\hat{A}, \hat{B}] \rangle \right|, \tag{14}$$

where it also follows that the linear UR is weaker than the variance product UR. In 1930 Schrödinger [54] improved the lower bound in Equation (12), so the new stronger UR reads:

$$\sigma_{P_A}^2 \sigma_{P_B}^2 \geq \frac{1}{4} \left| \langle [\hat{A}, \hat{B}] \rangle \right|^2 + \frac{1}{4} \left| \langle \{ \hat{A} - \langle \hat{A} \rangle, \hat{B} - \langle \hat{B} \rangle \} \rangle \right|^2, \tag{15}$$

where $\{\cdots, \cdots\}$ is the anti-commutator.

One interesting property of the *Heisenberg* UR in Equation (8) is that the lower bound is independent of the quantum state $\hat{\rho}$ under consideration. Another property is that it can be seen as a *bona fide* condition on the covariance matrix of an $n-$mode quantum state $\hat{\rho}$, *viz* the matrix of second moments of the CCOs, contained in the vector $\hat{\mathbf{x}}$, of the state $\hat{\rho}$:

$$\mathbf{V} := \frac{\langle \hat{\mathbf{x}}\hat{\mathbf{x}}^T \rangle + \langle \hat{\mathbf{x}}\hat{\mathbf{x}}^T \rangle^T}{2} - \langle \hat{\mathbf{x}} \rangle \langle \hat{\mathbf{x}}^\mathbf{T} \rangle. \tag{16}$$

Indeed, in [55,56] it was shown that the *bona fide* condition on the covariance matrix $\mathbf{V}$ of a quantum state $\hat{\rho}$ is,

$$\mathbf{V} + \frac{i\hbar}{2}\mathbf{J} \geq 0, \tag{17}$$

where the inequality means that the matrix on the left hand side is positive semi-definite, *viz.* all of its eigenvalues are greater or equal to zero. Applying the inequality in Equation (15) to the canonical conjugate operators $\hat{x}$ and $\hat{p}$, we have,

$$\sqrt{\det(\mathbf{V})} = \sqrt{\sigma_{P_x}^2 \sigma_{P_p}^2 - \frac{1}{4} |\langle \{\hat{x} - \langle \hat{x} \rangle, \hat{p} - \langle \hat{p} \rangle \} \rangle|} \geq \frac{\hbar}{2}. \tag{18}$$

For one mode systems, this inequality is equivalent to the *bona fide* condition in Equation (17). However, for multimode systems it is not enough. For multimode systems, a way to verify the *bona fide* of the covariance matrix was given in [57,58]. It was shown that testing the condition in Equation (17) is equivalent to verify the linear UR in Equation (13) for all the operators, $\hat{u}$ and $\hat{v}$, defined in Equation (3). Therefore, using Equation (14) we can write the series of implications:

$$\sigma_{P_u}^2 \sigma_{P_v}^2 \geq \frac{\hbar^2}{4} \gamma^2 \;\Rightarrow\; \sigma_{P_u}^2 + \sigma_{P_v}^2 \geq \hbar|\gamma| \Leftrightarrow \mathbf{V} + \frac{i\hbar}{2}\mathbf{J} \geq 0. \tag{19}$$

Thus, it is enough to verify the violation of the *Heisenberg* UR for some pair of operators $\hat{u}$ and $\hat{v}$ to confirm that the *bona fide* condition on the covariance matrix of some $n-$mode operator $\hat{\rho}$ is not satisfied.

2.2. Entropic URs

The use of entropy functions to quantify uncertainty of a probabilistic variable dates back to the early work of Shannon [59]. Since then, several different entropy functions have been defined, with distinct relations to meaningful characteristics of the probability distributions considered. A number of these entropy functions have found use in quantum mechanics and, in particular, in QIT [9]. Here we outline the application of these functions to uncertainty relations between non-commuting observables.

2.2.1. Shannon-entropy UR

The UR based on the differential Shannon entropy for operators defined in Equation (3) is:

$$h[P_u] + h[P_v] \geq \ln(\pi e \hbar |\gamma|), \tag{20}$$

where P_u and P_v are the marginal pdf defined in Equation (10) and the differential Shannon entropy of a pdf, P, is defined as [60]:

$$h[P] := -\int_{-\infty}^{\infty} dy\, P(y) \ln P(y). \tag{21}$$

For $\hat{u}$ and $\hat{v}$ as CCOs, this uncertainty relation was first proved in 1975 by Bialynicki-Birula and Mycielski [61]. In their derivation the authors used the L_p-L_q norm inequality for the Fourier transform operator obtained by Beckner [62]. Please note that in the literature this inequality is sometimes referred to as the Babenko-Beckner inequality (Equation 1.104 from [12] provides an extension of this inequality to the case of arbitrary mixed states, using two variants of the Minkowski inequality), because Babenko [63] had proved it before Beckner, but only for certain combinations of (p,q) parameters. For the sake of completeness, we should also mention that Hirschman [64] had derived a weaker version of (20) with the constant $e\pi$ inside the logarithm replaced by 2π. The extension of the validity for operators $\hat{u}$ and $\hat{v}$ that are not CCOs was provided very recently in References [58,65].

The *Shannon-entropy* UR is in general stronger than the *Heisenberg* UR as the former implies the latter. This can be seen by using the inequality for a pdf P [60]:

$$\ln\left(2\pi e \sigma_P^2\right) \geq 2h[P], \tag{22}$$

where σ^2 is the variance of P. Therefore, we can write the chain of inequalities:

$$\ln(2\pi e \sigma_{P_u} \sigma_{P_v}) \geq h[P_u] + h[P_v] \geq \ln(\pi e \hbar |\gamma|), \tag{23}$$

that compress the URs in Equations (8) and (20). It is clear from Equation (23) that the verification of the *Shannon-entropy* UR for any pair of the operators in Equation (3) is enough to guarantee the *bona fide* condition in Equation (17) [58].

When the quantum state $\hat{\rho}$ is Gaussian, *viz* when the Wigner function of $\hat{\rho}$ is a multivariate Gaussian probability distribution [11], the marginal pdfs, P_u and P_v, are also Gaussians. Remembering that the differential Shannon entropy of a Gaussian pdf P, with variance σ_P^2, is $h[P] = (1/2)\ln(2\pi e \sigma_P^2)$ [60], we can see that Gaussian states saturate the first inequality in Equation (23). Therefore, for Gaussian states the *Heisenberg* UR and the *Shannon-entropy* UR are completely equivalent. As we will see in Section 5 this is not the case for the coarse-grained versions of these URs.

2.2.2. Rényi-Entropy URs

The UR based on the differential Rényi entropy for the operators defined in Equation (3) that are CCOs is given by the inequality:

$$h_\alpha[P_u] + h_\beta[P_v] \geq \ln\left(\frac{\pi \hbar |\gamma|}{\alpha^{\frac{1}{(2-2\alpha)}} \beta^{\frac{1}{(2-2\beta)}}} \right), \tag{24}$$

where $1/\alpha + 1/\beta = 2$ with $1/2 \leq \alpha \leq 1$ and $\gamma = 1$ since we deal with CCO operators. As before, P_u and P_v are the marginal pdfs defined in Equation (10) and the differential Rényi entropy of order α relevant for an arbitrary pdf, P, is defined as [60]:

$$h_\alpha[P] := \frac{1}{1-\alpha} \ln\left(\int_{-\infty}^{\infty} dy \; [P(y)]^\alpha \right). \tag{25}$$

The *Rényi-entropy* UR was proved recently (in 2006) by Bialynicki-Birula [31] (see also [12]) again with the help of the powerful mathematical tools developed in [62]. Please note that in the limit $\alpha \to 1$ we also have $\beta \to 1$, and consequently $\alpha^{\frac{1}{(2-2\alpha)}} \beta^{\frac{1}{(2-2\beta)}} \to 1/e$. Therefore, in the limit $\alpha \to 1$ we have $h_\alpha[P_u] \to h[P_u]$ and $h_\beta[P_v] \to h[P_v]$, so the expression in Equation (24) reduces to the *Shannon-entropy* UR in Equation (20) for $\gamma = 1$. As far as we know, in contrast to the *Shannon-entropy* UR, the extension of the *Rényi-entropy* UR to the general case of operators that are not necessarily CCOs is still a challenge for the future. A first attempt in this direction was provided in Reference [65], where the authors show that the *Rényi* UR in Equation (24) is still valid when the eigenvectors of $\hat{u}$ and $\hat{v}$ are connected by a Fractional Fourier Transform [8], which corresponds to rotation in phase space.

All of the URs mentioned in this section (this is a general pattern though) can be cast in a general form

$$F(\hat{\rho}; \hat{u}, \hat{v}; P_u, P_v) \geq f(\hbar |\gamma|), \tag{26}$$

where F is an uncertainty functional [left hand side of inequalities Equations (8), (20) and (24) for example] and f represents its respective lower bounds. In particular, we do not pay much attention here to the Tsallis entropy and URs associated with it. Again such URs can be cast in the general form stated above and their derivation is usually very similar in spirit to the case of the Rényi entropy.

In Section 3 we will summarise the relevance of the *statistical* UR in general and in particular the URs in Equations (8), (20) and (24). In Section 5 we will present versions of the *Heisenberg*, *Shannon-entropy* and *Rényi-entropy* URs for coarse-grained measurements.

3. Utility of Uncertainty Relations in Quantum Physics

Uncertainty relations can be applied in several useful and interesting ways. First, they provide a way to test if experimental results are consistent with quantum mechanics, since data from the measurement of incompatible observables must verify any valid quantum UR. This is particularly helpful in identifying systematic errors in the measurement process, in testing the experimental reconstruction of density matrices, phase-space distributions (quantum state tomography), as well as the covariance matrix [66], or any other set of moments of the CCOs of the modes.

URs can also be used to characterize non-classical states of light, such as squeezed states [67]. In this case observation of the variance $\sigma_{\hat{P}_u}^2 \leq \hbar/4$ where $\hat{u}$ is a phase-space quadrature in Equation (3), indicates noise fluctuations in this quadrature that are smaller than the vacuum state. As a consequence of the Heisenberg UR, the noise fluctuations in the conjugate quadrature must be larger or equal to $\hbar/4\sigma_{\hat{P}_u}^2$. In a similar fashion, in Reference [68] it was shown that violation of one out of an infinite hierarchy of inequalities involving normally ordered quadrature moments is sufficient to demonstrate non-classicality. We note that $\sigma_{\hat{P}_u}^2 \leq \hbar/4$ corresponds to the lowest-order inequality of this set. Related techniques have been developed based on the quantum version of Bochner's theorem for the existence of a positive semi-definite characteristic function [69,70]. Both of these methods have been used experimentally in Reference [71]. More recently, these two techniques were unified into a single criteria involving derivatives of the characteristic function [72], and put to test on a squeezed vacuum state.

To our knowledge, the first application of URs to identify quantum correlations was described in Reference [73], in which the authors proposed a Heisenberg-like UR, similar to that in Equation (8), to identify non-classical correlations between both the phases and intensities of the fields produced by a non-degenerate parametric oscillator. It was shown by M. Reid [74] that these measurements provide a method to demonstrate correlations for which the seminal Einstein-Podolsky-Rosen (EPR) argument [75] is valid. An experiment using this UR-based method to demonstrate EPR-correlations between light fields was realized shortly therafter [76]. It was later shown by Wiseman et al. [77,78] that the Reid EPR-criterion was indeed a method to identify quantum states that violate a "local hidden state" model of correlations. This type of correlation has been called "EPR-steering", or just "steering" [79], as this was the terminology used by Schrödinger when he discussed EPR correlations in 1935 [80]. Since 2007, EPR-steering has been understood to make up part of a hierarchy of quantum correlations, situated between entanglement [81,82] and Bell non-locality [83]. In addition to methods utilizing variance-based URs [84], entropic URs, such as those in Section 2.2, can be used to identify EPR-steering [85,86] and to quantify high-dimensional entanglement [87,88]. Some of these URs can be used to test security in continuous variable quantum cryptography [89,90], and it has been shown that violation of entropic EPR-steering criteria are directly related to the secret key rate in one-sided device independent cryptography [91]. We also highlight techniques based on a matrix-of-moments approach [92]. Continuous-variable EPR-steering has been observed in intense fields [76,93,94] as well as photon pairs [85,95–97].

Perhaps one of the most important tasks in quantum information is identifying quantum entanglement. In this respect, URs have also found widespread use in simple and experimentally friendly entanglement detection methods, as we will now describe. Several early entanglement criteria for bipartite CV systems were developed using URs [98–101]. A particularly convenient method to construct entanglement criteria is to use the Peres-Horedecki positive partial transposition argument [102,103] (PPT), and apply it to uncertainty relations [82,104–107]. The PPT argument is as follows. A bipartite separable state $\hat{\sigma}_{12}$ can be written as [108]

$$\hat{\sigma}_{12} = \sum_i \lambda_i \hat{\rho}_{1i} \otimes \hat{\rho}_{2i}, \tag{27}$$

where $\hat{\rho}_{1i}$ and $\hat{\rho}_{2i}$ are *bona fide* density operators of subsystems 1 and 2, respectively. The transpose of the state $\hat{\rho}_{2i}$, here denoted $\hat{\rho}_{2i}^T$, is still a positive operator, since full transposition preserves the eigenspectrum. Thus, partial transposition (with respect to second subsystem) of $\hat{\sigma}_{12}$ gives the valid quantum state:

$$\hat{\sigma}_{12}^{T_2} = \sum_i \lambda_i \hat{\rho}_{1i} \otimes \hat{\rho}_{2i}^T. \tag{28}$$

On the other hand, partial transposition of an entangled state $\hat{\varrho}_{12}$, which cannot be written in the form (27), can lead to a non-physical density matrix since partial transposition may not preserve the positivity of the eigenspectrum. Thus, one can identify entanglement in a bipartite density operator by calculating the partial transposition and searching for negative eigenvalues, and even quantify the amount of entanglement via the negativity [109]. However, applications of this method in experiments requires quantum state tomography and reconstruction of the density operator, which involves a large number of measurements. A more experimentally friendly method to identify entanglement is to evaluate an UR applied to the partial transposition of $\hat{\varrho}_{12}$, which we describe in the next paragraph. The PPT-argument is only a sufficient entanglement criteria in a general bipartition of $m \times (n - m)$ modes, but is necessary and sufficient in the particular case of bipartitions of $1 \times (n - 1)$ modes in CV Gaussian states [10,110]. Thus, there are no Gaussian states which are PPT entangled states in bipartitions of the form $1 \times (n - 1)$. However, there do exist entangled CV Gaussian states that are PPT in general bipartitions of the type $m \times (n - m)$. These are called bound entangled states [111]. In Gaussian states, this set of bound entangled states coincides with the set of all states whose entanglement in a bipartition $m \times (n - m)$ cannot be distilled using local operations and classical communication [112–114]. However, to our knowledge, for non-Gaussian states it is conjectured that the set of bound entangled states in a given bipartition is only a sub-set of the set of undistillable states in that bipartition.

For continuous variables, Simon showed that transposition is equivalent to a momentum reflection, taking the single mode Wigner phase-space distribution $\mathcal{W}(\mathbf{x}, \mathbf{p}) \longrightarrow \mathcal{W}^{T_2}(\mathbf{x}, \mathbf{p}) = \mathcal{W}(\mathbf{x}, \mathbf{T}\mathbf{p})$ [57], where $\mathbf{T}$ is a diagonal matrix whose elements are $+1$ for non-transposed modes, and -1 for the transposed ones. Thus, evaluating the "transposed" Wigner function is the same as evaluating the original Wigner function with a sign change in the reflected p variables.

For simplicity, we consider now the particular example of global operators of a bipartite state:

$$\hat{u}_\pm = \hat{u}_1 \pm \hat{u}_2, \tag{29}$$

and

$$\hat{v}_\pm = \hat{v}_1 \pm \hat{v}_2. \tag{30}$$

We note that operators with the same sign satisfy the commutation relations $[\hat{u}_\pm, \hat{v}_\pm] = 2i\hbar\gamma$, so that these non-commuting operators after being an input to the uncertainty functionals fulfill the UR of the aforementioned form [note the factor of 2 in the argument of $f(\cdot)$]

$$F(\hat{\varrho}_{12}; \hat{u}_\pm, \hat{v}_\pm; P_{u_\pm}, P_{v_\pm}) \geq f(2\hbar|\gamma|). \tag{31}$$

Using the transformation of the Wigner function under partial transposition described above, one can evaluate the uncertainty functional of the partially transposed state $\hat{\varrho}_{12}^{T_2}$ via measurements on the actual state ϱ_{12} using the relation

$$F(\varrho_{12}^{T_2}; \hat{u}_\pm, \hat{v}_\pm; P_{u_\pm}, P_{v_\pm}) = F(\varrho_{12}; \hat{u}_\pm, \hat{v}_\mp; P_{u_\pm}, P_{v_\mp}), \tag{32}$$

which can be lower than $f(2\hbar|\gamma|)$ since the operators with different signs do commute. This possibility, when experimentally confirmed, indicates that $\varrho_{12}^{T_2}$ is not a *bona fide* density operator, and thus the bipartite quantum state ϱ_{12} is entangled.

Building on this general reasoning (PPT argument applied to an UR) several entanglement criteria have been developed. A comprehensive list of the criteria contains those based on the variances [115,116] and higher-order moments [117,118], Shannon entropy [105], Rényi entropy [106], characteristic function [119] as well as the triple product variance relation [120]. Particularly powerful is the formalism developed by Shchukin and Vogel, which provides an infinite set of inequalities involving moments of the bipartite state [121], such that violation of a single inequality indicates entanglement. We note that some of these criteria can be applicable to any non-commuting global operators. Uncertainty-based approaches (using the PPT method directly or not) have been developed for multipartite systems [122,123], and a general framework to construct entanglement criteria for multipartite systems based on the "PPT+UR" interrelation was presented in Reference [107]. The Shchukin-Vogel hierarchy of moment inequalities has also been applied to the multipartite case [124].

The PPT+UR approach has been used to identify continous variable entanglement experimentally in several systems, including entangled fields from parametric oscillators and amplifiers [94,125,126] as well as spatially entangled photon pairs produced from parametric down conversion [96,120,127], and time/frequency entangled photon pairs [128,129]. A higher-order inequality in the Shchukin-Vogel criteria [121] has been used to observe genuine non-Gaussian entanglement [130].

4. Realistic Coarse-Grained Measurements of Continuous Distributions

Coarse graining of observables with continuous spectra is a consequence of any realistic measurement process. In the laboratory, an experimentalist is given the task of designing projective measurements in order to recover information about probability densities of a continuous variable quantum system. Naturally, only partial information about the underlying continuous structure of the infinite-dimensional physical system is retrieved in a laboratory experiment. Whichever measurement design is chosen, the experimentalist is faced with two main difficulties, namely the finite *detector range* and finite *measurement resolution*, related to the size of the total region of possible detection events and the precision in which events are registered, respectively. The detector range problem [25,29] results from the finite amount of resource available to the experimentalist. For instance, consider a position discriminator based on a multi-element detector array. The array has a spatial reach (in a single spatial dimension) that increases linearly with the number of detectors. In a similar fashion, the sampling time of a single element detector used in raster scanning mode increases linearly with the chosen detection range. Continuous variables such as the position are also inevitably affected by the inherent finite resolution of the measurement apparatus [32], such as the size of each individual detector in the array, or the pixel size of a camera. Altogether, the finite detector range and measurement resolution restrict the capability to probe the detection position, limiting the experimentalist to a *coarse-grained* sample of the underlying CV degree of freedom.

The constraints imposed by the finite spatial reach and resolution of the measurement apparatus are then important features that must be considered in the experiment design. Ideally, the experimentalist would chose measurement settings producing the finest coarse-grained sample possible. As a trade-off, the increased resolution entails the sampling of a greater number of pixels (if the range of detection is preserved), increasing the amount of resources used in data acquisition and analysis. The compromise between the used resource and chosen resolution depends on the specific design and measurement technique. A single raster scanning detector is inherently inefficient and leads to acquisition times that grow with the number of scanned outcomes. On the other hand, the acquisition time is dramatically reduced by the use of multi-element detector arrays [131–134]. Other techniques such as position-to-time multiplexing [135,136] allow the sampling of multiple position outcomes with single element detectors, but at the expense of an increased dead-time between consecutive detections. We have exemplified the finite detector range and finite measurement resolution problems in terms of a detector that registers the position of a particle. However, similar considerations are valid for any detection system that registers a digitalized value of a continuous physical parameter.

Under constraints of resource utilisation—such as the number of detectors and/or sampling time—the experimentalist needs to set the number of possible detection outcomes for their coarse-grained measurements. Therefore, a natural question that arises regards the coarse-graining design allowing the extraction of the desired information. Naively, one might think that usual quantum mechanical features learnt from physics textbooks would be directly observable from the coarse-grained distributions obtained in the laboratory. The most prominent counter-example is the experimental observation of the *Heisenberg* UR in Equation (8). As shown in Reference [32], coarse-grained distributions of conjugate continuous variables do not necessarily satisfy the well known UR valid for continuous distributions. In order to accurately inspect the uncertainty product of the measured distributions in accordance with the *Heisenberg* UR, the latter must be modified to account for the detection resolution of the measurement apparatus. Another important quantum mechanical feature that one usually fails to observe from standard coarse-grained distributions is the mutual unbiasedness [137] relation between measurement outcomes of the incompatible observables. That is, eigenstates of-say-the coarse-grained position operator do not necessarily present a uniform distribution of outcomes for coarse-grained momentum measurements. In addition, some authors [138–141] have demonstrated that one can define functions of incompatible observables that indeed commute. Interestingly, it was shown in Reference [142] that one can indeed enjoy full quantum mechanical unbiasedness using a specific *periodic* coarse-graining design rather than the standard one. Other practical issues regarding false positives in entanglement detection [26,29] and cryptographic security [25,29] must also be reconsidered when one deals with realistic coarse-grained distributions.

In this section, we will introduce the projective measurement operators both for the *standard* and the *periodic* models of coarse graining. Practical features such as measurement resolution, detector range and positioning degrees of freedom in the measurement design will be discussed. We will also briefly discuss relations of mutual unbiasedness between coarse-grained measurement outcomes in domains of incompatibles observables. A detailed discussion of uncertainty relations for coarse-grained distributions will be presented in the next section.

4.1. Coarse-Graining Models

A laboratory experiment necessarily yields a discrete, finite set of measurement outcomes of any observable in any physical system. This is also the case for an experiment probing a continuous degree of freedom, $\hat{u}$, for which measurement outcomes $\{u_k\}$ labeled by the discrete integer index $k \in \mathbb{Z}$ relate to the underlying continuous real variable $u \in \mathbb{R}$ corresponding to the eigenspectra of $\hat{u}$. In the most general scenario, a coarse-graining model is obtained from an arbitrary partition of the set of real numbers $\mathbb{R}$, in intervals $\mathcal{R}_k$ with $u_k \in \mathcal{R}_k$. The orthogonality of the measurement outcomes requires the subsets to be mutually disjoint: $\mathcal{R}_k \cap \mathcal{R}_{k'} = \varnothing, \forall k \neq k'$. Even though the continuous variable can be formally discretised into an infinite number of outcomes (with k an unbounded integer), the experiment can only probe a finite range of the continuous variable. Thus, the detection range, $\mathcal{R}_{\text{range}}$, can be formally defined by the union of the disjoint subsets associated with the probed outcomes:

$$\bigcup_k \mathcal{R}_k = \mathcal{R}_{\text{range}} \subset \mathbb{R}. \tag{33}$$

This relation limits the set of possible values of k to a finite subset of integers $\mathcal{Z}_k \subset \mathbb{Z}$. Due to the finite range, $\mathcal{R}_{\text{range}}$, of the measurement process it is important to secure under reasonable experimental conditions that the underlying probability density is supported within the chosen range of detection [25,29]. Mathematically, a faithful coarse-grained measurement design should ensure that

$$\int_{\mathcal{R}_{\text{range}}} P_u(u) du \approx 1, \tag{34}$$

where P_u is the marginal pdf defined in Equation (10).

The probability $\mathrm{p}_k^{(u)}$ that the outcome u_k is produced writes as an integral of the marginal probability density, P_u, for the continuous variable:

$$\mathrm{p}_k^{(u)} = \int_{\mathcal{R}_k} P_u(u) du, \tag{35}$$

where the integration is performed in the interval $\mathcal{R}_k$. Due to the faithful coarse-grained condition in Equation (34) we have

$$\sum_{k \in \mathcal{Z}_k} \mathrm{p}_k^{(u)} \approx 1. \tag{36}$$

We can define projective operators associated with the coarse-grained measurements:

$$\hat{C}_k^{(u)} = \int_{\mathcal{R}_k} |u\rangle\langle u| du, \tag{37}$$

so that the probabilities (35) can be written as

$$\mathrm{p}_k^{(u)} = \mathrm{Tr}(\hat{\rho}\hat{C}_k^{(u)}), \tag{38}$$

with $P_u(u) = \langle u|\hat{\rho}|u\rangle$. In order to study mutual unbiasedness and uncertainty relations, we shall later in this and the following sections define coarse-grained operators like those in Equation (37) for conjugate variables of the quantum state, such as the position and the linear momentum of a quantum particle.

4.1.1. Standard Coarse Graining

The standard model of coarse graining describes, for example, the typical projective measurements performed with an array of adjacent, rectangular detectors. A conventional example of such an apparatus is the image sensor of a digital camera, for which the pixel size stands for the detection resolution whereas the length of the full sensor embodies the range of detection. In the current analysis, we shall consider a linear detector array along a single spatial dimension rather than the two-dimensional area of a typical image sensor, as illustrated in Figure 1. The coarse-graining interval representing the detection window of the k-th pixel of the linear array is then:

$$\mathcal{R}_k := \left(u_{\mathrm{cen}} + (k - \frac{1}{2})\Delta, u_{\mathrm{cen}} + (k + \frac{1}{2})\Delta \right], \tag{39}$$

where Δ is the detector or pixel size—also commonly referred to as the *coarse-graining width* or the *bin width*. Using the definition Equation (39), the discretised outcomes u_k represent the u value of the center of the corresponding bin:

$$u_k = u_{\mathrm{cen}} + k\Delta. \tag{40}$$

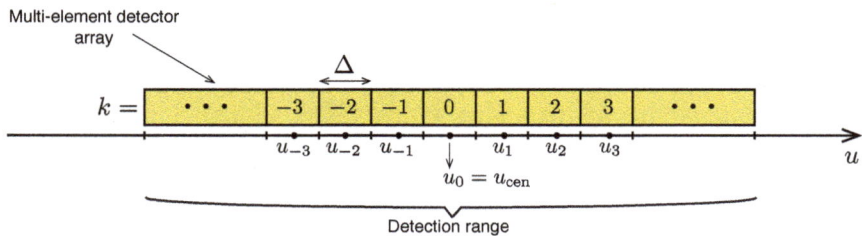

Figure 1. Multi-element detector array illustrating the standard coarse-graining geometry.

The parameter u_{cen} sets the position of the central bin of the array, whose outcome label is $k = 0$, yielding $u_0 = u_{\mathrm{cen}}$. To illustrate the effect of the coarse-graining design on measured distributions, we plot in Figure 2 coarse-grained distributions (blue bars) obtained using 3 different resolutions: $\Delta = 2$ (left colum), $\Delta = 1$ (central column) and $\Delta = 1/2$ (right column). For each resolution, we plot two distinct distributions obtained using $u_{\mathrm{cen}} = 0$ (top row) and $u_{\mathrm{cen}} = \Delta/2$ (bottom row). In other words, the coarse-graining bins of the distributions plotted at the bottom part of the figure are displaced by half a "pixel" in relation to the distributions at the top. Clearly, the distribution obtained using a fixed resolution is not unique, but the effect of small displacements (smaller than the bin width) gets less important as the resolution is increased. For comparison, the generating continuous distribution is plotted in red.

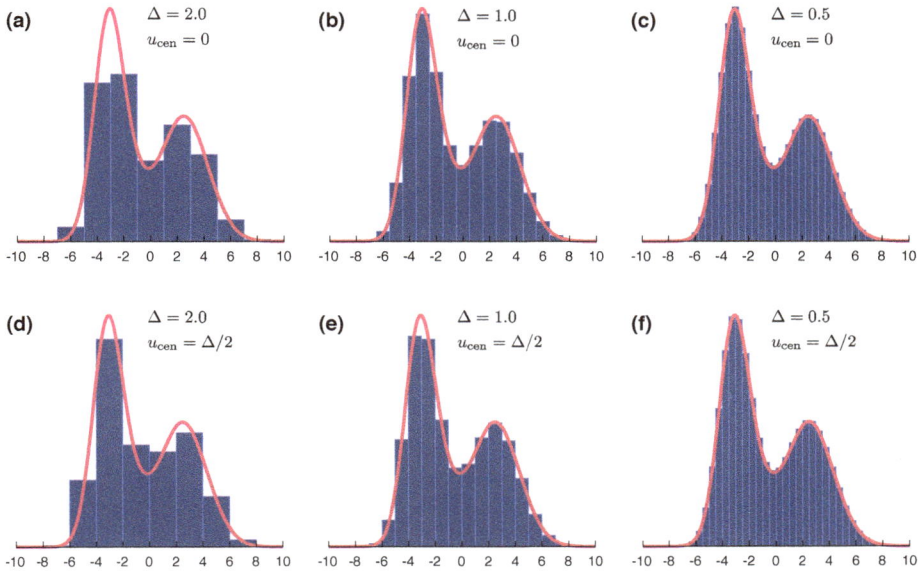

Figure 2. Coarse-grained distributions (blue bars) according to the standard model. The red solid line indicates the underlying continuous distribution used to generate the discretised versions. The used resolution Δ and positioning degree of freedom u_{cen} is indicated beside each distribution. For each resolution, two distinct distributions are shown, each of which associated with a different positioning of the coarse-graining bins.

We shall now use this model for standard coarse graining to explicitly define the discretised counterparts of the position and momentum operators given in Equation (3).

$$\hat{u}_\Delta = \sum_k u_k \hat{C}_k^{(u)}, \tag{41a}$$

$$\hat{v}_\delta = \sum_l v_l \hat{C}_l^{(v)}, \tag{41b}$$

where the projector $\hat{C}_k$ is defined in Equation (37) (with $\hat{C}_l^{(v)}$ having an equivalent definition for $\hat{v}$ measurements), and we used Δ (δ) as the detection resolution for $\hat{u}$ ($\hat{v}$) measurements. According to the definition in Equation (35), as a result of the the coarse-grained measurement of $\hat{u}$ and $\hat{v}$ we obtain the discrete probabilities, $p_{\Delta,k}^{(u)}$ and $p_{\delta,l}^{(v)}$. The discrete variances associate with these discrete probabilities are:

$$\sigma_{P_\Delta^{(u)}}^2 = \sum_k u_k^2\, p_{\Delta,k}^{(u)} - \left(\sum_k u_k\, p_{\Delta,k}^{(u)} \right)^2, \tag{42a}$$

$$\sigma_{P_\delta^{(v)}}^2 = \sum_l v_l^2\, p_{\delta,l}^{(v)} - \left(\sum_l v_l\, p_{\delta,l}^{(v)} \right)^2, \tag{42b}$$

where we define the set of discrete probabilities:

$$P_\Delta^{(u)} := \{p_{\Delta,k}^{(u)}\} \quad \text{and} \quad P_\delta^{(v)} := \{p_{\delta,k}^{(v)}\}. \tag{43}$$

One can see from the definitions (42) that if the bin widths Δ and δ are such that $p_{\Delta,k}^{(u)}$ and $p_{\delta,l}^{(v)}$ are sufficiently close to unity for some value of k and l, we have $\sigma_{P_\Delta^{(u)}}^2, \sigma_{P_\delta^{(v)}}^2 \longrightarrow 0$. Thus, naive application of any of the variance-based URs given in Section 2.1 would indicate a false violation of a UR. It has been shown in Reference [32] that the same argument applies to discretized versions of entropic URs, such as those of Section 2.2. Thus, proper treatment of standard coarse-grained measurements is essential in order to take advantage of the practical application of URs in QIT and quantum physics in general. In Section 5 we show how this can be done.

4.1.2. Periodic Coarse Graining

A distinct model of coarse graining discussed in the literature [142,143] is refereed to as *periodic coarse graining* (PCG). In this model, the partition of the whole set of real numbers $\mathbb{R}$ is performed in a periodic manner, leading to a finite number d of subsets $\mathcal{R}_k$, with $k = 0, \cdots, d-1$. The resulting discretization utilizes the index k as a direct label for the detection outcomes, in a similar fashion to what is usually defined for finite-dimensional quantum systems. The subsets $\mathcal{R}_k$ are defined as [142]:

$$\mathcal{R}_k := \{u \in \mathbb{R} \mid u_{\text{cen}} + k s_u \leqslant u (\text{mod } T_u) < u_{\text{cen}} + (k+1) s_u\}, \tag{44}$$

where s_u plays the role of a bin width similar to the resolution Δ used for the standard coarse graining. In the definition Equation (44), bins of size s_u are arranged periodically with the parameter T_u representing the period, as illustrated in Figure 3 for the particular design using $d = T_u/s_u = 5$ detection outcomes. It is important to notice that this coarse graining design do not distinguish detections in distinct bins associated with the same detection outcome k (ranging from 0 to 4 in Figure 3). For example, a detection within any bin colored in red in (44) would lead to the same detection outcome $k = 1$.

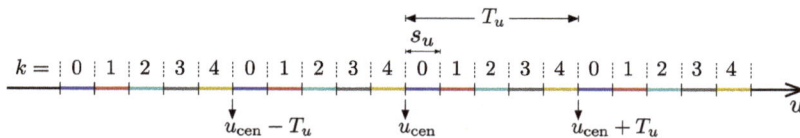

Figure 3. Periodic coarse-graining design with $d = T_u/s_u = 5$ detection outcomes. The parameter T_u is the periodicity in which bins of size s_u are arranged.

An interesting feature of the PCG model is that the number of detection outcomes is utterly adjustable by the choice of the parameters T_u and s_u, regardless of the chosen detection range. For instance, doubling the range of detection allows one to design PCG measurement using twice as much periods in its design, while maintaining the same number $d = T_u/s_u$ of detection outcomes. As with the standard model, the reference coordinate u_{cen} sets the center of the detection range also for the PCG design. Using the subset definition given in Equation (44), we can explicitly write the projector operators, Equation (37), for the PCG model as

$$\hat{\Pi}_k^{(u)} = \int_{\mathcal{R}_k} |u\rangle\langle u| du = \sum_{n \in \mathbb{Z}} \int_{u_{cen}+ks_u+nT_u}^{u_{cen}+(k+1)s_u+nT_u} |u\rangle\langle u| du, \tag{45}$$

where we extend the sum in n over $\mathbb{Z}$ without loss of generality, assuming that Equation (34) is satisfied. Analogously, we also define the PCG projective operators over the conjugate variable v:

$$\hat{\Pi}_l^{(v)} = \int_{\mathcal{R}_l} |v\rangle\langle v| dv = \sum_{n \in \mathbb{Z}} \int_{v_{cen}+ls_v+nT_v}^{v_{cen}+(l+1)s_v+nT_v} |v\rangle\langle v| dv, \tag{46}$$

where we define s_v and T_v as the bin width and periodicity used in the PCG measurements of v.

4.2. Mutual Unbiasedness in Coarse-Grained Measurements

If a quantum system , with finite dimension, is described as an eigenstate of a given observable, the measurement outcomes of complementary observables are completely unbiased: each one of them occurring with equal probability, $1/d$, where d is the dimension of the quantum system's Hilbert space. This unbiasedness relation is an important feature of quantum mechanics with no classical counterpart, and is usually cast in terms of the basis vectors constituting the eigenstates of two (or more) complementary observables. To be more precise, two orthonormal bases $\{|a_k\rangle\}$ and $\{|b_l\rangle\}$ are said to be *mutually unbiased* if and only if $|\langle a_k|b_l\rangle|^2 = 1/d$ for all $k,l = 0, \cdots, d-1$ [137]. The observation of unbiased measurement outcomes is customary in experiments with finite dimensional quantum systems. Not only routine, measurements in mutually unbiased bases (MUB) constitute a key procedure in several quantum information processing tasks, such as verification of cryptographic security [9], certification of quantum randomness [144], detection of quantum correlations [145–147] and tomographic reconstruction of quantum states [148,149].

Mutual unbiasedness is also extendable to continuous variables quantum systems [150], for which bases $\{|u\rangle\}$ and $\{|v\rangle\}$ such $[\hat{u}, \hat{v}] = i\hbar\gamma$, satisfy $|\langle u|v\rangle|^2 = 1/(2\pi\hbar|\gamma|)$, i.e., the overlap of the basis vectors $|u\rangle$ and $|v\rangle$ is independent (no bias) of their eigenvalues, u and v (note, however, that even though $\hat{u}$ and $\hat{v}$ are mutually unbiased observables, this does not imply that they are complementary, as would be the case for operators in a discrete quantum system [151]. In continuous variable quantum mechanics, mutual unbiasedness does not imply that $\hat{u}$ and $\hat{v}$ are maximally incompatible [152]. In this case, complementary observables are typically defined as CCOs, that is, forming a Fourier transform pair). For CV systems, nevertheless, this relation is rather a theoretical definition than an experimentally observable fact, since the experimentalist has neither the capability to prepare nor to measure the (infinitely squeezed) eigenstates of the $\hat{u}$ and $\hat{v}$. Instead, both the preparation and measurement procedures are limited to the finite resolution of the experimental apparatus. As discussed previously in this section, measurements of a CV degree of freedom render discretized, coarse-grained outcomes whose probabilities, Equation (35), are provided by a coarse-graining model described by the projective operators given in Equation (37). These coarse-grained probabilities obtained experimentally do not in general preserve the mutual unbiasedness complied by the underlying continuous variables.

To elaborate the issue, let us consider sets of projectors $\{\hat{C}_k^{(u)}\}$ and $\{\hat{C}_l^{(v)}\}$ defining coarse-graining measurements in the complementary domains u and v of a continuous variable quantum system $\hat{\rho}$. We assume measurement designs providing a number d of outcomes in each domain. In this scenario, the requirement for mutual unbiasedness is thus that the coarse-grained probabilities for measurements of one variable are evenly spread between all discretized outcomes whenever the quantum state is localized with respect to the coarse graining applied to its conjugate variable (and vice-versa). The subtlety in this requirement is the (infinite) degeneracy of normalizable quantum states that can be localized with respect to the chosen coarse graining. To emphasize this degeneracy, we refer to the outcome probabilities, Equation (35), with explicit dependency on the quantum state in order to mathematically phrase the condition for mutual unbiasedness in coarse-grained CV: the outcomes of $\{\hat{C}_k^{(u)}\}$ and $\{\hat{C}_l^{(v)}\}$ are mutually unbiased if *for all* quantum states $\hat{\rho}$ and $k_0, l_0 = 0, \cdots, d-1$ we have [142]:

$$\mathrm{p}_k^{(u)}(\hat{\rho}) = \delta_{k_0 k} \quad \Rightarrow \quad \mathrm{p}_l^{(v)}(\hat{\rho}) = d^{-1}, \tag{47a}$$

$$\mathrm{p}_l^{(v)}(\hat{\rho}) = \delta_{l_0 l} \quad \Rightarrow \quad \mathrm{p}_k^{(u)}(\hat{\rho}) = d^{-1}, \tag{47b}$$

where, again, we stress that $\mathrm{p}_k^{(u)}(\hat{\rho}) = \mathrm{Tr}(\hat{\rho}\hat{C}_k^{(u)})$ and $\mathrm{p}_l^{(v)}(\hat{\rho}) = \mathrm{Tr}(\hat{\rho}\hat{C}_l^{(v)})$, as in Equation (35).

Having formulated the conditions for mutual unbiasedness, Equation (47), it is easy to perceive that the adjacent, rectangular subsets defining the standard coarse graining [Equation (39)] will not lead to unbiased measurement outcomes. Any CV distribution localized in a single coarse-graining bin (for example in the u variable) generates a probability density that decays in the Fourier domain (the v variable) along the adjacent bins within the detection range. This decay generates a non constant coarse-grained distribution that, by definition, is biased. Furthermore, the number d of detection outcomes in the standard design depends directly on the selected detection range, as well as on the chosen resolution. As a consequence, even though a particular localized distribution could lead to approximately unbiased coarse-grained outcomes in the Fourier domain, an extended detection range would increase the number of outcomes, thus spoiling the unbiasedness.

It is thus evident that in order to retrieve unbiased outcomes from coarse-grained measurement, a more contrived coarse-graining design is needed. As it turns out, it was shown in Reference [142] that the PCG design exactly fulfils the requirements for unbiased measurements of finite cardinality stated in Equation (47). A relation between the periodicities T_u and T_v used in the PCG of the conjugate variables u and v was analytically derived as a single condition for unbiased coarse-grained measurements:

$$\frac{T_u T_v}{2\pi\hbar} = \frac{d}{m}, \quad m \in \mathbb{N} \quad s.t. \quad \forall_{n=1,\cdots,d-1} \quad \frac{mn}{d} \notin \mathbb{N}. \tag{48}$$

The unbiasedness condition stated in Equation (48) establishes infinite possibilities for the pair of periodicities T_u and T_v that can be used to design the mutually unbiased pair of PCG measurements defined in Equations (45) and (46), respectively. For instance, the simplest and most important case is the condition with $m = 1$, since it is valid for all d and provides the best trade-off between experimentally accessible periodicities: $T_u T_v = (2\pi\hbar)d$. Conditions with $m > 1$ are also possible but are not general since they depend on the chosen number of outcomes d [142]. For example, for $d = 4$, valid conditions are found using $m(\mathrm{mod}\, d) = 1, 3$ whereas for $d = 5$, valid conditions are found using $m(\mathrm{mod}\, d) = 1, 2, 3, 4$. Importantly, the case with $m(\mathrm{mod}\, d) = 0$ is always excluded, since in this case the PCG projectors describe commuting sets, $\left[\hat{\Pi}_k^{(u)}, \hat{\Pi}_l^{(v)}\right] = 0, \forall\, k, l$ [138–140]. In other words a joint eigenstate of the product $\hat{\Pi}_k^{(u)}\hat{\Pi}_l^{(v)}$ existis for all k and l whenever $T_u T_v = 2\pi\hbar/c$ with $c \in \mathbb{N}$ [153]. It is also interesting to note that using the periodicity definition from the PCG design ($T = ds$), it is possible to write the unbiasedness condition given in Equation (48) in alternative, equivalent ways:

$$(a)\ T_u T_v = \frac{2\pi\hbar}{m}d, \quad (b)\ T_u s_v = \frac{2\pi\hbar}{m}, \quad (c)\ s_u T_v = \frac{2\pi\hbar}{m}, \quad (d)\ s_u s_v = \frac{2\pi\hbar}{m}\frac{1}{d}. \tag{49}$$

Finally, in Reference [143] these results were generalized for PCG measurements applied to an arbitrary pair of phase space variables other than the conjugate pair formed by position and momentum. What is more, a triple of unbiased PCG measurements was also shown to exist for rotated phase space variables, along the same lines as the demonstration of a MUB triple in the continuous regime done in Reference [150]. Experimental demonstrations of unbiased PCG measurements were also carried out in References [142,143], both of them utilizing the transverse spatial variables of a paraxial light field.

5. UR for Coarse-Grained Observables

A kind of a paradigm shift in the theory of uncertainty relations was brought by the observation that everything can be efficiently characterized solely by means of probability distributions. As a result, tools known from information theory, such as information entropy, Fisher information and other measures, came into play. Additionally, the notion of uncertainty for discrete systems could better be captured that way. Since products of variances calculated for observables such as the spin are bounded in a state-dependent manner (so that the ultimate lower bound typically assumes the trivial value of 0), information entropies provide an attractive alternative [154]. Written already in the Rényi form,

$$H_\alpha \left[P\right] = \frac{1}{1-\alpha} \ln \sum_k [\mathrm{p}_k]^\alpha, \tag{50}$$

the above equation is a discrete counterpart of Equation (25), which corresponds to the discrete counterpart of Equation (21) when $\alpha = 1$.

In the finite-dimensional case given by an arbitrary state $\hat\rho$ acting on a d-dimensional Hilbert space $\mathcal{H}$, and a pair of non-degenerate, non-commuting observables, $\hat{A}$ and $\hat{B}$, one usually defines the probabilities associated to projective measurements:

$$\mathrm{p}_i^{(A)} = \langle a_i | \hat\rho | a_i \rangle, \qquad \mathrm{p}_j^{(B)} = \langle b_j | \hat\rho | b_j \rangle, \tag{51}$$

where $|a_i\rangle$ and $|b_j\rangle$, $i, j = 1, \dots, d$ are the eigenstates of the operators associated with both observables. Disctrete entropic URs for the above probability distributions are of the general form

$$H_\alpha \left[P^{(A)}\right] + H_\beta \left[P^{(B)}\right] \geq B_{\alpha\beta} \left(\mathbf{U}\right), \tag{52}$$

with $\mathbf{U} \in \mathcal{U}\left(d\right)$ being a unitary matrix with matrix elements $\mathbf{U}_{ij} = \langle a_i | b_j \rangle$. We denote $P^{(A)} := \{\mathrm{p}_i^{(A)}\}$ and $P^{(B)} := \{\mathrm{p}_j^{(B)}\}$ again with $i, j = 1, \dots, d$.

The first entropic uncertainty relation for discrete variables comes from Deutsch [154], who for $\alpha = 1 = \beta$ found the lower bound $B_{11}^D = -2 \ln C$, with $C = \left(1 + \sqrt{c_1}\right)/2$ and $c_1 = \max_{i,j} |U_{ij}|^2$. A substantially more renowned Maassen–Uffink (MU) bound [155] derived in 1988, is $B_{\alpha\beta}^{MU} = -\ln c_1$. This bound is however valid only for the conjugate parameters $1/\alpha + 1/\beta = 2$. Very recently, a plethora of new results [41,156–163] improving the celebrated MU bound has been obtained. In particular, an approach based on the notion of majorization (suitable from the perspective of resource theories and quantum thermodynamics [164]) provides a significant qualitative novelty [156,157,159,163], which will also be touched upon in this section.

In this review we are concerned with the case in which continuous probability distributions $P_u \left(u\right)$ and $P_v \left(v\right)$ are replaced (*viz.* they were measured this way) by their discrete counterparts $(k, l \in \mathbb{Z})$. According to the discussion in Section 4 we can use the definitions in Equations (35) and (39), and the condition in Equation (33), to write the discrete probabilities:

$$\mathrm{p}_{\Delta,k}^{(u)} = \int_{(k-1/2)\Delta}^{(k+1/2)\Delta} dy\, P_u \left(y\right), \qquad \mathrm{p}_{\delta,l}^{(v)} = \int_{(l-1/2)\delta}^{(l+1/2)\delta} dy\, P_v \left(y\right), \tag{53}$$

with $k \in \mathcal{Z}_k \subset \mathbb{Z}$. In the following we describe a series of URs for these discrete probabilities that are known as coarse-grained URs, derived in [24,30–32]. These are the coarse-grained counterpart of the *Heisenberg, Shannon entropy* and *Rényi entropy* URs in Equations (8), (20) and (24) respectively. Here, we will closely follow the treatment in [24,32]; however, before we start we give a short historical overview and discuss a path towards extensions going beyond CCOs.

The idea that generic quantum uncertainty could be quantified by the sum of Shannon entropies evaluated for discretized position and momentum probability distributions for the first time appeared in the contribution by Partovi [165]. He also derived the first coarse-grained UR which in the form is reminiscent to the Deutsch bound for finite-dimensional systems [154] (please note that both papers [154,165] have been published in 1983; however, Partovi in his first sentence refers to a "recent letter" by Deutsch). Both bounds [154,165] were obtained by means of a direct optimization, independently applied to every logarithmic contribution. Symmetry in developments of the URs for finite-dimensional and coarse-grained systems happened to be much deeper as the second coarse-grained result, by Bialynicki-Birula [30], is a counterpart of the MU bound [155]. The former result is an application of the continuous variant of the *Shannon entropy* UR (so the L_p-L_q norm inequality by Beckner [62]) supported by the Jensen inequality for convex functions, while the MU bound is a direct consequence of the Riesz theorem for the l_p-l_q norms. Please note that relatively often, integration limits in (53) were chosen as "from $k\Delta$ to $(k+1)\Delta$" and "from $l\delta$ to $(l+1)\delta$"; however this choice causes a formal pathology in the limit of infinite coarse graining [166]. Thus, sticking to terminology of Equation (39), in theory it is better to avoid borderline settings for the position of the central bin, i.e., $u_{cen} = \Delta/2$.

To briefly report later developments, one shall mention that Partovi reconsidered the problem he had posed several years ago, pioneering applications of majorizaiton techniques [167]. Also Schürmann and Hoffmann [168] discussed the *Shannon entropy* UR from the perspective of the integral equation associated to it, while the first author conjectured an improvement (later mentioned in detail) which agrees with his numerical tests [169]. Finally, we mention (without details) an erroneous improvement of [31] by Wilk and Wlodarczyk [170,171], mainly devoted to the case of the Tsallis entropy.

Although originally the URs were derived for CCOs, $\hat{u}$ and $\hat{v}$, here we show which of the URs in [24,32] can be valid also for operators $\hat{u}$ and $\hat{v}$ that are arbitrary linear combinations of all positions and momenta of the $n-$bosonic modes like the ones defined in Equation (3), *viz.* operators that are not necessarily CCOs. In the general case, we stress that there is always a unitary metaplectic transformation (so $\hat{U}_{\mathbf{S}}$ belongs to the metaplectic group $Mp(2n, \mathbb{R})$ and it is always associated with a matrix $\mathbf{S}$ that belongs the symplectic group $Sp(2n, \mathbb{R})$ [50]), $\hat{U}_{\mathbf{S}}$, that connects $\hat{u}$ and $\hat{v}$, *viz.* $\hat{v} = \hat{U}_{\mathbf{S}}^{\dagger}\hat{u}\hat{U}_{\mathbf{S}}$. However, this metaplectic transformation is not necessarily a $\pi/2$ rotation, which would be the case if $\hat{u}$ and $\hat{v}$ were CCOs. In order to see this, we first define two sets of operators $(\hat{\mathbf{u}}, \hat{\mathbf{u}}')^T = (\hat{u} = \hat{u}_1, \ldots, \hat{u}_n, \hat{u}'_1, \ldots, \hat{u}'_n)^T = \sqrt{\gamma}\,\tilde{\mathbf{S}}\,\hat{\mathbf{x}}$ and $(\hat{\mathbf{v}}, \hat{\mathbf{v}}')^T = (\hat{v} = \hat{v}_1, \ldots, \hat{v}_n, \hat{v}'_1 \ldots, \hat{v}'_n)^T = \sqrt{\gamma}\,\mathbf{S}'\,\hat{\mathbf{x}}$, where $\tilde{\mathbf{S}}$ and $\mathbf{S}'$ are some matrices belonging to the symplectic group $Sp(2n, \mathbb{R})$, with the only restriction that the first rows of $\tilde{\mathbf{S}}$ and $\mathbf{S}'$ correspond to the real coefficients $\mathbf{d}$ and $\mathbf{d}'$ in Equation (5), respectively, which define the operators $\hat{u}$ and $\hat{v}$ in Equation (3). Due to the properties of symplectic matrices, all the pairs $\hat{u}_i$ and $\hat{u}'_i$, and also $\hat{v}_i$ and $\hat{v}'_j$, satisfy CCRs, *viz.* $[\hat{u}_i, \hat{u}'_j] = i\hbar\gamma\delta_{ij}$ and $[\hat{v}_i, \hat{v}'_j] = i\hbar\gamma\delta_{ij}$ with $i, j = 1, \ldots, n$. However, it is immediate to see that $(\hat{\mathbf{v}}, \hat{\mathbf{v}}')^T = \mathbf{S}(\hat{\mathbf{u}}, \hat{\mathbf{u}}')^T$ where the matrix $\mathbf{S} := \mathbf{S}'\tilde{\mathbf{S}}^{-1}$ is a generic symplectic matrix. Then the Stone-von-Neumann theorem guarantees that the change $(\hat{\mathbf{u}}, \hat{\mathbf{u}}')^T \rightarrow (\hat{\mathbf{v}}, \hat{\mathbf{v}}')^T$ is unitarily implementable by a metaplectic transformation $\hat{U}_{\mathbf{S}}$ [50]. In particular we have $\hat{U}_{\mathbf{S}}^{\dagger}\,\hat{u}\,\hat{U}_{\mathbf{S}} = (\mathbf{S}\,\hat{\mathbf{x}})_1 =: \hat{v}$.

5.1. URs Proved Only for CCOs

The key concept behind the treatment of coarse-grained URs in [24,32] is the introduction of the piece-wise continuous probability density functions:

$$Q_{\Delta,u}(u) := \sum_{k \in \mathcal{Z}_k} \mathrm{p}_{\Delta,k}^{(u)}\, D_\Delta(u, u_k) \quad \text{and} \quad Q_{\delta,v}(v) := \sum_{l \in \mathcal{Z}_l} \mathrm{p}_{\delta,l}^{(v)}\, D_\delta(v, v_l), \tag{54}$$

where $D_\Delta(u, u_k)$ and $D_\delta(v, v_l)$ are called the histogram functions (HF) with u_k (and v_l in an analogous way) defined in Equation (40). Generically, these functions are defined such that they are normalized in each bin:

$$\int_{(k-1/2)\Delta}^{(k+1/2)\Delta} D_\Delta(u, u_k)\, du = 1 \quad \text{and} \quad \int_{(l-1/2)\delta}^{(l+1/2)\delta} D_\delta(v, v_l)\, dv = 1, \tag{55}$$

and approach the Dirac delta distribution for infinitesimal bin size:

$$\lim_{\Delta \to 0} D_\Delta(u, u_k) = \delta(u - u_k) \quad \text{and} \quad \lim_{\delta \to 0} D_\delta(v, v_l) = \delta(v - v_l). \tag{56}$$

Therefore, in the limit $\mathcal{Z}_k, \mathcal{Z}_l \to \mathbb{Z}$ and $\Delta, \delta \to 0$ we have $Q_{\Delta,u}(u) \to P_u(u)$ and $Q_{\delta,v}(v) \to P_v(v)$. We shall stress here that the HF can, in general, have any functional form as long as it is non-negative, normalized and fulfills Equation (56). However, the most common histogram function is the rectangular HF:

$$D_\Delta^R(u, u_k) := \begin{cases} 1/\Delta & \text{for } u \in \left((k - \frac{1}{2})\Delta, (k + \frac{1}{2})\Delta \right], \\ 0 & \text{otherwise.} \end{cases} \tag{57}$$

with an equivalent definition for $D_\delta^R(v, v_l)$. In Figure 2 we show an example of coarse-grained probability distributions functions $Q_{\Delta,u}(u)$ (the area beneath these functions are displayed in full) using rectangular histogram functions and for different size bins Δ.

Here, we generalise the results in [24,32] through the following expression that will be justified later:

$$h_\alpha[Q_{\Delta,u}] + h_\beta[Q_{\delta,v}] \geq \ln \left(\frac{\pi \hbar |\gamma|\ e^{h_\alpha[D_\Delta] - \ln \Delta + h_\beta[D_\delta] - \ln \delta}}{\varepsilon_\alpha(\Gamma/4)} \right), \tag{58}$$

with $1/\alpha + 1/\beta = 2$ and $1/2 \leq \alpha \leq 1$. To simplify the notation we define the function:

$$\varepsilon_\alpha(x) := \min \left\{ \alpha^{\frac{1}{2-2\alpha}} \beta^{\frac{1}{2-2\beta}}, \frac{1}{2} R_{00}^2(x, 1) \right\}, \tag{59}$$

where $R_{00}(x, y)$ denotes one of the radial prolate spheroidal wave functions of the first kind [172], and introduce the joint coarse-graining parameter $\Gamma = \Delta\delta/(\hbar|\gamma|)$. We stress that Equation (58) involves the differential Rényi entropies of the piece-wise continuous distributions defined in Equation (54).

Let us see how the results in [12,24,30–32] can be derived from Equation (58). First, we observe that the Rényi entropies of rectangular HFs, for every values of α and β, are:

$$h_\alpha[D_\Delta^R] = \ln \Delta \quad \text{and} \quad h_\beta[D_\delta^R] = \ln \delta, \tag{60}$$

so Equation (58) reduces to:

$$h_\alpha[Q_{\Delta,u}] + h_\beta[Q_{\delta,v}] \geq \ln \left(\frac{\pi \hbar |\gamma|}{\varepsilon_\alpha(\Gamma/4)} \right). \tag{61}$$

If we perform the limit $\Gamma/4 \to 0$ in Equation (61), we have $(1/2)R_{00}^2(\Gamma/4, 1) \to 1/2$, and considering that $1/e < \alpha^{\frac{1}{2-2\alpha}} \beta^{\frac{1}{2-2\beta}} \leq 1/2$ when $1/2 < \alpha \leq 1$ (see Figure 4) we recover the *Rényi-entropy* UR in Equation (24) and when $\alpha = 1$ the *Shannon* UR in Equation (20).

Now, we can decompose the differential Rényi entropies in the left hand side of Equation (58) as (see Appendix A):

$$h_\alpha[Q_{\Delta,u}] = H_\alpha\left[P_\Delta^{(u)}\right] + h_\alpha[D_\Delta] \quad \text{and} \quad h_\beta[Q_{\delta,v}] = H_\beta\left[P_\delta^{(v)}\right] + h_\beta[D_\delta], \tag{62}$$

where we denote the set of discrete probabilities appearing in Equation (53) as $P_\Delta^{(u)} := \{p_{\Delta,k}^{(u)}\}$ and $P_\delta^{(v)} := \{p_{\delta,k}^{(v)}\}$, respectively. Please note that, for pdfs with bounded support, the Rényi entropy is maximized for the uniform distribution [173], so we always have: $h_\alpha[D_\Delta] \leq \ln(\Delta)$ and $h_\beta[D_\delta] \leq \ln(\delta)$.

If we apply the result Equation (62) to the inequality Equation (58) we recover the result proved in Reference [24] for the discrete entropies:

$$H_\alpha[P_\Delta^{(u)}] + H_\beta[P_\delta^{(v)}] \geq \ln\left(\frac{\pi}{\varepsilon_\alpha(\Gamma/4)\Gamma}\right). \tag{63}$$

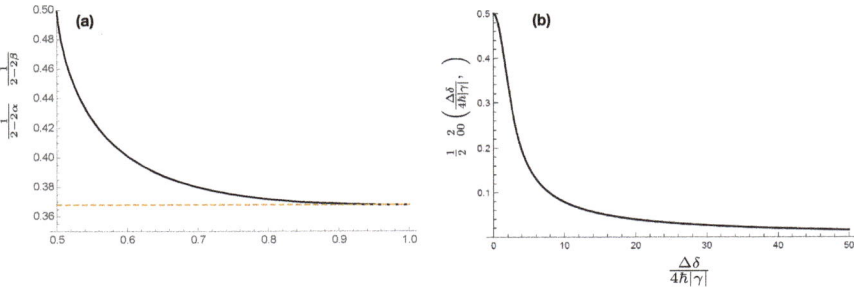

Figure 4. In panel (**a**) the full line is the graph of the function $f(\alpha) = \alpha^{\frac{1}{2-2\alpha}}\beta^{\frac{1}{2-2\beta}}$, with $0 < \alpha \leq 1$, and where $\beta(\alpha) = \alpha/(2\alpha-1)$ that stems from the condition $1/\alpha + 1/\beta = 2$. The horizontal dashed line is drawn to indicate the limit $\lim_{\alpha\to 1} f(\alpha) = 1/e$. In panel (**b**) we plot the behaviour of $g(y) = (1/2)R_{00}(y,1)$ as a function of $y := \Delta\delta/(4\hbar|\gamma|)$. Although $g(y)$ is shown in the range $0 \leq y \leq 50$, it is important to note that $g(y)$ is continuous monotonically decreasing function in the positive real axis such $\lim_{y\to\infty} g(y) = 0$.

This is the coarse-grained version of the *Rényi entropy* UR in Equation (24) (Schürmann conjectured [169] that $\varepsilon_1(z)$ defined in (59), in the context of Equation (63) could be replaced by $e^{-1}R_{00}^2(2z/e,1)$). We shall also emphasize, as the title of this subsection suggests, that the demonstration of the URs (63) presented in Reference [24] uses explicitly the fact that $\hat{u}$ and $\hat{v}$ form a CCO pair. Therefore, the UR in Equation (58) is, in principle, valid only for CCO pairs, since it can be obtained from Equation (63) by adding $h_\alpha[D_\Delta] + h_\beta[D_\delta]$ to both sides, and using Equation (62).

The discrete Rényi entropy is always positive, and we have

$$\lim_{\Gamma\to+\infty} \ln\left(\frac{\pi}{\Gamma\varepsilon_\alpha(\Gamma/4)}\right) = \lim_{\Gamma\to+\infty} \ln\left(\frac{\pi}{\frac{1}{2}\Gamma R_{00}^2\left(\frac{\Gamma}{4},1\right)}\right) = 0, \tag{64}$$

with the last line being valid because $\lim_{x\to\infty}(2x/\pi)R_{00}^2(x,1) = 1$ (Equation (28) in [174] reads: $\frac{z}{2\pi}R_{00}^2(z/4,1) \sim 1 - 2\sqrt{\pi z}e^{-z/2}$. This result is based on the appropriate asymptotic expansion [175] valid for $z \to \infty$). This results show that the coarse-grained UR in Equation (63) is non-trivially satisfied for an arbitrary (even very large) values of the coarse-graining widths. However, this desired property is not enjoyed by the UR

$$H_\alpha[P_\Delta^{(u)}] + H_\beta[P_\delta^{(v)}] \geq \ln\left(\frac{\pi}{\alpha^{\frac{1}{2-2\alpha}}\beta^{\frac{1}{2-2\beta}}\Gamma}\right), \tag{65}$$

first derived in [31]. This UR corresponds to Equation (63) in the coarse-grained regime $\Gamma/4 \lesssim 1.79$ in which $\varepsilon_1(\Gamma/4) = 1/e$. Obviously, this is not a mere coincidence, as Equation (63) subsumes (65). This is clearly visible inside the definition of ε which involves the minimum of two different bounds.

When $\Gamma/4 > 1.79$ the lower bound in Equation (65) is negative so this UR is trivially satisfied, since the discrete entropy is always non-negative.

From the above considerations we can obtain an UR for the variances, $\sigma^2_{Q_{\Delta,u}}$ and $\sigma^2_{Q_{\delta,v}}$, if we set $\alpha = 1$ in Equation (58) and use the inequality (22):

$$\ln\left(2\pi e \sigma_{Q_{\Delta,u}}\sigma_{Q_{\delta,v}}\right) \geq h[Q_{\Delta,u}] + h[Q_{\delta,v}] \geq \ln\left(\frac{\pi\hbar|\gamma|\ e^{h[D_\Delta]-\ln\Delta+h[D_\delta]-\ln\delta}}{\varepsilon_1(\Gamma/4)}\right), \tag{66}$$

where $h[\cdot]$ stands for the Shannon entropy. Now, we can use the decompositions:

$$\sigma^2_{Q_{\Delta,u}} = \sigma^2_{P^{(u)}_\Delta} + \sigma^2_{D_\Delta} \quad \text{and} \quad \sigma^2_{Q_{\delta,v}} = \sigma^2_{P^{(v)}_\delta} + \sigma^2_{D_\delta}, \tag{67}$$

where the variances of the discrete probability distributions were defined in Equation (42), while $\sigma^2_{D_\Delta}$ and $\sigma^2_{D_\delta}$, are the variances of the generic HFs. Therefore, applying the above splitting to Equation (66) we arrive at the lower bound [24]:

$$\left(\sigma^2_{P^{(u)}_\Delta} + \sigma^2_{D_\Delta}\right)\left(\sigma^2_{P^{(v)}_\delta} + \sigma^2_{D_\delta}\right) \geq \frac{\hbar^2\gamma^2}{4}\frac{e^{2(h[D_\Delta]-\ln\Delta+h[D_\delta]-\ln\delta-1)}}{\varepsilon_1^2(\Gamma/4|\gamma|)}. \tag{68}$$

When the HF are rectangular, and in the coarse-grained regime $\Gamma/(4|\gamma|) \lesssim 1.79$ where $\varepsilon_1(\Gamma/4|\gamma|) = 1/e$, we recover the UR [32]:

$$\left(\sigma^2_{P^{(u)}_\Delta} + \frac{\Delta^2}{12}\right)\left(\sigma^2_{P^{(v)}_\delta} + \frac{\delta^2}{12}\right) \geq \frac{\hbar^2\gamma^2}{4}, \tag{69}$$

where we have used the fact that in this case

$$\sigma^2_{D_\Delta^R} = \frac{\Delta^2}{12} \quad \text{and} \quad \sigma^2_{D_\delta^R} = \frac{\delta^2}{12}. \tag{70}$$

Both Equations (68) and (69) are the coarse-grained versions of the *Heisenberg* UR in Equation (8). It is important to emphasize that Equation (69) cannot be obtained by the simple substitution $\sigma^2_{P_u} \to \sigma^2_{P^{(u)}_\Delta}$ and $\sigma^2_{P_v} \to \sigma^2_{P^{(v)}_\delta}$ done inside the *Heisenberg* UR.

Although both $\sigma^2_{D_\Delta}$ and $\sigma^2_{D_\delta}$ are the variances of a generic HF, *viz.* $D_\Delta(u, u_k)$ and $D_\delta(v, v_k)$ for any value of k, it is interesting to associate them to the respective central bins, namely those that contain the mean value of the probability distributions P_u and P_v. By doing this, together choosing the origins of the coordinates in the middle of the central bin, we can see that the variances $\sigma^2_{P^{(u)}_\Delta}$ and $\sigma^2_{P^{(v)}_\delta}$ are free from contributions associated with the statistics relevant for the central bins. Thus, if the widths of the coarse graining increase in the measurement of $\hat{u}$ and $\hat{v}$, the respective central bin-widths grow, so that the variances $\sigma^2_{P^{(u)}_\Delta}$ and $\sigma^2_{P^{(v)}_\delta}$ only involve contributions from the tails of the probability distributions $Q_{\Delta,u}$ and $Q_{\delta,v}$. Therefore, for large coarse grainings, the variances $\sigma^2_{D_\Delta}$ and $\sigma^2_{D_\delta}$ become more important in the inequalities Equations (68) and (69). Thus, in the regime when:

$$\Gamma \geq \pi e \Rightarrow \Gamma \geq \frac{\pi}{\varepsilon_1(\Gamma/4|\gamma|)} \Rightarrow \Gamma^2 \geq \frac{1}{4}\frac{e^{2(h[D_\Delta]+h[D_\delta])}}{e^2\sigma^2_{D_\Delta}\sigma^2_{D_\delta}\varepsilon_1^2(\Gamma/4|\gamma|)} \Rightarrow$$

$$\Rightarrow \quad \sigma^2_{D_\Delta}\sigma^2_{D_\delta} \geq \frac{\hbar^2|\gamma|^2}{4}\frac{e^{2(h[D_\Delta]-\ln\Delta+h[D_\delta]-\ln\delta-1)}}{\varepsilon_1^2(\Gamma/4|\gamma|)} \tag{71}$$

both Equations (68) and (69) are satisfied trivially. Note, that in Equation (71) we have used the relation $4\pi^2 \geq e^{2(h[D_\Delta]+h[D_\delta])}/e^2\sigma^2_{D_\Delta}\sigma^2_{D_\delta} > 0$ which can be obtained from the inequality in Equation (22).

However, Equation (68) is only the starting point for the second construction, proposed in [24], that is free from the above limitation, and cannot be trivially satisfied. This improved UR reads:

$$K\left(\frac{\sigma^2_{P^{(u)}_\Delta}}{\Delta^2}\right) K\left(\frac{\sigma^2_{P^{(v)}_\delta}}{\delta^2}\right) \geq \frac{\pi^2}{\Gamma^2 \varepsilon_1^2(\Gamma/4)}, \tag{72}$$

where $K(t)$ is implicitly defined as

$$K(t) := \frac{\exp\left[2t\mathcal{M}^{-1}(t)\right]}{\text{erf}^2\left(\sqrt{\mathcal{M}^{-1}(t)}/2\right)},$$

with $\text{erf}(x) := (2/\sqrt{\pi}) \int_0^x e^{-y^2} dy$ being the error function and $\mathcal{M}^{-1}(t)$ denoting the inverse of the invertible function

$$\mathcal{M}(y) := \frac{\exp(-y/4)}{2\sqrt{\pi y}\, \text{erf}(\sqrt{y}/2)}.$$

The idea behind derivation of the coarse-grained UR in Equation (72) is the following. Let us rewrite Equation (68) in the form:

$$\xi(h[D_\Delta], \sigma^2_{D_\Delta}, h[D_\delta], \sigma^2_{D_\delta}) := \frac{\left(\sigma^2_{P^{(u)}_\Delta} + \sigma^2_{D_\Delta}\right)\left(\sigma^2_{P^{(v)}_\delta} + \sigma^2_{D_\delta}\right)}{e^{2(h[D_\Delta]+h[D_\delta]-1)}} \geq \frac{1}{4\Gamma^2\varepsilon_1^2(\Gamma/4)}.$$

Now the function ξ is supposed to be minimized; however, because the Shannon entropy $h[D_\Delta]$ ($h[D_\delta]$) is interrelated with (bounded by a function of) the variance $\sigma^2_{D_\Delta}$ ($\sigma^2_{D_\delta}$) the minimization needs to be performed in two steps. For fixed values of the variances $\sigma^2_{D_\Delta}$ and $\sigma^2_{D_\delta}$, the function ξ achieves its minimum when the Shannon entropies $h[D_\Delta]$ and $h[D_\delta]$ are maximized with respect to the functional form of the HFs, D_Δ and D_δ. As already stated, the HFs are constrained by the requirement of the fixed value for both variance. The form of the HF with maximum Shannon entropy [24] is a Gaussian with support restricted to the central bin and whose variance is an appropriate function of $\sigma^2_{D_\Delta}$ ($\sigma^2_{D_\delta}$) (for details see [24].) Therefore, for this optimal HF its Shannon entropy $h[D_\Delta^{op}]$ ($h[D_\delta^{op}]$) is only a function of the variance $\sigma^2_{D_\Delta}$ ($\sigma^2_{D_\delta}$), thus we have $G(\sigma^2_{D_\Delta}, \sigma^2_{D_\delta}) = \xi(h[D_\Delta^{op}], \sigma^2_{D_\Delta}, h[D_\delta^{op}], \sigma^2_{D_\delta})$. The second step is a direct minimization of $G(\sigma^2_{D_\Delta}, \sigma^2_{D_\delta})$, which results in the left hand side product in Equation (72).

According to the discussion above Equation (71) the coarse-grianed UR in Equation (72) has no contributions from the statistics corresponding to the central bin. In the limit when $\Delta, \delta \to 0$ we recover the *Heisenberg* UR in Equation (8) thanks to the identities [24]

$$\lim_{\Delta \to 0} \Delta^2 K\left(\frac{\sigma^2_{P^{(u)}_\Delta}}{\Delta^2}\right) = \sigma^2_{P_u} \lim_{y \to 0} \frac{1}{\mathcal{M}(y)} \frac{\exp(2y\mathcal{M}(y))}{\text{erf}^2\left(\sqrt{y}/2\right)} = 2\pi e \sigma^2_{P_u}. \tag{73}$$

In the opposite limit of infinite coarse graining, *viz* $\Delta, \delta \to \infty$, we have $\sigma^2_{P^{(u)}_\Delta}, \sigma^2_{P^{(v)}_\delta} \longrightarrow 0$ and

$$\overbrace{\lim_{\sigma^2_{P^{(u)}_\Delta} \to 0} K\left(\frac{\sigma^2_{P^{(u)}_\Delta}}{\Delta^2}\right)}^{=1} \overbrace{\lim_{\sigma^2_{P^{(v)}_\delta} \to 0} K\left(\frac{\sigma^2_{P^{(v)}_\delta}}{\delta^2}\right)}^{=1} \geq \overbrace{\lim_{\Gamma \to \infty} \frac{\pi^2}{\Gamma^2 \varepsilon_1^2(\Gamma/4)}}^{=1}. \tag{74}$$

It is important to note that since

$$\frac{\pi^2}{\Gamma^2 \varepsilon_1^2(\Gamma/4)} > 1,$$

(75)

whenever both Δ and δ are finite, it is forbidden to set $\sigma^2_{P_\Delta^{(u)}}$ and $\sigma^2_{P_\delta^{(v)}}$ as simultaneously equal to zero, as it would contradict the coarse-grained UR (72). This means that any quantum state (pure or mixed) cannot be localised in both observables $\hat{u}$ and $\hat{v}$ that are CCOs. In other words, the associated probability distributions cannot simultaneously have compact support.

This remarkable conclusion somehow threatens the scientific program to recover classical mechanics solely from coarse-grained averaging, physically originating from the finite-precision of the observations [19,176,177]. Indeed, quantum features can be observed in the measurement of $\hat{u}$ and $\hat{v}$ irrespective of the precision of the detectors. However, for very large coarse-graining widths the variances $\sigma^2_{P_\Delta^{(u)}}$ and $\sigma^2_{P_\delta^{(v)}}$ are dominated by the contributions from the tails of the $P_\Delta^{(u)}$ and $P_\delta^{(v)}$. Thus, as these probabilities are likely very small, they would be particularly susceptible to statistical fluctuations and it would in general require very long acquisition times to collect the sufficient amount of data necessary to verify the UR (72) in the regime of extremely large coarse graining.

5.2. URs Valid for General Observables, $\hat{u}$ and $\hat{v}$, Defined in Equation (3)

If we let $\alpha = 1$ in Equation (58), use rectangular HFs such that Equation (60) is valid and restrict the size of the involved bins such that $\varepsilon_1(\Gamma/4|\gamma|) = 1/e$—this is the regime of the coarse graining when $\Gamma/4 \lesssim 1.79$—we obtain the simplified coarse-grained UR of the form:

$$h[Q_{\Delta,u}] + h[Q_{\delta,v}] \geq \ln\left(\pi e \hbar |\gamma|\right).$$

(76)

Because the coarse-grained UR in Equation (58) was derived only for CCOs, $\hat{u}$ and $\hat{v}$, a priori it is not clear why the above UR could remain valid also for generalized observables defined in Equation (3). This fact, however, can be proved with the help of the *Shannon-entropy* UR (20), that has properly been extended to the desired observables, and the inequalities:

$$h[Q_{\Delta,u}] \geq h[P_u] \quad \text{and} \quad h[Q_{\delta,v}] \geq h[P_v],$$

(77)

whose detailed derivation based on the Jensen inequality is relegated to Appendix B. Passing to the discrete entropies we find the coarse-grained UR:

$$H[P_\Delta^{(u)}] + H[P_\delta^{(v)}] \geq \ln\left(\frac{\pi e}{\Gamma}\right),$$

(78)

which looks the same as the one derived in [30] for CCOs. Here, the validity of this UR has been extended for any observables $\hat{u}$ and $\hat{v}$ as defined in Equation (3). Also, following the same arguments that lead from Equation (66) to the UR in Equation (69) we can see that the UR for the discrete variances is also valid for general $\hat{u}$ and $\hat{v}$ as defined in Equation (3).

To briefly summarize, entropic uncertainty relations for coarse-grained probability distributions were almost only considered for position and momentum variables. As far as we know, the only exceptions are given in References [58,65]. However, as we have shown here, the generalization of entropic URs for differential probabilities associated with general observables $\hat{u}$ and $\hat{v}$, which are linear combinations of position and momentum, can be done in many cases. However, in each case a careful analysis should be carried out to verify that the related coarse-grained URs are also valid for these generalised operators. Here, we have done this only in the simple cases.

5.3. Coarse-Grained URs Merged with the Majorization Approach

In [174] the coarse-grained scenario has been discussed with the help of the results obtained in [156,157,159], namely the majorization-based approach to quantification of uncertainty. To say it briefly, a majorization relation $x \prec y$ between two arbitrary d-dimensional probability distributions means that for every $n \leq d$ the inequality $\sum_{k=1}^{n} x_k^{\downarrow} \leq \sum_{k=1}^{n} y_k^{\downarrow}$ holds, with an equality (normalization) for $n = d$. Traditionally, by "$\downarrow$" we denote the decreasing order, so that $(x^{\downarrow})_k \geq (x^{\downarrow})_l$, for all $k \leq l$. The Rényi entropy (and also others, such as the Tsallis entropy) is Schur-concave, which implies $H_{\alpha}[x] \geq H_{\alpha}[y]$ whenever $x \prec y$.

In the context of coarse-grained probability distributions it was conceptually simpler to consider the so-called direct-sum majorization introduced in [159]. An advantage of the majorization approach is that it covers a regime of (α, β) parameters, $\beta = \alpha$ to be precise, which in some way is perpendicular to the conjugate choice $1/\alpha + 1/\beta = 2$. In [174] an infinite hierarchy of majorization vectors, depending on a single parameter $\Gamma = \Delta\delta/\hbar$, has been derived. The discussion is conducted for CCOs, thus one can easily recognize the dimensionless Γ parameter as those which appears in all previous URs with $\gamma = 1$.

The main result, namely a family of lower bounds denoted as $\mathcal{R}_{\alpha}^{(n)}(\Delta\delta/\hbar)$ for $n = 2, \dots, \infty$, has been presented in Equation (27) from [174], however, we refrain from providing its detailed construction here. It seems enough to say that the bound in question is a function of $R_{00}^2(j_0\Gamma/4, 1)$ with j_0 being certain positive integers. In other words, in spirit, the majorization bound is close to that derived in [24] and extensively discussed above. A comparison of the new bound and (63) for $\alpha = 1 = \beta$—the only value of both parameters for which the involved bounds describe the same situation—showed that $\mathcal{R}_1^{(3)}$ outperforms (63) in the regime when the R_{00}-term does contribute to ε_1.

Asymptotic behavior of the new and previous coarse-grained bounds shows that for $\alpha = 1 = \beta$ and large Γ, all $\mathcal{R}_1^{(n)}$ bounds improve (63) by a divergent factor $\Gamma/4$. Moreover, the typical behavior of discrete majorization bounds has been confirmed in the coarse-grained setting. In the discrete case, the majorization relations almost surely dominate the MU bound, with an exception being a small neighborhood of the point for which the unitary matrix **U** is the Fourier matrix. The analog of the Fourier matrix in the coarse-grained scenario is the continuous limit $\Gamma \to 0$. This probably intuitive fact has been rigorously shown by means of the asymptotics of $\mathcal{R}_1^{(\infty)}$ for small Γ, which is equal to $-\frac{1}{2}\ln\Gamma$.

5.4. Other Coarse-Grained URs

At the very end of this long section we would like to touch on a few coarse-grained URs which go beyond the standard position-momentum conjugate pair. First of all, Bialynicki-Birula also provided his major *Shannon entropy* UR in the case of angle and angular momentum [30], as well as (together with Madajczyk) to the variables on the sphere [178]. Coarse graining in these physical settings is only relevant for the periodic CVs (angle on a circle and two angles on a sphere), as the conjugate variables are discrete (though infinite dimensional).

Also, the coarse-grained scenario has been developed [179] in relation to the memory-assisted UR [180] relevant for quantum key distribution. The result, even though non-trivial, differs from Equation (63) in a similar fashion as the MU bound differs from the UR in the presence of quantum memory by Berta et al [180].

Going in a completely different direction, Rastegin [181] in his recent contribution proposed an extension of (65) to the case of a modified CCR, which assumes the form $[\hat{x}, \hat{p}] = i\hbar(1 + \beta\hat{p}^2)$. The parameter β is related to the so-called minimal length predicted by certain variants of string theory and similar approaches (not to be confused with β playing the role of a conjugate parameter in the MU bound and similar URs for the Rényi entropies).

Last but not least, some of us have very recently derived an inequality (see Equations 9–12 from [182]), which could be understood as an UR (valid for CCOs) in the setting relevant for periodic coarse graining discussed in Section 4.1.2. As this UR involves additional averaging of $\mathrm{p}_k^{(x)}(\hat{\rho})$ and

$p_l^{(p)}(\hat{\rho})$ defined below Equation (47) with respect to the positioning degrees of freedom, we do not provide further details of this construction encouraging the interested reader to consult [182].

6. Applications of Coarse-Grained Measurements and Coarse-Grained Uncertainty Relations

As discussed above, when detecting the position and momentum of particles such as photons or individual atoms, coarse-grained measurements are not just necessary but can be much more practical. In this regard, URs that deal with coarse-grained measurements can be useful for several applications, such as those discussed in Section 3.

Section 3 discussed the use of URs along with the PPT arguement for the convenient detection of quantum entanglement in continous variable quantum systems. However, sufficient care must be taken in regards to coarse-grained measurements. The pitfalls of applying the usual entanglement criteria for continuous variables to coarse-grained measurements was discussed in Reference [26], where it was argued that this can lead to false-positive identifications of entanglement, such that the entanglement criteria based on uncertainty relations discussed in Section 3 can be (falsely) violated even for separable states. For a simple illustration of this, consider the most trivial example of a separable continuous variable state, the two-mode vacuum state [10]. Even though the state is separable, improper application of entanglement criteria without correctly taking account of coarse graining can lead to erroneous results. A demonstration of this is shown in Figure 5. We consider the results from coarse grained measurements, and apply entanglement criteria based on an ideal continuous variable UR and its coarse-grained version. The red circles show a variance-based entanglement criteria based on the variance product UR Equation (8), using the global operators defined in Equations (29) and (30), as developed in Reference [99]. Here we have subtracted the lower bound from the product of variances, so that a negative value indicates entanglement. As can be seen, when the coarse graining is large, one would erroneously conclude that the quantum state is entangled. On the other hand, the coarse-grained variance product UR (69) applied to the global operators (29) and (30) never indicates that the state is entangled, as indicated by the blue squares in Figure 5. Similar results hold for other UR-based entanglement criteria.

To show how coarse-grained data should be properly handled to identify entanglement, an experimental study was performed in a system of spatially-entangled photons [26]. In particular, the same variance criteria based on (69) was tested for the global operators defined in Equations (29) and (30), in which case entanglement was identified for a wide range of coarse-graining widths. It was also shown that coarse-grained entropic entanglement criteria, for example based on inequality Equation (58) ($\alpha = \beta = 1$) applied to operators (29) and (30), can be superior to coarse-grained variance-based criteria, identifying entanglement when variance criteria do not, even for the case of Gaussian states. This is due to the fact that the coarse-grained probability distributions functions such as those shown in Figure 2 are not Gaussian functions, even when the quantum state under investigation is Gaussian.

An advantage of coarse-graining is that the measurement time can be drastically reduced. In Reference [86] EPR-steering was tested for discrete distributions of measurements made from standardized binning on the two-photon state produced from spontaneous parametric down-conversion, using a coarse-grained version of the EPR-steering criteria of Reference [85]. Bi-dimensional steering was observed for sample sizes ranging from 8×8 to 24×24, representing a considerable reduction in measurement overhead when compared with the quasi-continuous measurements reported in Reference [85], which sampled about 100 data points per cartesian direction (about 10^4 total measurements) to evaluate entropic EPR-steering criteria of continuous variables.

Standard coarse graining has been studied in the context of quantum state reconstruction of single and two-mode Gaussian states, and the quantum to classical transition [183]. Two scenarios were considered: direct reconstruction of the covariance matrix alone, and full reconstruction of the state using maximum likelihood estimation. The reconstructed coarse-grained functions were compared to those of Gaussian states subject to thermal squeezed reservoirs, indicating that in this context coarse graining does not produce a thermalized (decohered) Gaussian state.

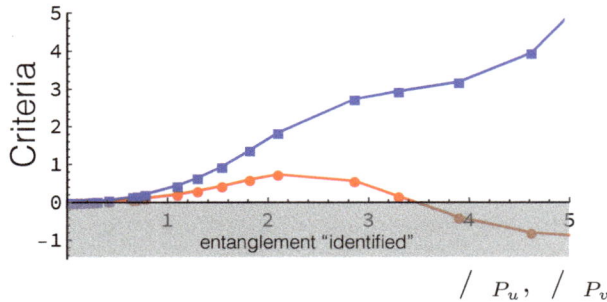

Figure 5. Numerical results testing entanglement criteria for the two-mode vacuum state, a separable pure state. The entanglement criteria are based on URs following the PPT argument outlined in Section 3. The criteria are evaluated as a function of the bin widths $\Delta = \delta$, which are given in units of the standard deviations σ_{P_u} and σ_{P_v}. We note that $\sigma_{P_u} = \sigma_{P_v}$ for the two-mode vacuum state. The red circles show the variance product UR Equation (8), where we apply the naive approach in which the variances of the continous variables are calculated from the discretized data using Equation (42). One can see that in this case we obtain a false-positive for entanglement when the coarse-graining widths are large. The blue squares show the coarse-grained variance product UR Equation (69), both applied to the global operators Equations (29) and (30). Here the lower bounds for both inequalities have been subtracted, so that a negative value indicates entanglement. The lines are merely guides for the eye.

The work mentioned above considered standard coarse graining, as described in Section 4.1. In some cases it is interesting to consider different models, such as that of periodic coarse graining described in Section 4.1.2. The mutual unbiasedness of periodic coarse graining described in Section 4.2 has been tested experimentally for two [142] and even three [143] phase-space directions. It was shown that mutual unbiasedness appears when the appropriate bin widths of the two or three conjugate variables are chosen. Periodic coarse graining has also been used in the detection of spatial correlations of photon pairs from SPDC [182]. Using a novel entanglement criteria based on the UR for characteristic functions [153], it was possible to identify entanglement with as few as 2×2 measurements in position and momentum (8 total), representing a considerable reduction in measurement overhead.

Simple binary binning of homodyne measurements has been proposed as a means to test dichotomic Bell's inequalities in CV systems, while allowing for high detection efficency [184–187]. Other types of non-standard coarse graining have been proposed as a means to violate Bell's inequality using homodyne measurements on non-Gaussian states [188]. Though it was shown that one could achieve maximal violation in principle, exotic non-Gaussian states are required. In Reference [27] it was shown that imperfect binning could result in false violations of Bell's inequalities, and even in violations of Cirelson's bound for quantum Bell correlations.

A closely related subject to periodic coarse graining of CVs is that of the so called modular variables [189–191], for which phase-space variables u are rewritten as $u = n_u \ell + \bar{u}$, where n_u is the integer component and $\bar{u}$ the modular component, such that $0 \le \bar{u} < \ell$. Here ℓ is a scaling parameter of appropriate dimension. For two CCOs, such as $\hat{x}$ and $\hat{p}$ for example, the integer operator of one observable-say-$\hat{n}_x$ and the modular operator of the other observable-$\bar{\hat{p}}$ satisfy URs that closely resemble those of the angle and angular momentum variables [30]. The modular variable construction was first introduced by Aharanov et al. [138,189] as a method to identify non-locality in quantum mechanics. Since then, several interesting applications have been developed. Variance-based URs for the modular variable construction were proposed as a method to identify a novel type of squeezing, as well as entanglement in pairs of atoms [192]. This entanglement criteria was used in References [193], along with one based on entropic uncertainty relations, to identify spatial entanglement of photon pairs that have passed through multiple slit apertures. Application to multiple-photon states was studied in Reference [194]. It is worth noting that in this case the usual CV entanglement criteria as discussed in Section 3 are incapable of detecting entanglement. Modular variables have been proposed as a way to test for the Greenberger-Horne-Zeilinger paradox in CV systems [195], as well as quantum contextuality [196–198] and as a method to construct algebras resembling that of discrete systems [190,191,199].

Finally, we briefly mention that URs play an important role in the attempt to unify quantum theory with general relativity. In this case, the Heisenberg uncertainty principle is modified to become a generalized uncertainty principle, taking into account Planck scale effects, which impose coarse-graining that is a fundamental part of nature, leading to minimum and maximum length quantum mechanics. An extensive amount of literature exists on the subject, for two recent reviews, see References [200,201].

7. Conclusions

Uncertainty relations play an important role in quantum physics, which is two-fold: on the one hand they have historically represented the difference between classical and quantum physics, while on the other hand they are a tool that can be used to identify and even quantify interesting quantum properties. Beginning with the seminal work of Heisenberg in 1927, several uncertainty relations have been developed for continuous variable quantum systems. However, in a realistic experimental setting, one never has access to the infinite dimensional spectrum associated to these observables. Thus, coarse graining is imposed by the detection apparatus to account for the measurement precision and range.

Here we have provided a review of several quantum mechanical uncertainty relations tailored specifically to coarse-grained measurement of continuous quantum observables. Our aim was to survey the state-of-the-art of the subject, from both the theoretical advances to experimental application of coarse-grained uncertainty relations. We also extend the validity of some of the coarse-grained URs, already in the literature, to general linear combinations of canonical observables in n-mode bosonic systems.

Several interesting open questions remain. First, it would be interesting to see the generalization of all the coarse-grained URs presented here for pairs of observables that are connected by general unitary metaplectic transformations. Second, one can consider applying coarse graining to URs not mentioned explicitly here, such as the triple variance product criteria [120,150], UR for characteristic functions [153], among others, as well the plethora of moment inequalities arising from tests for non-classicality [68,72] and entanglement [117,118,121]. Third, and more important, a deep discussion of the role of coarse-grained URs within the scientific program to recover classical mechanics solely from coarse-grained averaging should be developed. We hope that this review encourage this discussion.

Author Contributions: All authors contributed equally to this work.

Acknowledgments: The authors acknowledge financial support from the Brazilian Funding Agencies CNPq, CAPES (PROCAD2013 project) and FAPERJ, and the National Institute of Science and Technology—Quantum Information. Ł.R. acknowledges financial support by Grant number 2015/18/A/ST2/00274 of the National Science Center, Poland.

Conflicts of Interest: The authors declare no conflict of interest. The founding sponsors had no role in the design of the study, in the collection, analyses, or interpretation of data, in the writing of the manuscript, and in the decision to publish the results.

Abbreviations

The following abbreviations are used in this manuscript:

CV   Continuous variable
UR   Uncertainty relation
QIT   Quantum information theory
CCR   Canonical commutation relation
CCO   Canonically conjugate operators
pdf   probability distribution function
EPR   Einstein-Podolsky-Rosen
PPT   Positive partial transposition
PCG   Periodic coarse graining
MU   Maassen-Uffink
HF   Histogram function

Appendix A

Following [24] we aim to prove the decomposition in Equation (62). To this end it is enough to discuss the case of $\hat{u}$ since the proof for $\hat{v}$ looks the same. We can write:

$$
\begin{aligned}
h_\alpha[Q_{\Delta,u}] &= \frac{1}{1-\alpha} \ln\left(\int_{-\infty}^{\infty} du \, [Q_{\Delta,u}(u)]^\alpha \right) = \frac{1}{1-\alpha} \ln\left(\sum_{k\in\mathcal{Z}_k} \int_{\mathcal{R}_k} du \, [Q_{\Delta,u}(u)]^\alpha \right) \\
&= \frac{1}{1-\alpha} \ln\left(\sum_{k\in\mathcal{Z}_k} \left[p_{\Delta,k}^{(u)} \right]^\alpha \int_{\mathcal{R}_k} du \, [D_\Delta(u,u_k)]^\alpha \right),
\end{aligned}
\tag{A1}
$$

where we use the fact that the function $Q_{\Delta,u}(u)$ in the interval $\mathcal{R}_k$ is equal to $p_{\Delta,k}^{(u)} D_\Delta(u,u_k)$. Now, because the shape of the HF, $D_\Delta(u,u_k)$, is the same for all values of k, the integral $\int_{\mathcal{R}_k} du \, [D_\Delta(u,u_k)]^\alpha$ does not depend on k. Therefore, we can write:

$$
h_\alpha[Q_{\Delta,u}] = \frac{1}{1-\alpha} \ln\left(\sum_{k\in\mathcal{Z}_k} \left[p_{\Delta,k}^{(u)} \right]^\alpha \right) + \frac{1}{1-\alpha} \ln\left(\int_{\mathcal{R}_k} du \, [D_\Delta(u,u_k)]^\alpha \right),
\tag{A2}
$$

that corresponds to the decomposition in Equation (62).

Appendix B

Here, we prove the inequalities in Equation (77). As before, is enough to consider the single case relevant for the variable u. In the next few lines, we actually closely follow the treatment presented in [12]. First we define the mean value within the kth histogram bin:

$$
\langle \ldots \rangle_k := \frac{1}{\Delta} \int_{(k-1/2)\Delta}^{(k+1/2)\Delta} \ldots du.
\tag{A3}
$$

Then, because the function $f(x) = x\ln(x)$ is convex we can apply Jensen's inequality [60] to obtain,

$$
\langle P_u \ln P_u \rangle_k \geq \langle P_u \rangle_k \ln\langle P_u \rangle_k.
\tag{A4}
$$

Now we can use the definition in Equation (53), multiply both sides by -1 and sum over $k \in \mathcal{Z}_k$:

$$
-\sum_{k\in\mathcal{Z}_k} p_{\Delta,k}^{(u)} \ln p_{\Delta,k}^{(u)} + \left(\sum_{k\in\mathcal{Z}_k} p_{\Delta,k}^{(u)} \right) \ln(\Delta) \geq -\sum_{k\in\mathcal{Z}_k} \int_{(k-1/2)\Delta}^{(k+1/2)\Delta} P_u(u) \ln P_u(u) \, du.
\tag{A5}
$$

After using the condition in Equation (36), the definition of the discrete Shannon entropy $H[P_\Delta^{(u)}] := -\sum_{k\in\mathcal{Z}_k} p_{\Delta,k}^{(u)} \ln p_{\Delta,k}^{(u)}$, the decomposition in Equation (62) with $\alpha = 1$ and $h[D_\Delta^R] = \ln\Delta$, and the definition of the differential Shannon entropy in Equation (21) we obtain:

$$h[Q_{\Delta,u}] := H[P_\Delta^{(u)}] + \ln\Delta \geq h[P_u], \tag{A6}$$

which is the desired result.

References

1. Wheeler, J.A.; Zurek, W.H. (Eds.) *Quantum Theory and Measurement*; Princeton University Press: Princeton, NJ, USA, 1983.
2. Scully, M.O.; Englert, B.G.; Walther, H. Quantum optical tests of complementarity. *Nature* **1991**, *351*, 111–116. [CrossRef]
3. Kim, Y.H.; Kulik, S.; Shih, Y.; Scully, M. Delayed Choice Quantum Eraser. *Phys. Rev. Lett.* **2000**, *84*, 1. [CrossRef] [PubMed]
4. Bertet, P.; Osnaghi, S.; Rauschenbeutel, A.; Nogues, G.; Auffeves, A.; Brune, M.; Raimond, J.M.; Haroche, S. A complementarity experiment with an interferometer at the quantum-classical boundary. *Nature* **2001**, *411*, 166–170. [CrossRef] [PubMed]
5. Walborn, S.P.; Cunha, M.O.T.; Pádua, S.; Monken, C.H. Double-slit quantum eraser. *Phys. Rev. A* **2002**, *65*, 0338. [CrossRef]
6. Mandel, L. Coherence and indistinguishability. *Opt. Lett.* **1991**, *16*, 1882–1883. [CrossRef] [PubMed]
7. Englert, B.G. Fringe Visibility and Which-Way Information: An Inequality. *Phys. Rev. Lett.* **1996**, *77*, 2154. [CrossRef] [PubMed]
8. Ozaktas, H.M.; Zalevsky, Z.; Kutay, M.A. *The Fractional Fourier Transform: with Applications in Optics and Signal Processing*; John Wiley and Sons Ltd.: New York, NY, USA, 2001.
9. Coles, P.J.; Berta, M.; Tomamichel, M.; Wehner, S. Entropic uncertainty relations and their applications. *Rev. Mod. Phys.* **2017**, *89*, 848–858. [CrossRef]
10. Braunstein, S.L.; van Loock, P. Quantum information with continuous variables. *Rev. Mod. Phys.* **2005**, *77*, 513. [CrossRef]
11. Adesso, G.; Ragy, S.; Lee, A.R. Continuous Variable Quantum Information: Gaussian States and Beyond. *Open Syst. Inf. Dyn.* **2014**, *21*, 1440001. [CrossRef]
12. Bialynicki-Birula, I.; Rudnicki, Ł. Entropic Uncertainty Relations in Quantum Physics. In *Statistical Complexity: Applications in Electronic Structure Chapter 1*; Sen, K., Ed.; Springer: Dordrecht, The Netherlands, 2011; pp. 1–34.
13. Wehner, S.; Winter, A. Entropic uncertainty relations—A survey. *New J. Phys.* **2010**, *12*, 025009. [CrossRef]
14. Sperling, J.; Vogel, W. Verifying continuous-variable entanglement in finite spaces. *Phys. Rev. A* **2009**, *79*, 052313. [CrossRef]
15. Willard, J. *Elementary Principles in Statistical Mechanics*; Scribner's sons: New York, NY, USA, 1902.
16. Ehrenfest, P.; Ehrenfest, T. *Begriffliche Grundlagen der Statistischen Auffassung in der Mechanik*; B. G. Teubner: Leipzig, Germany, 1912.
17. Ehrenfest, P.; Ehrenfest, T. *The Conceptual Foundations of the Statistical Approach in Mechanics*; Dover: New York, NY, USA, 1990.
18. Mackey, M. *Time's Arrow: The Origins of Thermodynamic Behavior*; Springer: New York, NY, USA, 1992.
19. Kofler, J.; Brukner, C. Classical World Arising out of Quantum Physics under the Restriction of Coarse-Grained Measurements. *Phys. Rev. Lett.* **2007**, *99*, 180403. [CrossRef] [PubMed]
20. Kofler, J.; Brukner, C.V. Conditions for Quantum Violation of Macroscopic Realism. *Phys. Rev. Lett.* **2008**, *101*, 090403. [CrossRef] [PubMed]
21. Raeisi, S.; Sekatski, P.; Simon, C. Coarse Graining Makes It Hard to See Micro-Macro Entanglement. *Phys. Rev. Lett.* **2011**, *107*, 250401. [CrossRef] [PubMed]
22. Wang, T.; Ghobadi, R.; Raeisi, S.; Simon, C. Precision requirements for observing macroscopic quantum effects. *Phys. Rev. A* **2013**, *88*, 062114. [CrossRef]

23. Jeong, H.; Lim, Y.; Kim, M.S. Coarsening Measurement References and the Quantum-to-Classical Transition. *Phys. Rev. Lett.* **2014**, *112*, 010402. [CrossRef] [PubMed]
24. Rudnicki, L.; Walborn, S.P.; Toscano, F. Optimal uncertainty relations for extremely coarse-grained measurements. *Phys. Rev. A* **2012**, *85*, 042115. [CrossRef]
25. Ray, M.R.; van Enk, S.J. Missing data outside the detector range. II. Application to time-frequency entanglement. *Phys. Rev. A* **2013**, *88*, 062327. [CrossRef]
26. Tasca, D.S.; Rudnicki, L.; Gomes, R.M.; Toscano, F.; Walborn, S.P. Reliable Entanglement Detection under Coarse-Grained Measurements. *Phys. Rev. Lett.* **2013**, *110*, 210502. [CrossRef] [PubMed]
27. Tasca, D.S.; Walborn, S.P.; Toscano, F.; Souto Ribeiro, P.H. Observation of tunable Popescu-Rohrlich correlations through postselection of a Gaussian state. *Phys. Rev. A* **2009**, *80*, 030101. [CrossRef]
28. Semenov, A.A.; Vogel, W. Fake violations of the quantum Bell-parameter bound. *Phys. Rev. A* **2011**, *83*, 032119. [CrossRef]
29. Ray, M.R.; van Enk, S.J. Missing data outside the detector range: Continuous-variable entanglement verification and quantum cryptography. *Phys. Rev. A* **2013**, *88*, 042326. [CrossRef]
30. Bialynicki-Birula, I. Entropic Uncertainty Relations. *Phys. Lett.* **1984**, *103*, 253–254. [CrossRef]
31. Bialynicki-Birula, I. Formulation of the uncertainty relations in terms of the Rényi entropies. *Phys. Rev. A* **2006**, *74*, 52101. [CrossRef]
32. Rudnicki, Ł.; Walborn, S.P.; Toscano, F. Heisenberg uncertainty relation for coarse-grained observables. *EPL* **2012**, *97*, 38003.
33. Heisenberg, W. Über den anschaulichen Inhalt der quantentheoretischen Kinematik und Mechanik. *Z. Phys.* **1927**, *43*, 172–198. [CrossRef]
34. Busch, P.; Heinonen, T.; Lahti, P. Heisenberg's uncertainty principle. *Phys. Rep.* **2007**, *452*, 155–176. [CrossRef]
35. Ozawa, M. Universally valid reformulation of the Heisenberg uncertainty principle on noise and disturbance in measurement. *Phys. Rev. A* **2003**, *67*, 042105. [CrossRef]
36. Ozawa, M. Uncertainty relations for noise and disturbance in generalized quantum measurements. *Ann. Phys.* **2004**, *311*, 350–416. [CrossRef]
37. Ozawa, M. Universal uncertainty principle in the measurement operator formalism. *J. Opt. B Quantum Semiclass. Opt.* **2005**, *7*, S672–S681. [CrossRef]
38. Werner, R.F. The Uncertainty Relation for Joint Measurement of Postion and Momentum. *Quantum Inf. Comput.* **2004**, *4*, 546–562.
39. Busch, P.; Heinonen, T.; Lahti, P. Noise and disturbance in quantum measurement. *Phys. Lett. A* **2004**, *320*, 261–270. [CrossRef]
40. Busch, P.; Lahti, P.; Werner, R.F. Proof of Heisenberg's Error-Disturbance Relation. *Phys. Rev. Lett.* **2013**, *111*, 160405. [CrossRef] [PubMed]
41. Korzekwa, K.; Lostaglio, M.; Jennings, D.; Rudolph, T. Quantum and classical entropic uncertainty relations. *Phys. Rev. A* **2014**, *89*, 042122. [CrossRef]
42. Arthurs, E.; Kelly, J.L. On the Simultaneous Measurement of a Pair of Conjugate Observables. *Bell Syst. Tech. J.* **1965**, *44*, 725–729. [CrossRef]
43. Davies, E.B. *Quantum Theory of Open Systems*; Academic Press London ; New York, NY, USA, 1976; 171p.
44. Busch, P. Indeterminacy relations and simultaneous measurements in quantum theory. *Int. J. Theor. Phys.* **1985**, *24*, 63–92. [CrossRef]
45. Arthurs, E.; Goodman, M.S. Quantum correlations: A generalized Heisenberg uncertainty relation. *Phys. Rev. Lett.* **1988**, *60*, 2447–2449. [CrossRef] [PubMed]
46. Ishikawa, S. Uncertainty relations in simultaneous measurements for arbitrary observables. *Rep. Math. Phys.* **1991**, *29*, 257–273. [CrossRef]
47. Raymer, M. Uncertainty principle for joint measurement of noncommuting variables. *Am. J. Phys.* **1994**, *62*, 986. [CrossRef]
48. Ozawa, M. Uncertainty relations for joint measurements of noncommuting observables. *Phys. Lett. A* **2004**, *320*, 367–374. [CrossRef]
49. Tasca, D.S.; Gomes, R.M.; Toscano, F.; Souto Ribeiro, P.H.; Walborn, S.P. Continuous-variable quantum computation with spatial degrees of freedom of photons. *Phys. Rev. A* **2011**, *83*, 052325. [CrossRef]
50. Dutta, B.; Mukunda, N.; Simon, R. The real symplectic groups in quantum mechanics and optics. *Pramana* **1995**, *45*, 471–497.

51. Kennard, E.H. Zur Quantenmechanik einfacher Bewegungstypen. *Z. Phys.* **1927**, *44*, 326–352. [CrossRef]
52. Weyl, H. *Gruppentheorie und Quantenmechanik (Leipzig: S Hirzel) Weyl H 1950 The Theory of Groups and Quantum Mechanics*; Dover: New York, NY, USA, 1928.
53. Robertson, H. The uncertainty principle. *Phys. Rev.* **1929**, *34*, 163–164. [CrossRef]
54. Schrödinger, E. On Heisenberg's Uncertainty Principle. *Phys. Math.* **1930**, *19*, 296–303.
55. Simon, R.; Mukunda, N.; Dutta, B. Quantum-noise matrix for multimode systems: U (n) invariance, squeezing, and normal forms. *Phys. Rev. A* **1994**, *49*, 1567. [CrossRef] [PubMed]
56. Solomon Ivan, J.; Sabapathy, K.K.; Mukunda, N.; Simon, R. Invariant theoretic approach to uncertainty relations for quantum systems. *arXiv* **2012**, arXiv:1205.5132v1.
57. Simon, R. Peres-Horodecki Separability Criterion for Continuous Variable Systems. *Phys. Rev. Lett.* **2000**, *84*, 2726–2729. [CrossRef] [PubMed]
58. Huang, Y. Entropic uncertainty relations in multidimensional position and momentum spaces. *Phys. Rev. A* **2011**, *83*, 052124. [CrossRef]
59. Shannon, C.E. A Mathematical Theory of Communication. *Bell Syst. Tech. J.* **1948**, *27*, 379–423. [CrossRef]
60. Cover, T.M.; Thomas, J.A. *Elements of Information Theory*; John Wiley and Sons: New York, NY, USA, 2006.
61. Bialynicki-Birula, I.; Mycielski, J. Uncertainty Relations for Information Entropy in Wave Mechanics. *Commun. Math. Phys.* **1975**, *44*, 129. [CrossRef]
62. Beckner, W. Inequalities in Fourier Analysis. *Ann. Math.* **1975**, *102*, 159–182. [CrossRef]
63. Babenko, K.I. IAn inequality in the theory of Fourier integrals. *Izv. Akad. Nauk SSSR Ser. Mater.* **1961**, *25*, 531–542.
64. Hirschman, I.I. A Note on Entropy. *Am. J. Math.* **1957**, *79*, 152–156. [CrossRef]
65. Guanlei, X.; Xiaotong, W.; Xiaogang, X. Generalized entropic uncertainty principle on fractional Fourier transform. *Signal Process.* **2009**, *89*, 2692–2697. [CrossRef]
66. Narcowich, F.J. Geometry and uncertainty. *J. Math. Phys.* **1990**, *31*, 354–364. [CrossRef]
67. Slusher, R.E.; Hollberg, L.W.; Yurke, B.; Mertz, J.C.; Valley, J.F. Observation of Squeezed States Generated by Four-Wave Mixing in an Optical Cavity. *Phys. Rev. Lett.* **1985**, *55*, 2409–2412. [CrossRef] [PubMed]
68. Shchukin, E.; Richter, T.; Vogel, W. Nonclassicality criteria in terms of moments. *Phys. Rev. A* **2005**, *71*, 011802. [CrossRef]
69. Vogel, W. Nonclassical States: An Observable Criterion. *Phys. Rev. Lett.* **2000**, *84*, 1849–1852. [CrossRef] [PubMed]
70. Richter, T.; Vogel, W. Nonclassicality of Quantum States: A Hierarchy of Observable Conditions. *Phys. Rev. Lett.* **2002**, *89*, 283601. [CrossRef] [PubMed]
71. Kiesel, T.; Vogel, W.; Hage, B.; DiGuglielmo, J.; Samblowski, A.; Schnabel, R. Experimental test of nonclassicality criteria for phase-diffused squeezed states. *Phys. Rev. A* **2009**, *79*, 022122. [CrossRef]
72. Ryl, S.; Sperling, J.; Agudelo, E.; Mraz, M.; Köhnke, S.; Hage, B.; Vogel, W. Unified nonclassicality criteria. *Phys. Rev. A* **2015**, *92*, 011801. [CrossRef]
73. Reid, M.D.; Drummond, P.D. Quantum Correlations of Phase in Nondegenerate Parametric Oscillation. *Phys. Rev. Lett.* **1988**, *60*, 2731–2733. [CrossRef] [PubMed]
74. Reid, M.D. Demonstration of the Einstein-Podolsky-Rosen paradox using nondegenerate parametric amplification. *Phys. Rev. A* **1989**, *40*, 913–923. [CrossRef]
75. Einstein, A.; Podolsky, D.; Rosen, N. Can Quantum-Mechanical Description of Physical Reality Be Considered Complete? *Phys. Rev.* **1935**, *47*, 777. [CrossRef]
76. Ou, Z.Y.; Pereira, S.F.; Kimble, H.J.; Peng, K.C. Realization of the Einstein-Podolsky-Rosen paradox for continuous variables. *Phys. Rev. Lett.* **1992**, *68*, 3663–3666. [CrossRef] [PubMed]
77. Wiseman, H.M.; Jones, S.J.; Doherty, A.C. Steering, Entanglement, Nonlocality, and the Einstein-Podolsky-Rosen Paradox. *Phys. Rev. Lett.* **2007**, *98*, 140402. [CrossRef] [PubMed]
78. Jones, S.J.; Wiseman, H.M.; Doherty, A.C. Entanglement, Einstein-Podolsky-Rosen correlations, Bell nonlocality, and steering. *Phys. Rev. A* **2007**, *76*, 052116. [CrossRef]
79. Cavalcanti, D.; Skrzypczyk, P. Quantum steering: a review with focus on semidefinite programming. *Rep. Prog. Phys.* **2017**, *80*, 024001. [CrossRef] [PubMed]
80. Schrödinger, E. The Present Status of Quantum Mechanics. *Naturwissenschaften* **1935**, *23*, 807. [CrossRef]
81. Horodecki, R.; Horodecki, P.; Horodecki, M.; Horodecki, K. Quantum entanglement. *Rev. Mod. Phys.* **2009**, *81*, 865–942. [CrossRef]

82. Gühne, O.; Tóth, G. Entanglement detection. *Phys. Rep.* **2009**, *474*, 1–75. [CrossRef]
83. Brunner, N.; Cavalcanti, D.; Pironio, S.; Scarani, V.; Wehner, S. Bell nonlocality. *Rev. Mod. Phys.* **2014**, *86*, 419–478. [CrossRef]
84. Ji, S.W.; Lee, J.; Park, J.; Nha, H. Steering criteria via covariance matrices of local observables in arbitrary-dimensional quantum systems. *Phys. Rev. A* **2015**, *92*, 062130. [CrossRef]
85. Walborn, S.P.; Salles, A.; Gomes, R.M.; Toscano, F.; Souto Ribeiro, P.H. Revealing Hidden Einstein-Podolsky-Rosen Nonlocality. *Phys. Rev. Lett.* **2011**, *106*, 130402. [CrossRef] [PubMed]
86. Schneeloch, J.; Dixon, P.B.; Howland, G.A.; Broadbent, C.J.; Howell, J.C. Violation of Continuous-Variable Einstein-Podolsky-Rosen Steering with Discrete Measurements. *Phys. Rev. Lett.* **2013**, *110*, 130407. [CrossRef] [PubMed]
87. Schneeloch, J.; Howland, G.A. Quantifying high-dimensional entanglement with Einstein-Podolsky-Rosen correlations. *Phys. Rev. A* **2018**, *97*, 042338. [CrossRef]
88. Schneeloch, J.; Tison, C.C.; Fanto, M.L.; Alsing, P.M.; Howland, G.A. Quantifying entanglement in a 68-billion dimensional quantum system. *arXiv* **2018**, arXiv:1804.04515.
89. Reid, M.D. Quantum cryptography with a predetermined key, using continuous-variable Einstein-Podolsky-Rosen correlations. *Phys. Rev. A* **2000**, *62*, 062308. [CrossRef]
90. Grosshans, F.; Cerf, N.J. Continuous-Variable Quantum Cryptography is Secure against Non-Gaussian Attacks. *Phys. Rev. Lett.* **2004**, *92*, 047905. [CrossRef] [PubMed]
91. Branciard, C.; Cavalcanti, E.G.; Walborn, S.P.; Scarani, V.; Wiseman, H.M. One-sided device-independent quantum key distribution: Security, feasibility, and the connection with steering. *Phys. Rev. A* **2012**, *85*, 010301. [CrossRef]
92. Kogias, I.; Skrzypczyk, P.; Cavalcanti, D.; Acín, A.; Adesso, G. Hierarchy of Steering Criteria Based on Moments for All Bipartite Quantum Systems. *Phys. Rev. Lett.* **2015**, *115*, 210401. [CrossRef] [PubMed]
93. Silberhorn, C.; Lam, P.K.; Weiß, O.; König, F.; Korolkova, N.; Leuchs, G. Generation of Continuous Variable Einstein-Podolsky-Rosen Entanglement via the Kerr Nonlinearity in an Optical Fiber. *Phys. Rev. Lett.* **2001**, *86*, 4267–4270. [CrossRef] [PubMed]
94. Bowen, W.P.; Schnabel, R.; Lam, P.K.; Ralph, T.C. Experimental Investigation of Criteria for Continuous Variable Entanglement. *Phys. Rev. Lett.* **2003**, *90*, 043601. [CrossRef] [PubMed]
95. D'Angelo, M.; Kim, Y.H.; Kulik, S.P.; Shih, Y. Identifying Entanglement Using Quantum Ghost Interference and Imaging. *Phys. Rev. Lett.* **2004**, *92*, 233601. [CrossRef] [PubMed]
96. Howell, J.C.; Bennink, R.S.; Bentley, S.J.; Boyd, R.W. Realization of the Einstein-Podolsky-Rosen Paradox Using Momentum- and Position-Entangled Photons from Spontaneous Parametric Down Conversion. *Phys. Rev. Lett.* **2004**, *92*, 210403. [CrossRef] [PubMed]
97. Tasca, D.S.; Walborn, S.P.; Souto Ribeiro, P.H.; Toscano, F.; Pellat-Finet, P. Propagation of transverse intensity correlations of a two-photon state. *Phys. Rev. A* **2009**, *79*, 033801. [CrossRef]
98. Duan, L.M.; Giedke, G.; Cirac, J.I.; Zoller, P. Inseparability Criterion for Continuous Variable Systems. *Phys. Rev. Lett.* **2000**, *84*, 2722–2725. [CrossRef] [PubMed]
99. Mancini, S.; Giovannetti, V.; Vitali, D.; Tombesi, P. Entangling Macroscopic Oscillators Exploiting Radiation Pressure. *Phys. Rev. Lett.* **2002**, *88*, 120401. [CrossRef] [PubMed]
100. Giovannetti, V.; Mancini, S.; Vitali, D.; Tombesi, P. Characterizing the entanglement of bipartite quantum systems. *Phys. Rev. A* **2003**, *67*, 022320. [CrossRef]
101. Zhang, C.J.; Nha, H.; Zhang, Y.S.; Guo, G.C. Entanglement detection via tighter local uncertainty relations. *Phys. Rev. A* **2010**, *81*, 012324. [CrossRef]
102. Peres, A. Separability Criterion for Density Matrices. *Phys. Rev. Lett.* **1996**, *77*, 1413. [CrossRef] [PubMed]
103. Horedecki, M.; Horodecki, P.; Horodecki, R. Separability of mixed states: necessary and sufficient conditions. *Phys. Lett. A* **1996**, *223*, 1–8.
104. Nha, H.; Zubairy, M.S. Uncertainty Inequalities as Entanglement Criteria for Negative Partial-Transpose States. *Phys. Rev. Lett.* **2008**, *101*, 130402. [CrossRef] [PubMed]
105. Walborn, S.P.; Taketani, B.G.; Salles, A.; Toscano, F.; de Matos Filho, R.L. Entropic Entanglement Criteria for Continuous Variables. *Phys. Rev. Lett.* **2009**, *103*, 160505. [CrossRef] [PubMed]
106. Saboia, A.; Toscano, F.; Walborn, S.P. Family of continuous-variable entanglement criteria using general entropy functions. *Phys. Rev. A* **2011**, *83*, 032307. [CrossRef]

107. Toscano, F.; Saboia, A.; Avelar, A.T.; Walborn, S.P. Systematic construction of genuine-multipartite-entanglement criteria in continuous-variable systems using uncertainty relations. *Phys. Rev. A* **2015**, *92*, 052316. [CrossRef]

108. Werner, R.F. Quantum states with Einstein-Podolsky-Rosen correlations admitting a hidden-variable model. *Phys. Rev. A* **1989**, *40*, 4277–4281. [CrossRef]

109. Vidal, G.; Werner, R.F. Computable measure of entanglement. *Phys. Rev. A* **2002**, *65*, 032314. [CrossRef]

110. Werner, R.F.; Wolf, M.M. Bound Entangled Gaussian States. *Phys. Rev. Lett.* **2001**, *86*, 3658–3661. [CrossRef] [PubMed]

111. Horodecki, M.; Horodecki, P.; Horodecki, R. Mixed-State Entanglement and Distillation: Is there a "Bound" Entanglement in Nature? *Phys. Rev. Lett.* **1998**, *80*, 5239–5242. [CrossRef]

112. Bennett, C.H.; Bernstein, H.J.; Popescu, S.; Schumacher, B. Concentrating partial entanglement by local operations. *Phys. Rev. A* **1996**, *53*, 2046. [CrossRef] [PubMed]

113. Giedke, G.; Kraus, B.; Duan, L.M.; Zoller, P.; Cirac, I.J.; Lewenstein, M. Separability and Distillability of bipartite Gaussian States–the Complete Story. *Fortschr. Phys.* **2001**, *49*, 973–980. [CrossRef]

114. Giedke, G.; Kraus, B.; Duan, L.M.; Lewenstein, M.; Cirac, I.J. Entanglement Criteria for All Bipartite Gaussian States. *Phys. Rev. Lett.* **2001**, *87*, 167904. [CrossRef] [PubMed]

115. Hyllus, P.; Eisert, J. Optimal entanglement witnesses for continuous-variable systems. *New J. Phys.* **2006**, *8*, 51. [CrossRef]

116. Nha, H. Entanglement condition via su(2) and su(1,1) algebra using Schrödinger-Robertson uncertainty relation. *Phys. Rev. A* **2007**, *76*, 014305. [CrossRef]

117. Agarwal, G.S.; Biswas, A. Inseparability inequalities for higher order moments for bipartite systems. *New J. Phys.* **2005**, *7*, 211. [CrossRef]

118. Hillery, M.; Zubairy, M.S. Entanglement Conditions for Two-Mode States. *Phys. Rev. Lett.* **2006**, *96*, 050503. [CrossRef] [PubMed]

119. Paul, E.C.; Tasca, D.S.; Rudnicki, L.; Walborn, S.P. Detecting entanglement through direct measurement of biphoton characteristic functions. **2018**, submitted for publication.

120. Paul, E.C.; Tasca, D.S.; Rudnicki, L.; Walborn, S.P. Detecting entanglement of continuous variables with three mutually unbiased bases. *Phys. Rev. A* **2016**, *94*, 012303. [CrossRef]

121. Shchukin, E.; Vogel, W. Inseparability Criteria for Continuous Bipartite Quantum States. *Phys. Rev. Lett.* **2005**, *95*, 230502. [CrossRef] [PubMed]

122. Van Loock, P.; Furusawa, A. Detecting genuine multipartite continuous-variable entanglement. *Phys. Rev. A* **2003**, *67*, 052315. [CrossRef]

123. Sun, Q.; Nha, H.; Zubairy, M.S. Entanglement criteria and nonlocality for multimode continuous-variable systems. *Phys. Rev. A* **2009**, *80*, 020101. [CrossRef]

124. Shchukin, E.; Vogel, W. Conditions for multipartite continuous-variable entanglement. *Phys. Rev. A* **2006**, *74*, 030302. [CrossRef]

125. Villar, A.S.; Cruz, L.S.; Cassemiro, K.N.; Martinelli, M.; Nussenzveig, P. Generation of Bright Two-Color Continuous Variable Entanglement. *Phys. Rev. Lett.* **2005**, *95*, 243603. [CrossRef] [PubMed]

126. Coelho, A.S.; Barbosa, F.A.S.; Cassemiro, K.N.; Villar, A.S.; Martinelli, M.; Nussenzveig, P. Three-Color Entanglement. *Science* **2009**, *6*, 823–826. [CrossRef] [PubMed]

127. Tasca, D.S.; Walborn, S.P.; Ribeiro, P.H.S.; Toscano, F. Detection of transverse entanglement in phase space. *Phys. Rev. A* **2008**, *78*, 010304. [CrossRef]

128. Shalm, L.K.; Hamel, D.R.; Yan, Z.; Simon, C.; Resch, K.J.; Jennewein, T. Three-photon energy-time entanglement. *Nat. Phys.* **2012**, *9*, 19–22. [CrossRef]

129. MacLean, J.P.W.; Donohue, J.M.; Resch, K.J. Direct Characterization of Ultrafast Energy-Time Entangled Photon Pairs. *Phys. Rev. Lett.* **2018**, *120*, 053601. [CrossRef] [PubMed]

130. Gomes, R.M.; Salles, A.; Toscano, F.; Ribeiro, P.H.S.; Walborn, S.P. Quantum Entanglement Beyond Gaussian Criteria. *Proc. Natl. Acad. Sci. USA* **2009**, *106*, 21517–21520. [CrossRef] [PubMed]

131. Edgar, M.; Tasca, D.; Izdebski, F.; Warburton, R.; Leach, J.; Agnew, M.; Buller, G.; Boyd, R.; Padgett, M. Imaging high-dimensional spatial entanglement with a camera. *Nat. Commun.* **2012**, *3*, 984. [CrossRef] [PubMed]

132. Aspden, R.S.; Tasca, D.S.; Boyd, R.W.; Padgett, M.J. EPR-based ghost imaging using a single-photon-sensitive camera. *New J. Phys.* **2013**, *15*, 073032. [CrossRef]

133. Moreau, P.A.; Devaux, F.; Lantz, E. Einstein-Podolsky-Rosen Paradox in Twin Images. *Phys. Rev. Lett.* **2014**, *113*, 160401. [CrossRef] [PubMed]

134. Tentrup, T.B.H.; Hummel, T.; Wolterink, T.A.W.; Uppu, R.; Mosk, A.P.; Pinkse, P.W.H. Transmitting more than 10 bit with a single photon. *Opt. Express* **2017**, *25*, 2826–2833. [CrossRef] [PubMed]

135. Warburton, R.E.; Izdebski, F.; Reimer, C.; Leach, J.; Ireland, D.G.; Padgett, M.; Buller, G.S. Single-photon position to time multiplexing using a fiber array. *Opt. Express* **2011**, *19*, 2670–2675. [CrossRef] [PubMed]

136. Leach, J.; Warburton, R.E.; Ireland, D.G.; Izdebski, F.; Barnett, S.M.; Yao, A.M.; Buller, G.S.; Padgett, M.J. Quantum correlations in position, momentum, and intermediate bases for a full optical field of view. *Phys. Rev. A* **2012**, *85*, 013827. [CrossRef]

137. Durt, T.; Englert, B.G.; Bengtsson, I.; Życzkowski, K. On Mutually Unbiased Bases. *Int. J. Quant. Inf.* **2010**, *8*, 535–640. [CrossRef]

138. Aharonov, Y.; Pendleton, H.; Petersen, A. Modular variables in quantum theory. *Int. J. Theor. Phys.* **1969**, *2*, 213–230. [CrossRef]

139. Busch, P.; Lahti, P.J. To what extent do position and momentum commute? *Phys. Lett. A* **1986**, *115*, 259–264. [CrossRef]

140. Reiter, H.; Thirring, W. Are x and p incompatible observables? *Found. Phys.* **1989**, *19*, 1037–1039. [CrossRef]

141. Ylinen, K. Commuting functions of the position and momentum observables on locally compact abelian groups. *J. Math. Anal. Appl.* **1989**, *137*, 185–192. [CrossRef]

142. Tasca, D.S.; Sánchez, P.; Walborn, S.P.; Rudnicki, L. Mutual Unbiasedness in Coarse-Grained Continuous Variables. *Phys. Rev. Lett.* **2018**, *120*, 040403. [CrossRef] [PubMed]

143. Paul, E.C.; Walborn, S.P.; Tasca, D.S.; Rudnicki, L. Mutually Unbiased Coarse-Grained Measurements of Two or More Phase-Space Variables. *Phys. Rev. A* **2018**, *97*, 052103. [CrossRef]

144. Vallone, G.; Marangon, D.G.; Tomasin, M.; Villoresi, P. Quantum randomness certified by the uncertainty principle. *Phys. Rev. A* **2014**, *90*, 052327. [CrossRef]

145. Spengler, C.; Huber, M.; Brierley, S.; Adaktylos, T.; Hiesmayr, B.C. Entanglement detection via mutually unbiased bases. *Phys. Rev. A* **2012**, *86*, 022311. [CrossRef]

146. Krenn, M.; Huber, M.; Fickler, R.; Lapkiewicz, R.; Ramelow, S.; Zeilinger, A. Generation and confirmation of a (100×100)-dimensional entangled quantum system. *Proc. Natl. Acad. Sci. USA* **2014**, *111*, 6243–6247. [CrossRef] [PubMed]

147. Erker, P.; Krenn, M.; Huber, M. Quantifying high dimensional entanglement with two mutually unbiased bases. *Quantum* **2017**, *1*, 22. [CrossRef]

148. Fernández-Pérez, A.; Klimov, A.B.; Saavedra, C. Quantum process reconstruction based on mutually unbiased basis. *Phys. Rev. A* **2011**, *83*, 052332. [CrossRef]

149. Giovannini, D.; Romero, J.; Leach, J.; Dudley, A.; Forbes, A.; Padgett, M.J. Characterization of High-Dimensional Entangled Systems via Mutually Unbiased Measurements. *Phys. Rev. Lett.* **2013**, *110*, 143601. [CrossRef] [PubMed]

150. Weigert, S.; Wilkinson, M. Mutually unbiased bases for continuous variables. *Phys. Rev. A* **2008**, *78*, 020303. [CrossRef]

151. Kraus, K. Complementary observables and uncertainty relations. *Phys. Rev. D* **1987**, *35*, 3070–3075. [CrossRef]

152. Grassl, M.; McNulty, D.; Mišta, L.; Paterek, T. Small sets of complementary observables. *Phys. Rev. A* **2017**, *95*, 823–826. [CrossRef]

153. Rudnicki, L.; Tasca, D.S.; Walborn, S.P. Uncertainty relations for characteristic functions. *Phys. Rev. A* **2016**, *93*, 022109. [CrossRef]

154. Deutsch, D. Uncertainty in Quantum Measurements. *Phys. Rev. Lett.* **1983**, *50*, 631–633. [CrossRef]

155. Maassen, H.; Uffink, J.B.M. Generalized entropic uncertainty relations. *Phys. Rev. Lett.* **1988**, *60*, 1103–1106. [CrossRef] [PubMed]

156. Friedland, S.; Gheorghiu, V.; Gour, G. Universal Uncertainty Relations. *Phys. Rev. Lett.* **2013**, *111*, 230401. [CrossRef] [PubMed]

157. Puchała, Z.; Rudnicki, Ł.; Życzkowski, K. Majorization entropic uncertainty relations. *J. Phys. A Math. Theor.* **2013**, *46*, 272002. [CrossRef]

158. Coles, P.J.; Piani, M. Improved entropic uncertainty relations and information exclusion relations. *Phys. Rev. A* **2014**, *89*, 022112. [CrossRef]

159. Rudnicki, L.; Puchała, Z.; Życzkowski, K. Strong majorization entropic uncertainty relations. *Phys. Rev. A* **2014**, *89*, 052115. [CrossRef]

160. Bosyk, G.M.; Zozor, S.; Portesi, M.; Osán, T.M.; Lamberti, P.W. Geometric approach to extend Landau-Pollak uncertainty relations for positive operator-valued measures. *Phys. Rev. A* **2014**, *90*, 052114. [CrossRef]

161. Zozor, S.; Bosyk, G.M.; Portesi, M. General entropy-like uncertainty relations in finite dimensions. *J. Phys. A Math. Theor.* **2014**, *47*, 495302. [CrossRef]

162. Kaniewski, J.M.K.; Tomamichel, M.; Wehner, S. Entropic uncertainty from effective anticommutators. *Phys. Rev. A* **2014**, *90*, 012332. [CrossRef]

163. Puchała, Z.; Łukasz Rudnicki.; Krawiec, A.; Życzkowski, K. Majorization uncertainty relations for mixed quantum states. *J. Phys. A Math. Theor.* **2018**, *51*, 175306. [CrossRef]

164. Brandão, F.; Horodecki, M.; Ng, N.; Oppenheim, J.; Wehner, S. The second laws of quantum thermodynamics. *Proc. Natl. Acad. Sci. USA* **2015**, *112*, 3275–3279. [CrossRef] [PubMed]

165. Partovi, M.H. Entropic Formulation of Uncertainty for Quantum Measurements. *Phys. Rev. Lett.* **1983**, *50*, 1883–1885. [CrossRef]

166. Rudnicki, Ł. Shannon entropy as a measure of uncertainty in positions and momenta. *J. Russ. Laser Res.* **2011**, *32*, 393.

167. Partovi, M.H. Majorization formulation of uncertainty in quantum mechanics. *Phys. Rev. A* **2011**, *84*, 052117. [CrossRef]

168. Schürmann, T.; Hoffmann, I. A Closer Look at the Uncertainty Relation of Position and Momentum. *Found. Phys.* **2009**, *39*, 958–963. [CrossRef]

169. Schürmann, T. A note on entropic uncertainty relations of position and momentum. *J. Russ. Laser Res.* **2012**, *33*, 52–54. [CrossRef]

170. Wilk, G.; Włodarczyk, Z. Uncertainty relations in terms of the Tsallis entropy. *Phys. Rev. A* **2009**, *79*, 062108. [CrossRef]

171. Bialynicki-Birula, I.; Rudnicki, L. Comment on "Uncertainty relations in terms of the Tsallis entropy". *Phys. Rev. A* **2010**, *81*, 026101. [CrossRef]

172. Abramowitz, M.; Stegun, I. *Handbook of Mathematical Functions*; Dover: New York, NY, USA, 1964.

173. Lassance, N. Optimal RRnyi Entropy Portfolios. *SSRN Electron. J.* **2017**, 1–15. [CrossRef]

174. Rudnicki, L. Majorization approach to entropic uncertainty relations for coarse-grained observables. *Phys. Rev. A* **2015**, *91*, 032123. [CrossRef]

175. Fuchs, W. On the eigenvalues of an integral equation arising in the theory of band-limited signals. *J. Math. Anal. Appl.* **1964**, *9*, 317–330. [CrossRef]

176. Ballentine, L. *Quantum Mechanics: A Modern Development*; World Scientific: Singapore, 1998.

177. Kofler, J.; Brukner, Č. *A Coarse-Grained Schrödinger Cat*; IOS Press: Amsterdam, The Netherlands, 2007.

178. Bialynicki-Birula, I.; Madajczyk, J. Entropic uncertainty relations for angular distributions. *Phys. Lett. A* **1985**, *108*, 384–386. [CrossRef]

179. Furrer, F.; Berta, M.; Tomamichel, M.; Scholz, V.B.; Christandl, M. Position-momentum uncertainty relations in the presence of quantum memory. *J. Math. Phys.* **2014**, *55*, 122205. [CrossRef]

180. Berta, M.; Matthias Christandl and, R.C.; Renes, J.M.; Renner, R. The uncertainty principle in the presence of quantum memory. *Nat. Phys.* **2010**, *6*, 659. [CrossRef]

181. Rastegin, A.E. On entropic uncertainty relations in the presence of a minimal length. *Ann. Phys.* **2017**, *382*, 170–180. [CrossRef]

182. Tasca, D.S.; Rudnicki, L.; Aspden, R.S.; Padgett, M.J.; Souto Ribeiro, P.H.; Walborn, S.P. Testing for entanglement with periodic coarse graining. *Phys. Rev. A* **2018**, *97*, 042312. [CrossRef]

183. Park, J.; Ji, S.W.; Lee, J.; Nha, H. Gaussian states under coarse-grained continuous variable measurements. *Phys. Rev. A* **2014**, *89*, 042102. [CrossRef]

184. Gilchrist, A.; Deuar, P.; Reid, M.D. Contradiction of Quantum Mechanics with Local Hidden Variables for Quadrature Phase Amplitude Measurements. *Phys. Rev. Lett.* **1998**, *80*, 3169–3172. [CrossRef]

185. Gilchrist, A.; Deuar, P.; Reid, M.D. Contradiction of quantum mechanics with local hidden variables for quadrature phase measurements on pair-coherent states and squeezed macroscopic superpositions of coherent states. *Phys. Rev. A* **1999**, *60*, 4259–4271. [CrossRef]

186. Munro, W.J. Optimal states for Bell-inequality violations using quadrature-phase homodyne measurements. *Phys. Rev. A* **1999**, *59*, 4197–4201. [CrossRef]

Entropy **2018**, *20*, 454

187. García-Patrón, R.; Fiurášek, J.; Cerf, N.J.; Wenger, J.; Tualle-Brouri, R.; Grangier, P. Proposal for a Loophole-Free Bell Test Using Homodyne Detection. *Phys. Rev. Lett.* **2004**, *93*, 130409. [CrossRef] [PubMed]

188. Wenger, J.; Hafezi, M.; Grosshans, F.; Tualle-Brouri, R.; Grangier, P. Maximal violation of Bell inequalities using continuous-variable measurements. *Phys. Rev. A* **2003**, *67*, 012105. [CrossRef]

189. Aharanov, Y.; Rohrlich, D. *Quantum Paradoxes*; Wiley: Berlin, Germany, 2005.

190. Vernaz-Gris, P.; Ketterer, A.; Keller, A.; Walborn, S.P.; Coudreau, T.; Milman, P. Continuous discretization of infinite-dimensional Hilbert spaces. *Phys. Rev. A* **2014**, *89*, 052311. [CrossRef]

191. Ketterer, A.; Keller, A.; Walborn, S.P.; Coudreau, T.; Milman, P. Quantum information processing in phase space: A modular variables approach. *Phys. Rev. A* **2016**, *94*, 022325. [CrossRef]

192. Gneiting, C.; Hornberger, K. Detecting Entanglement in Spatial Interference. *Phys. Rev. Lett.* **2011**, *106*, 210501. [CrossRef] [PubMed]

193. Carvalho, M.A.D.; Ferraz, J.; Borges, G.F.; de Assis, P.L.; Pádua, S.; Walborn, S.P. Experimental observation of quantum correlations in modular variables. *Phys. Rev. A* **2012**, *86*, 032332. [CrossRef]

194. Barros, M.R.; Farías, O.J.; Keller, A.; Coudreau, T.; Milman, P.; Walborn, S.P. Detecting multipartite spatial entanglement with modular variables. *Phys. Rev. A* **2015**, *92*, 022308. [CrossRef]

195. Massar, S.; Pironio, S. Greenberger-Horne-Zeilinger paradox for continuous variables. *Phys. Rev. A* **2001**, *64*, 062108. [CrossRef]

196. Plastino, A.R.; Cabello, A. State-independent quantum contextuality for continuous variables. *Phys. Rev. A* **2010**, *82*, 022114. [CrossRef]

197. Asadian, A.; Budroni, C.; Steinhoff, F.E.S.; Rabl, P.; Gühne, O. Contextuality in phase space. *arXiv* **2015**, arXiv:1502.05799.

198. Laversanne-Finot, A.; Ketterer, A.; Barros, M.R.; Walborn, S.P.; Coudreau, T.; Keller, A.; Milman, P. General conditions for maximal violation of non-contextuality in discrete and continuous variables. *J. Phys. A Math. Theor.* **2017**, *50*, 155304. [CrossRef]

199. Asadian, A.; Erker, P.; Huber, M.; Klöckl, C. Heisenberg-Weyl Observables: Bloch vectors in phase space. *Phys. Rev. A* **2016**, *94*, 010301. [CrossRef]

200. Chang, L.N.; Lewis, Z.; Minic, D.; Takeuchi, T. On the Minimal Length Uncertainty Relation and the Foundations of String Theory. *Adv. High Energy Phys.* **2011**, *2011*, 493514. [CrossRef]

201. Tawfik, A.N.; Diab, A.M. Review on Generalized Uncertainty Principle. *Rep. Prog. Phys.* **2015**, *78*, 126001. [CrossRef] [PubMed]

MDPI

St. Alban-Anlage 66

4052 Basel

Switzerland

Tel. +41 61 683 77 34

Fax +41 61 302 89 18

www.mdpi.com

Entropy Editorial Office

E-mail: entropy@mdpi.com

www.mdpi.com/journal/entropy

www.ingramcontent.com/pod-product-compliance
Lightning Source LLC
Chambersburg PA
CBHW051855210326
41597CB00033B/5904